Hydrogen Energy for Beginners

Hydrogen Energy for Beginners

Alexander Gavrilyuk

Pan Stanford Publishing

Published by

Pan Stanford Publishing Pte. Ltd.
Penthouse Level, Suntec Tower 3
8 Temasek Boulevard
Singapore 038988

Email: editorial@panstanford.com
Web: www.panstanford.com

British Library Cataloguing-in-Publication Data
A catalogue record for this book is available from the British Library.

Hydrogen Energy for Beginners

ISBN 978-981-4364-77-5 (Hardcover)
ISBN 978-981-4364-78-2 (eBook)

Printed in the USA

Contents

Preface

Why and how did this book appear? What were the reasons?

For many years I have been dealing with research related to hydrogen. The more I learn about this ubiquitous element, the more I understand that our knowledge about hydrogen and its properties suffers from one-sidedness.

These days one speaks of hydrogen very often; every day we hear about hydrogen energy, hydrogen infrastructure, hydrogen economy, and sometimes even on hydrogen society. Hydrogen is proclaimed to be the fuel of the future; it has been nominated to substitute fossil fuels since their use has become a threat to the environment and, in turn, to human existence.

While considering the importance of hydrogen, the only unforgivable mistake would be its underestimation. Hydrogen is extremely important for many processes, phenomena, and reactions that have been long investigated in the various and different branches of natural sciences, such as biology, chemistry, and physics. Hydrogen is a very important player in metabolism; it always works for life. Hydrogen is the most abundant element in the universe. Hydrogen powers stars; the energy delivered to the earth from the sun is due to thermonuclear reaction, in which hydrogen is the main player. The importance of hydrogen for medical purposes is also growing day by day. Recently it was discovered that hydrogen can be an exclusive antioxidant in medicine, which makes it an eventual element for medical treatment against various diseases. On the other hand, hydrogen is an inalienable element of electronics technology.

This book is devoted to the fundamentals, concepts, and ideas relating to hydrogen rather than the technical problems associated with it. In my opinion, it is important to present the bigger picture of the hydrogen world to explain the important processes arising with the inalienable participation of hydrogen. It makes sense to cover important generalizations that elucidate the unique features of this mysterious element and to present the keys to understanding its functions in various processes: hydrogen unites all natural sciences.

The most important feature that makes hydrogen unique is its crucial role in processes involving the conversion, transfer, and storage of energy. In many cases, to perform these functions, hydrogen atoms are split into electrons and protons, and this mobile couple of small, electrically charged particles with the opposite charge performs various and unique functions in the energy processes.

Hydrogen atoms create hydrogen bonds, which are enormously important for the whole of nature and especially for vital processes. Hydrogen atoms are able to create special pathways for protons, facilitating their displacements and creating a one-dimensional chain of hydrogen bonds along which proton solitons can be transferred to large distances. Hydrogen bonds are especially important in processes such as enzyme catalysis and the creation of supramolecules.

Hydrogen is used to cure dangling bonds in amorphous silicon.

Hydrogen atoms can be easily inserted in great quantities in many solids to drastically change their optical, electrical, structural, and magnetic parameters.

Hydrogen may act as a fuel, dopant, or catalyst, sometimes combining these functions. Sometimes it acts latently, and it takes a lot of effort to discover its role in various processes.

An amazing and intriguing fact is that cosmic hydrogen is engaged to provide energy to the earth, whereas its counterpart, terrestrial hydrogen, plays the main role in its conversion via photosynthesis, providing vital power—enormous range of activities!

The objective of this book is to present in full width a realm of processes related to hydrogen that occur exclusively due to the peculiarities of hydrogen species. The book collects "hydrogen knowledge" belonging to different natural sciences, and I hope that it would be interesting for students and postgraduates in various disciplines who would like to elaborate their "hydrogen world outlook", "the hydrogen Weltanschauung".

And I think that this enigmatic, exceptional, ubiquitous, unique, wonderful, uncanny hydrogen deserves thorough consideration of all its various properties.

I have tried to cover the common features of hydrogen species in various phenomena, which make its activity so multisided. The book highlights many different issues from various scientific trends. It is for those beginners who want to learn about this magic hydrogen world. I think that the main task for a young researcher is to realize the bewildering amount of scientific ideas connected with hydrogen processes in nature and hydrogen technologies. The prime aim of the book is to inspire interest in the hydrogen world and in hydrogen science.

But the problems of hydrogen energy seem to be even more important. Natural resources are distributed between countries nonuniformly. There are countries that are terrifically rich in resources and whose economy depends entirely on the output and sale of fossil fuels. Since the prices of fossils at present are very high, there is no necessity to develop new technologies in those countries: all financial problems can be solved by the great amount of money coming from oil and gas production. And since only new technologies power the development of societies, those "fossil" societies are stagnating in all aspects of social life. Having an ugly, monopolistic economy based on the output and sale of raw materials, they also usually possess monopolistic, authoritarian, or totalitarian political regimes that are often very aggressive both inside their own country and abroad. Using the high prices for the fossil fuels, they get enormous financial power used not to achieve high living standards for their own people, but for militarization, political repressions, and corruption, which they try to spread all over the world to achieve their primitive targets. In these countries, human rights are brutally suppressed. These states are enemies of the open society.

Thus, creating a hydrogen energy economy by making energy available for the whole humankind without restrictions and blackmailing from countries rich in energy resources would be an important step not only from the ecological and economical but also from a political and human point of view.

I think it would not be an overestimation to say that it is a challenge for the civilized world to develop energy resources that are cheap and available to all countries and peoples in the world. All said, stressing the enormous importance of the breakthroughs in science and technology could create a world of hydrogen energy.

This book comprises three parts.

The first part covers the main properties of hydrogen species that isolate hydrogen from all other elements and that make them so exclusively important in energy processes. Atomic hydrogen is an element that is the smallest in mass and size, has poor electron density and has the potential to get rid of it, facilely turning into a couple consisting of two particles with elementary charges of opposite signs (proton and electron). Hydrogen is a particle with small spatial restrictions; it is very reactive and capable of participating in tunneling reactions. In this part, special attention has been given to the hydrogen bond (the first step of the hydrogen transfer reactions) and its importance. The possibility of the hydrogen bonds forming one-dimensional structures where the solitons can spread is underlined.

The second part highlights a series of outstanding natural processes of hydrogen. The first is hydrogen fusion, which powers the stars and the sun, the energy of which is delivered to the earth via radiation. This energy is converted into chemical energy during photosynthesis, where hydrogen is an inalienable player again and again. The chemical energy is then, again with the irreplaceable help of hydrogen, used in the oxidation process, powering the life of animals and human beings. The last chapter of this part covers the biological production of hydrogen, which is crucial for the production of dihydrogen, which has the potential to become the dream fuel for technological purposes. The research directed toward the technological improvement of this attractive natural process is also elucidated here.

The third part is devoted to the technologies of hydrogen production and storage, which are inalienable parts of future hydrogen energetics. Along with conventional methods of hydrogen production, such as cracking and electrolysis of water, new methods based on higher technologies, including artificial photosynthesis and photocatalytic degradation of organic substances, are discussed. In addition, photoinjection of hydrogen in solids is covered, because this process is connected with abstraction of hydrogen atoms under the action of light from organic molecules adsorbed preliminarily on the surface of transition metal oxides. Readers will find especially interesting the technologies related to the double-layered structures that can transport hydrogen atoms between various materials.

All these processes present the total picture of hydrogen behavior in energy processes and make it possible to understand the total manifestation of hydrogen science.

To read and understand this book is not an easy job, as it requires knowledge of many different branches of science. However, such reading may prove very useful to make the entire picture of the natural world and to develop scientific imagination. I have tried to make the book as lucid as possible, and I wish success to all those who would like to learn about the splendid and wide field of hydrogen science and its importance as a source of energy that humankind would be able to use.

Alexander Gavrilyuk
Autumn 2013

Chapter 1

Unique Properties of Hydrogen

1.1 Introduction

Hydrogen is one of the most important and mysterious elements of the periodic table of chemical elements. It possesses a whole series of unique properties that predetermine its exclusive importance in various processes in physics, chemistry, and biology. However, before considering the importance of hydrogen and its specific features, it makes sense to make a short excurse into the history of hydrogen. When and how did it appear? What is its origin?

1.2 History of Hydrogen

In fact we all live in the universe that evolved after the Big Bang which is the instant when the universe is supposed to have started rapidly expanding from infinitively dense and high-energy state named as singular state. According to the best available measurements, the Big Bang occurred between 13.3 and 13.9 billion years ago [1,2].

The earliest instance when the theoretical description was possible is the moment of the so-called Planck epoch with the temperature $T = 10^{32}$ K (Planck temperature) and the density $d = 10^{93}$ g cm^{-3} (Planck density) [3,4]. The early universe appeared to be highly homogeneous medium with enormously high energy density, temperature, and pressure.

Hydrogen Energy for Beginners
Alexander Gavrilyuk

ISBN 978-981-4364-77-5 (Hardcover), 978-981-4364-78-2 (eBook)
www.panstanford.com

During further expansion and cooling the universe had overcome the phase transitions in elementary particles system. The very early universe occurred to be enormously hot and elementary particles had energies higher than those currently accessible in particle accelerators on the earth. Approximately 10^{-35} s after the Planck epoch the phase transition triggered an exponential expansion of the universe called the "period of cosmic inflation" [3,4].

After the end of this period the building material for the universe comprised the quark–gluon plasma. With time the temperature fell down to the values when the next phase transition called bariogenesis occurred. At this stage the quarks and gluons fused into barions such as protons and neutrons. Thus, just at this stage the first ever known hydrogen species, namely, protons were born.

Further cooling led to the next phase transition: formation of physical forces and elementary particles in its modern form.

Then the epoch of nuclear synthesis came when protons fused with neutrons and formed the nuclei of deuterium, helium-4, and some other light isotopes. On further cooling and expansion of the universe the moment occurred when gravitation became the dominant force. Some 380,000 years after the Big Bang the temperature became so low that the appearance of hydrogen atoms became possible. Prior to this the processes of ionization of hydrogen atoms and recombination of protons and electrons were in equilibrium.

With the formation of neutral hydrogen the decoupling of substance and radiation occurred and the cosmic microwave background was emitted.

The color temperature of the decoupled radiation has continued to decrease ever since; now down to 2.725 K, its temperature will continue to drop as the universe expands. According to the Big Bang model, the radiation from the sky measured today comes from a spherical surface called the "surface of last scattering" [5].

Figure 1.1 shows the above-mentioned three stages of the universe evolution after the moment of the Big Bang. So, the hydrogen atoms appeared on the second stage whereas hydrogen atoms were formed only on the third stage, thus 380,000 years later than protons.

Mother Nature celebrated the birth of Big Bang's children, that is, hydrogen atoms, by emitting of cosmic microwave background that contained a sophisticated problem to solve for future generations of researchers. After 13.7 billion years, this splendid problem was solved; in 1965, Arno Penzias and Robert Woodrow Wilson from Bell Telephone Laboratories registered the cosmic background and were awarded the Nobel Prize in physics in 1978.

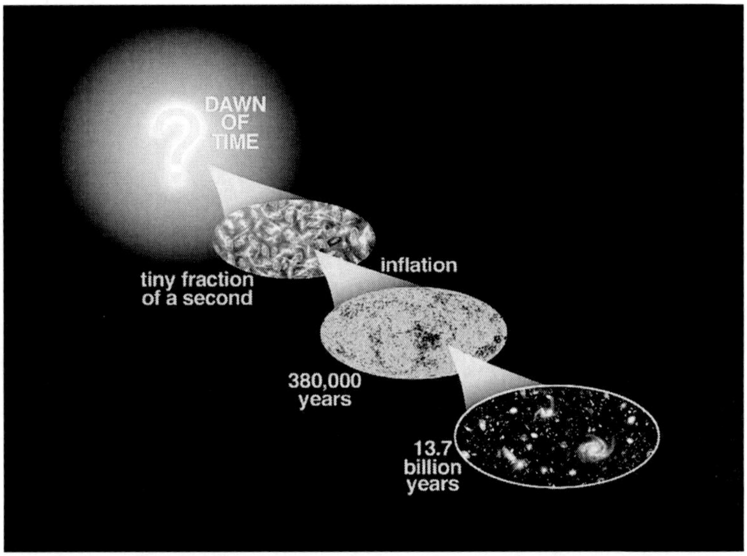

Figure 1.1 Universe evolution after the Big Bang. (Image: NASA/WMAP Science Team) [6].

Meanwhile the decoupling of substance and radiation made it possible to observe another microwave irradiation which is called "hydrogen line" [7]. This radiation is related directly to hydrogen atoms in the universe and we will consider it when elucidating the hydrogen atom electron structure.

After appearance of hydrogen atoms, substance then continued to aggregate into the first stars and ultimately galaxies, quasars, clusters of galaxies, and superclusters. Hydrogen dominates in our universe being the most abundant element in the cosmos, of which it makes up 88.6% of the composition by weight and 93% of all atoms in the universe; stars consisting predominantly of hydrogen in their plasma state [8]. The chemical evolution of the stars is crucially determined upon the hydrogen fusion.

In the interstellar space hydrogen exists as separate molecules, atoms, and ions and can form molecular clouds essentially different in size, density, and temperature.

Thus, hydrogen atom is the direct and closest ancestor of the Big Bang; it being one of the oldest creations in the universe and this is a right time to consider its structure.

1.3 Hydrogen Atom Electron Structure and Hydrogen Line

Hydrogen atom has a simplest electron structure consisting of a single proton orbited by a single electron.

Hydrogen atom is a quantum particle. The electron may orbit around the proton only on stationary orbits; a transition between the orbits is accompanied either by absorption or emittance of a photon. For this reason a hydrogen atom has its characteristic absorption spectrum where several bands appear due to electron transition under the action of photons between different energy levels corresponding to the electron orbits. These bands are named after the scientists that discovered them. Possible electron transitions are depicted in Fig. 1.2.

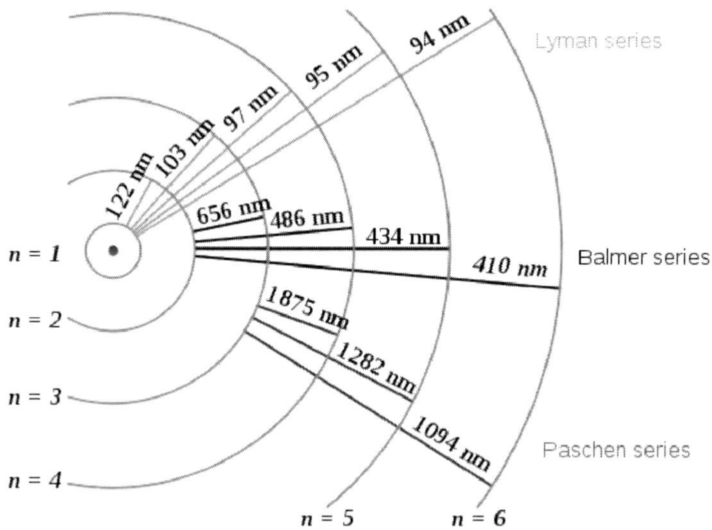

Figure 1.2 Electron transitions and their resulting wavelengths for hydrogen. Energy levels are not to scale (Image: A_hidrogen_szinkepei.jpg) [7].

1.4 Origin of Neutral Hydrogen Emission

Like their orbital motion, the protons and electrons also have spin. They may have their spin oriented in the same or in the opposite direction. Because of magnetic interactions between the particles, a hydrogen atom that has the spins of the electron and proton aligned in the same direction (parallel) has slightly more energy than one where the spins of the electron and proton are in opposite directions (antiparallel). The fact that the lowest-energy configuration arises in the antiparallel spin configuration is an inherently quantum-mechanical result [7].

The proton has a nuclear spin $I = 1/2$. The associated magnetic dipole moment interacts with the spin of the electron via a magnetic coupling. This interaction gives rise to a splitting of the ground state of hydrogen $n = 0$, $l = 0$, $j = 1/2$ into a $F = 1$ level and a $F = 0$ level, F being the total angular momentum including the nuclear spin. It can be shown that the splitting, named the hyperfine splitting, is 1420 MHz. (Fig. 1.3)

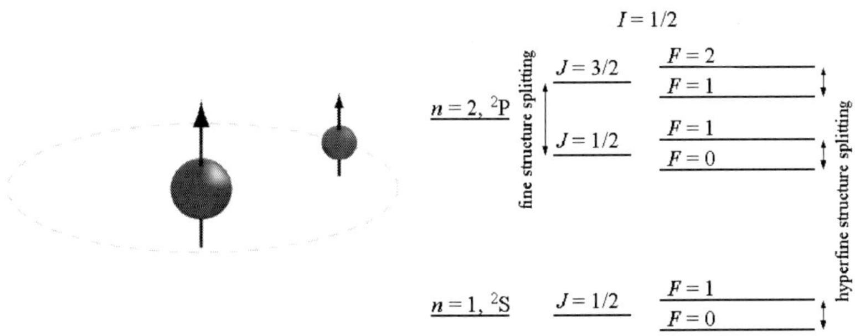

Figure 1.3 (Left) An electron orbiting a proton with parallel spins (pictured) has higher energy than if the spins were antiparallel (Image: JabberWok) [9]. (Right) Fine and hyperfine structure in hydrogen. The hyperfine splitting of the ground ^2S state is the source of the 21 cm hydrogen line. (Image: DJIndica) [10].

The lowest orbital energy state of atomic hydrogen has hyperfine splitting arising from the spins of the proton and electron changing from a parallel to antiparallel configuration. This transition is highly forbidden with an extremely small probability of 2.9×10^{-15} s^{-1}. This means that the time for a single isolated atom of neutral hydrogen to undergo this transition is around 10 million (10^7) years and so is unlikely to be seen in a laboratory on the earth.

However, as the total number of atoms of neutral hydrogen in the interstellar medium is very large, this emission line is easily observed by radio telescopes. Also, the lifetime can be considerably shortened by collisions with other hydrogen atoms and interaction with the cosmic microwave background already mentioned.

It is a rare event for H atoms in the interstellar medium to switch from the parallel to the antiparallel configuration, but when it happens radio waves are emitted with a wavelength of 21 cm and a corresponding frequency of 1420 MHz. The abundance of H atoms in the universe is so high that the 21 cm spectral feature is readily observable with radio telescopes. Sky surveys of the 21 cm line can be made, hence the hydrogen in the universe can be mapped out.

The line has an extremely small natural width because of its long lifetime, so most broadening is due to Doppler shifts caused by the motion of the emitting regions relative to the observer. It is believed that this frequency should act as a universal marker for other technologically advanced civilizations to transmit at 1420 MHz. One of the sky surveys obtained with the help of radio telescope is presented in Fig. 1.4.

This giant "spur" is interstellar gas and dust emerging from disk of our Milky Way. Seen against a backdrop of distant quasars, Lynds Bright Nebula 679 is over 200 light years away and apparently driven out of the disk of the galaxy by giant interstellar shock waves. Lying in the constellation Perseus, the spur is over 2° in angular extent, about four moon diameters. At the surface of the spur lies an interface between the cool and hot matter forming the diffuse interstellar medium. On the left side, facing the plane of the galaxy the material in the spur consists of cold hydrogen gas and warm dust. On the side facing away from the plane of galaxy the spur is being evaporated by ultraviolet radiation. The hot gas being driven off the spur can be seen as a faint green halo arising from hot, fluorescing hydrogen gas.

This map displays the stuff between stars in the Perseus Spiral Arm (6500 light years distant) of our Milky Way by assigning different colors to radiation not detected by the human eye. Invisible to optical telescopes, hydrogen gas can only be revealed by observations of faint radio waves emitted at a wavelength of 21 cm [11].

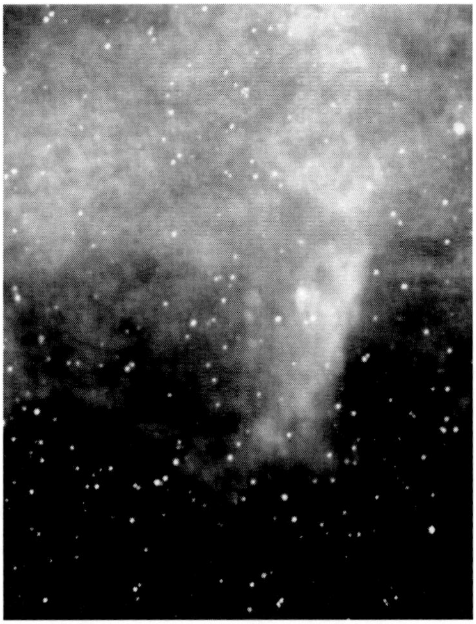

Figure 1.4 Interstellar gas clouds collide: Canadian Galactic Plane Survey maps a bright "spur" in our Milky Way. Reproduced with permission from Jayanne English (University of Manitoba), using data acquired by the Canadian Galactic Plane Survey (NRC/NSERC) and the Infrared Astronomical Satellite, and produced with the support of Russ Taylor (University of Calgary) [11].

1.5 Unique Properties of Hydrogen Species

The origin of hydrogen was covered in the previous section and now the properties of hydrogen species will be elucidated, especially those properties that discriminate hydrogen from all other elements of the periodic table of elements and predetermine its functional abilities in many reactions and processes.

Hydrogen has three naturally occurring isotopes, protium, deuterium, and tritium denoted as ^{1}H, ^{2}H, and ^{3}H; protium and deuterium being stable. Deuterium contains one proton and one neutron in its nucleus. Essentially all deuterium in the universe is thought to have been produced at the time of the Big Bang, and has endured since that time. Deuterium is not radioactive, and does not represent a significant toxicity hazard. Water enriched in molecules that include deuterium instead of normal hydrogen is called "heavy water". Deuterium and its compounds are used as a non-radioactive label in chemical experiments and in solvents for ^{1}H-nuclear magnetic spectroscopy. Heavy water is used as a neutron moderator and coolant for nuclear reactors. Deuterium is also a potential fuel for commercial nuclear fusion.

^{3}H is known as tritium and contains one proton and two neutrons in its nucleus. It is radioactive, decaying into helium-3 through beta decay with a half-life of 12.32 years.

$$^3_1T \rightarrow {}^3_2He^{+} + e^{-} + \nu_e \qquad (1.1)$$

Hydrogen is the lightest chemical element with atomic number 1 (the first element in the periodic system of elements); an average atomic mass being of 1.00794 u (1.007825 u for the most common isotope protium H^0). Atomic H^0 has the smallest covalent radius, ~0.37 Å, among all of the atoms of the chemical elements. Hydrogen is nature's lightest and smallest atom: its mass and size are the smallest among all chemical elements [8]. This is a very important property which predetermines mobility of hydrogen atoms and more than this, as we will see later, under definite conditions the low mass gives a chance for hydrogen to participate in tunneling reactions.

The simplicity of hydrogen yields fantastic chemical and physical properties of the element and its species. Being strongly inclined to form either a totally empty or a totally filled 1s electron shell, hydrogen exhibits three common oxidation states: +1, 0, and –1 (reflecting a $1s^0$, $1s^1$, or $1s^2$ electronic configuration, respectively). The relative change in the number of electrons surrounding the nucleus associated with the transformations $H^0 \rightarrow H^-$ and $H^0 \rightarrow H^+$ comprises ±1e, or ±100%; this is thus the largest possible change among the chemical elements of the periodic table, which in turn provides the relative changes of many key chemical and physical properties of these three unitary species [8].

The most common isotope of hydrogen is protium with a single proton thus without neutrons; its abundance being of more than 99.98%. Atomic hydrogen (H^0) is a highly unstable arrangement for it often occurs energetically beneficial either to fill completely its 1s shell or to split off the electron turning into H^- or H^+ ion, correspondingly. Thus, being in contact with other atoms or molecules, it is readily ionized. The bare proton H^+ has a radius close to 10^{-5} Å, incredibly low value, especially if to compare with another single-charged ions where this value is about 1 Å.

In case of losing its sole electron (what happens by chemical reactions) hydrogen atom turns into a tiny proton that immediately strives to acquire aggressively electron density from other atoms and molecules perturbing the electron density map.

With incredibly small size and the total absence of an electron shell a proton provides very strong polarization of the medium around itself. Being a particle with low steric restrictions, it is accommodated readily in numerous mediums, and is called "ubiquitous" because of this property.

For this reason a wide variety of phenomena might be considered as proton transfer reactions that are thought to be simple since they represent the motion of a tiny nucleus deprived of the electron shell [12]. The peculiarities of the proton transfer reactions are that they are carried out without essential reconstruction of the bonding electrons system and without participation of repulsion forces between non-bonding electrons.

However, some reactions include hydrogen atom transfer and not proton transfer, but those reactions are usually carried out under the action of irradiation or bombardment by high-energy particles [12]. Furthermore this along with a small mass is revealed in mobility of hydrogen in many solids and liquids. The proton is an unusually strong Lewis acid which has a great affinity to accept an electron pair. For this reason, the bare proton is never found in solution or inside of solids, as it always spontaneously extracts a degree of electron density from (even very poor) Lewis acid, still remaining a moderately strong Lewis acid even after partial salvation [8].

At the same time having acquired some electron density a proton nevertheless has an effective positive charge being capable of reacting upon the action of electric field. It participates facilely in the process of the charge separation which makes it an outstanding energy element and an inalienable partner of an electron in all energy cycles. The proton property of being hard electron density acceptor makes hydrogen-containing compounds very reactive: energy required to break a bond with a hydrogen atom usually being compensated due to possible strong polarization of the surrounding medium by the detached proton which in turn enhances the reaction probability.

A proton is also a unique partner for an electron in the processes which need charge separation that is necessary in many photochemical reactions where excitation yielded by photon is stabilized via the exchange of the photogenerated hole for a proton. Hydrogen abstraction reactions are good examples of bio-energy conversion: the solar energy is transformed into chemical energy. These processes will be covered in the next chapters.

It is customary to suppose that nuclear mass does not play the main role in chemical reactions since the interaction forces are determined rather by electron and nuclear charges than by masses. However there are a lot of cases where the light weight of the proton has become an important factor in reactions.

Hydrogen forms strong covalent bonds with many elements. The combination of the small mass along with high force constants yields the values of the vibration frequencies for these bonds, which are much higher than those for all other types of the bonds. In quantum physics, this means high energy for a vibration quantum and the zero-point vibration energy (½ hν).

For a long time the tunneling of the atomic particles was considered as a rare and exotic phenomenon. However, it was shown in [13] that tunneling of hydrogen species must be often taken into account. From very general considerations it is obvious that in order to reveal its wave properties, an atomic particle with mass m and velocity V (so, with the energy $E = 0.5\ \mathrm{mV}^2$) must have the de Broglie wavelength λ comparable with the width of the energy barrier d. Thus:

$$\lambda = \hbar / \sqrt{2mE}\ , \tag{1.2}$$

where $\hbar$ is the Planck constant divided by 2π.

At the same time, to make the classical over-barrier transitions not as fast as to diminish all possibilities to observe the tunneling, the barrier height E must be substantially higher than $k_b T$, where k is the Boltzmann constant and T the temperature.

So, to achieve the prevalence of the tunneling transitions of a particle through the energy barrier against the Arrhenius transitions over the barrier it is necessary to fulfill the inequality:

$$\lambda/d > k_b T/E \tag{1.3}$$

Thus,

$$T < \hbar / k_b \sqrt{2} d \sqrt{E/m} \tag{1.4}$$

So, the dominance of tunneling transitions against the Arrhenius ones is possible at temperatures lower than a certain temperature T_t, which is called "tunneling temperature". At temperatures lower than T_t the tunneling transitions prevail over the Arrhenius ones whereas at temperatures higher than T_t there is the opposite correlation. Goldanskii derived the criterion for the tunneling temperature as the highest limit for the exponential prevalence of the tunneling transitions over the Arrhenius ones as [13]:

$$T_t = (\sqrt{2}\hbar / k_b d)\sqrt{E/m} \quad (1.5)$$

With the typical chemical reaction values E = 1 eV and d = 1 Å, one gets the value T_t = 320 K for proton and T_t = 240 K for deuterium. Since these temperatures are rather high, one can see that it is always necessary to take into account the possibility for tunneling when dealing with hydrogen: the tunneling is not an exotic phenomenon for the hydrogen processes.

Hydrogen abstraction reactions often contain the tunneling channel, which makes their investigations especially exciting. The same could be said for quantum diffusion. Tunneling makes it possible to provide the research at very low temperatures; the isotope exchange of protium for heavier hydrogen isotope deuterium is a powerful tool in the research since the mass of the tunneling particle is changed twice and this is the highest possible change of mass via isotope exchange, if not to consider the exchange the other hydrogen species, namely, of protium to tritium.

Free H^- has a substantially larger Pauling ionic radius of 2.08 Å, comparable to those of highly polarizable bromide (1.95 Å) and iodide (2.16 Å) anions. Indeed, the static electric dipole polarizability (R) of free H^- (10.17 Å^3) is even larger than that of I^{-1} (7.16 Å^3) [14]. Electron density is relatively loosely bound to the nucleus in the hydride anion, and H^- is recognized as an avid electron donating species, a soft but strong Lewis base, and a potent reducing agent [8].

The vast difference in chemical properties between H^+ and H^- is further emphasized and explained by the values of the first ionization energy, I_P = 13.6 eV, and electron affinity, E_A = 0.75 eV, of an H atom (see Fig. 1.5). The I_P of H is smaller than that of Kr (14 eV) and larger than that of Cl (12.97 eV) [8]. It is thus characteristic of prototypical nonmetals. Interestingly, however, the E_A of H is much less than those of nonmetals and is rather close to those of typical metals (for Li, E_A = 0.61 eV). The result is a fascinating duality in the physical and chemical properties of H: it behaves as a hard cation toward nonmetals and as a (usually) soft anion toward metals. The very large energy difference, some 14.35 eV, between H^+ and H^- in a vacuum is thus among the largest values of the (I_P + E_A) parameter of all chemical elements if the common oxidation states are considered. And the vast difference in the chemical properties between H^+, a hard electron density acceptor, and H^-, a soft electron density donor, is probably the largest absolute and relative variation of properties among the (n – 1) and (n + 1) oxidation states for any element available to chemistry [8].

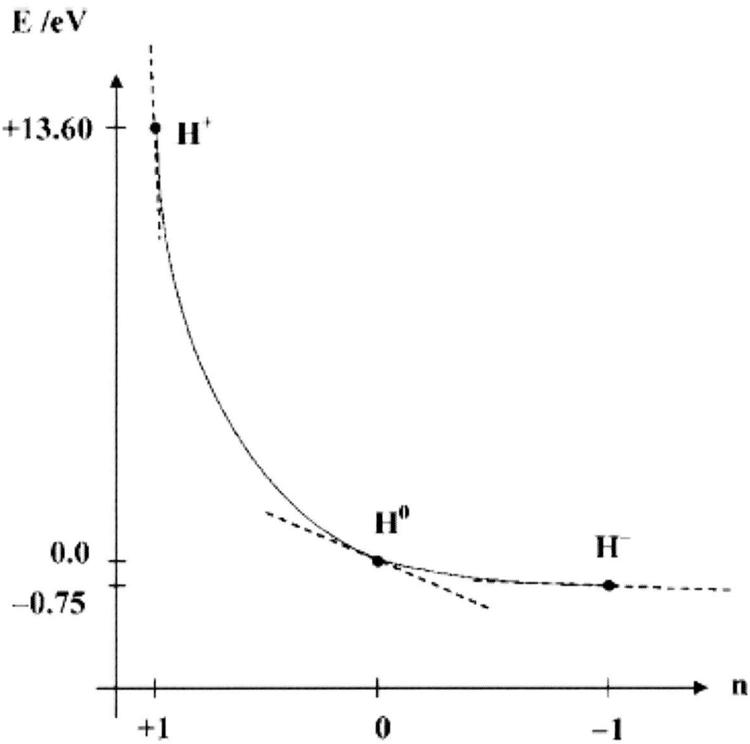

Figure 1.5 Dependence of the electronic energy of the H_n species on the oxidation state of hydrogen, n. The hardness (the derivative of energy on the electron density) is shown with the dotted lines. Reprinted with permission from Ref. [8]. Copyright (2004) American Chemical Society.

Summarizing the above, it is possible to make very important conclusions concerning the unique properties of hydrogen as a chemical element, which make it the main player in various processes related to conversion and transport of energy:

(i) Hydrogen is the smallest particle.
(ii) Hydrogen is the lightest particle.
(iii) Hydrogen has a poor electron shell that it loses easily and turns into a tiny proton with a positive charge.

These properties make hydrogen very suitable for energy conversion. The energy of light or chemical energy can be converted directly into the electrical energy or the energy of charge separation and then further to the chemical energy, that is, the energy of the rupture or creation of chemical bonds. Hydrogen takes active place in all energy cycles on the earth that will be considered below.

Now it is the time to consider other facets of hydrogen that attract the interest of researchers.

1.6 The Properties of Molecular Hydrogen

Pure hydrogen in nature exists as a H_2 molecule. Molecular hydrogen H_2 is a highly combustible diatomic gas. The enthalpy of the H–H bond in H_2 (436 kJmol^{-1} or 4.52 eV) is huge, much larger than the respective values for the molecules Li_2 (110.2 kJ mol^{-1}, 1.14 eV) and Cl_2 (242.6 kJmol^{-1}, 2.52 eV), and rather close to that of the strongly bound O_2 molecule (498.4 kJmol^{-1}, 5.17 eV) [8]:

$$2H_{2(gas)} + O_{2(gas)} \rightarrow 2H_2O_{(liquid)} + 572\ \text{kJ} \quad (1.6)$$

(286 kJmol^{-1} per one hydrogen atom)

Equation 1.6 clearly demonstrates why hydrogen has undisputable backgrounds for being a "dream" material for energy production: first, combustion of hydrogen yields a great amount of energy; second, the reaction product is water which is environment friendly. Hydrogen has the highest combustion energy release per unit weight compared to any commonly occurring material. This property makes it the fuel of choice for upper stages of multi-stage rockets [15].

However, it is not easy to organize the hydrogen fuel infrastructure. One should overcome numerous obstacles to create one that is convenient and cheap. These problems will be discussed in detail in the chapter on hydrogen storage.

The work with hydrogen gas requires special caution: hydrogen gas forms explosive mixtures with air in the concentration range 4–74% (volume per cent of hydrogen in air). The mixtures spontaneously detonate by spark, heat, or sunlight.

As in the case of atomic hydrogen H^0 there exist two different spin isomers of hydrogen diatomic molecules (H_2) that differ by the total spin of their nuclei (Fig. 1.6). In the orthohydrogen form, the spins of the two protons are parallel and form a triplet state with a molecular spin quantum number of 1 (½ + ½); in the parahydrogen form the spins are antiparallel and form a singlet with a molecular spin quantum number of 0 (½ – ½) (Fig. 1.6). At standard temperature and pressure, hydrogen gas contains ~25% of the para form and 75% of the ortho form, also known as the "normal form" [16]. The equilibrium ratio of orthohydrogen to parahydrogen depends on temperature; the ortho form being an excited state having a higher energy than the para form. At very low temperatures, the equilibrium state is composed almost exclusively of the para form. The uncatalyzed interconversion between para and ortho H_2 increases with an increase in temperature; therefore, rapidly condensed H_2 contains large quantities of high-energy ortho form that converts to the para form very slowly [17]. The ortho/para ratio in condensed H_2 is an important consideration in the preparation and storage of liquid hydrogen: the conversion from ortho to para is exothermic and produces enough heat to evaporate some of the hydrogen liquid, leading to loss of liquefied material. Catalysts for the ortho–para interconversion, such as ferric oxide, activated carbon, platinized asbestos, rare earth metals, uranium compounds, chromic oxide, or some nickel [18] compounds, are used during hydrogen cooling [19].

The liquid and gas phase thermal properties of pure parahydrogen differ significantly from those of the normal form because of differences in rotational heat capacities [20]. The ortho/para distinction also occurs in other hydrogen-containing molecules or functional groups, such as water and methylene, but is of little significance for their thermal properties [21].

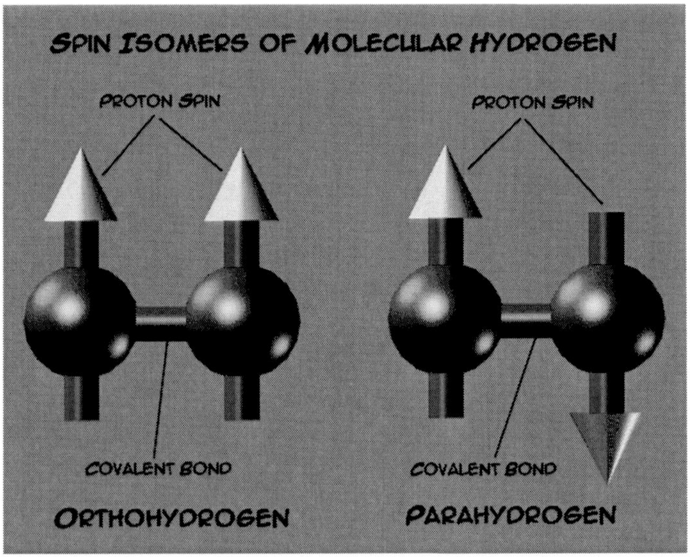

Figure 1.6 Spin isomers of molecular hydrogen (Image: Jim Farris) [22].

1.7 Hydrogen and Life

Huge amounts of life-giving solar rays deliver energy to the earth in the form of heat and light to our plant world via the remarkable process of photosynthesis. It is worthwhile to stress that in processes related to creation and adoption of light energy hydrogen plays an outstanding role. First, the light energy from the sun's irradiation is released by the thermonuclear reaction taking place in the sun whereas hydrogen is a fuel for that. Second, as we will see later, hydrogen is an extremely important participant in the photosynthesis that adopts the photon energy from the sun in living organisms, namely, plants and photosynthetic bacteria.

What a unique combination: hydrogen acts in cosmos to provide the energy for the earth and, at the same time it plays the main role in converting the sun's energy into chemical energy on the earth via photosynthesis! Indeed, hydrogen is an uncanny element.

Hydrogen is pivotal to life; it would be different to find another element capable of competing for this priority. H_2 is a product of some types of anaerobic metabolism and is produced by several microorganisms, usually via reactions catalyzed by iron- or nickel-containing enzymes called hydrogenases. These enzymes catalyze the reversible redox reaction between H_2 and its components two protons and two electrons. Creation of hydrogen gas occurs in the transfer of reducing equivalents produced during pyruvate fermentation into water (see Chapter 4).

Water cleavage, in which water is decomposed into its component protons, electrons, and oxygen, occurs in the light reactions taking place in all photosynthetic organisms. Some such organisms, including the alga *Chlamydomonas reinhardtii* and cyanobacteria, have evolved a second step in the dark reactions in which protons and electrons are reduced to form H_2 gas by specialized hydrogenases in the chloroplast. Efforts are being made to genetically modify cyanobacterial hydrogenases to efficiently synthesize H_2 gas even in the presence of oxygen. Efforts have also been made to genetically modified alga in a bioreactor (see Chapter 6).

1.8 Hydrogen in Chemistry

On the other hand, hydrogen is also one of the most important elements in the chemistry of energy materials, such as hydrocarbons (mineral oil and methane gas).

As it was already said, the contemporary challenges such as continuous fuel crises, the increasing pollution of our environment, and the problems of environment protection against greenhouse effect have been the driving forces for the utilization of clean and renewable energy resources. Hydrogen is undoubtedly, if not the most important, one of the main candidates to substitute petroleum products as a clean energy carrier for both transportation and stationary applications

Hydrogen forms compounds with most elements and is present in water and most organic compounds. It plays an important role particularly in the acid–base chemistry, with many reactions exchanging protons between soluble molecules.

Large quantities of H_2 are needed in the petroleum and chemical industries, especially for the processing of fossil fuels and in the production of ammonia. H_2 is consumed in great amounts in such widely used chemical technologies as hydrodealkylation, hydrodesulfurization, and hydrocracking. It is used also as a hydrogenating agent to increase the level of saturation of unsaturated fats and oils and in the production of methanol, and in the manufacture of hydrochloric acid. H_2 is also used as a reducing agent for metallic ores.

1.9 Hydrogen in Physics and Engineering

Apart from its use as a reactant, H_2 has wide applications in physics and engineering. It is used as a shielding gas in welding methods, such as atomic hydrogen welding. H_2 is used as the rotor coolant in electrical generators at power stations, because it has the highest thermal conductivity of any gas. Liquid H_2 is used in cryogenic research, including superconductivity studies. Because of its very low density H_2 can be used as a lifting gas in balloons and airships.

Hydrogen is used pure or mixed with nitrogen (sometimes called forming gas) as a tracer gas for minute leak detection. Its applications can be found in the automotive, chemical, power generation, aerospace, and telecommunications industries. Hydrogen is an authorized food additive (E 949) that allows leak testing in food packages, among other antioxidizing properties [23].

Due to their small size, hydrogen atoms can be accommodated in great quantities in the transition metal oxides, which yield formation of hydrogen metal bronzes with general formula $H_xMe_mO_n$, where Me the transition metal. In hydrogen tungsten bronzes H_xWO_3 x may reach the value of 0.5, whereas in hydrogen molybdenum bronzes H_xMoO_3 and in hydrogen vanadium bronzes $H_xVO_{2.5}$; x may reach the value close to 2. Furthermore, hydrogen doping can occur in these materials at room or even at very low temperatures, and under the action either of electric field or light. Hydrogen atoms implanted into the oxide structure donate their electrons to the host; released protons remain localized in the vicinity of the nearest oxygen anions, from where they extract some degree of electron density. Thus, the inserted hydrogen atoms are turned into the electron–proton plasma. Such in situ doping provides great changes in the various parameters of the oxide. It gives a possibility to achieve and observe in detail the semiconductor–metal transition in situ in the WO_3 films. The hydrogen doping in situ is the background for the electrochromism and photochromism in the transition metal oxides since it yields great changes in the optical parameters of the latter [24,25]. Furthermore the electron–proton plasma can be transported to other materials that are in contact with the transition metal oxide which creates the intriguing possibilities for the engineering of the various energy conversion systems.

Hydrogen is employed to saturate broken ("dangling") bonds of amorphous silicon and amorphous carbon that helps in stabilizing material properties. It is also a potential electron donor in various oxide materials.

Very often hydrogen ions have high diffusivity not only in liquid solutions, but also in solids, where there is a special class of materials, namely, superionic conductors or simply superionics. From this point of view it is interesting to observe the diffusion of the protons in the ice lattice, where it has specific features inherent only to the protons.

Hydrogen is highly soluble in many rare earth and transition metals and is soluble in both nanocrystalline and amorphous metals. Hydrogen solubility in metals is influenced by local distortions or impurities in the crystal lattice. These properties may be useful when hydrogen is purified by passage through hot palladium disks, but the gas serves as a metallurgical problem as hydrogen solubility contributes in an unwanted way to embrittle many metals, complicating the design of pipelines and storage tanks.

1.10 Applications of Hydrogen Isotopes

Hydrogen's rarer isotopes also have specific applications each. Deuterium is used in nuclear fission applications as a moderator to slow neutrons and in nuclear fusion reactions. Deuterium compounds have applications in chemistry and biology, for example, in studies of reaction isotope effects. Tritium, produced in nuclear reactors, is used in the production of hydrogen bombs, as an isotopic label in biosciences, and as a radiation source in luminous paints.

The radioactivity of tritium can be used for many purposes, for example, for investigations of hydrogen diffusion.

Small amounts of tritium occur naturally because of the interaction of cosmic rays with atmospheric gases; tritium has also been released during nuclear weapons tests. It is used in nuclear fusion reactions, as a tracer in isotope geochemistry, and especially in self-powered lighting devices. Tritium has also been used in chemical and biological labeling experiments as a radiolabel.

Thus the atomic mass ratio for the isotopes is 1:2:3. The big atomic mass ratio makes it possible to investigate tunneling processes via isotope exchange.

1.11 Conclusions

In this chapter, we made our very first acquaintance with the basic properties of hydrogen. The unique properties that isolate this element from others are its small size, mass, and the ease with which it loses its poor electronic shell. All these properties predetermine its exclusive reactive ability. The excurse made covered far less functions and possibilities of hydrogen, but is still enough to feel its exclusive importance for the universe and human beings. We live in the world of "hydrogenicity", where the motion and action of hydrogen determines our life in various numerous aspects.

Chapter 2

Hydrogen Bond

2.1 What Is a Hydrogen Bond?

Because of its exceptional qualities, especially the tiny size, the hydrogen atom is able to form a fine pattern of an intermolecular interaction called a hydrogen bond, that is, the interaction of a hydrogen atom with an electronegative atom having a lone electron pair, such as F, O, N, Cl, and S, that belongs to another molecule or chemical group.

If there is a molecule containing a common covalent –X-H bond, where X is an atom with a high electronegativity (at least higher than hydrogen) then the hydrogen atom can form another bond with another electronegative atom Y belonging to another molecule or chemical group. Usually this bond is depicted by a stroke line as –X-H···Y-Z where Z is another molecule or its fragment.

It makes sense to present here an abstract from the current definition of the hydrogen bond officially adopted by the International Union of Pure and Applied Chemistry (IUPAC) that claims that the hydrogen bond is "a form of association between an electronegative atom and a hydrogen atom attached to a second, relatively electronegative atom. It is best considered as an electrostatic interaction, heightened by the small size of hydrogen, which permits proximity of the interacting dipoles or charges."

Hydrogen Energy for Beginners
Alexander Gavrilyuk

ISBN 978-981-4364-77-5 (Hardcover), 978-981-4364-78-2 (eBook)
www.panstanford.com

One can see that even in this official definition the unique property of hydrogen is underlined, namely, its small size that in turn provides the proximity of interacting dipoles.

The energy for the typical hydrogen bonds ranges between 0.13–0.31 eV [26]. A classical example is the formation of hydrogen bonds between two water molecules (see Fig. 2.1), which results in the formation of a water dimer $(H_2O)_2$ with bond energy of 0.2 eV [26]. The hydrogen bond is depicted by the stroke line. If there are two water molecules one of the hydrogen would tend to interact with both oxygens (Fig. 2.1); as a result a new compound is formed where the dots denote a new type of intermolecular interaction namely hydrogen bond. The hydrogen bond is related to weak bonds, since its energy contains approximately one twentieth of the covalent O–H bond energy. The distance between two oxygen atoms in the complex is approximately equal to 2.76 Å which is less than the sum of the van der Waals radii of oxygen atoms equal to 3.06 Å [26]. The van der Waals radius, r_w, of an atom is the radius of an imaginary hard sphere which can be used to model the atom for many purposes.

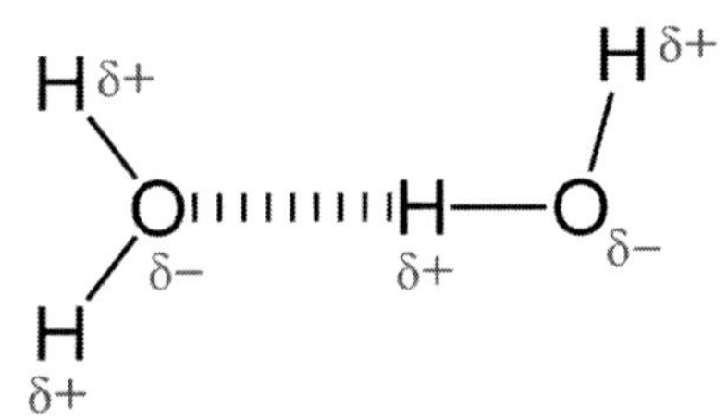

Figure 2.1 Formation of a water dimer via the hydrogen bond.

Since the covalent bond of the hydrogen atom is saturated, the lone electron pair is donated by the atom Y to form the hydrogen bond with it; thus the hydrogen bond being a case of donor–acceptor or coordination bond.

As it was mentioned in Chapter 1, a hydrogen atom is always inclined to play the role of a Lewis acid, thus of an acceptor of electron density. For this reason the hydrogen bond is more intensive if the hydrogen atom is bonded by the covalent bond to another atom Y with high electronegativity (e.g., oxygen or fluorine); then the hydrogen atom acquires partial positive charge that enhances its Lewis acidity. Due to this fact the electron density on the hydrogen atom is reduced as compared with the electron density of an isolated hydrogen atom and the effective charge of such an atom occurs to be positive which makes it possible for the hydrogen atom to interact with one more electronegative atom; this interaction, as a rule, being weaker as that with the first atom is achieved with the help of lone electron pairs. Electron pairs are considered lone pairs if two electrons are paired but are not used in chemical bonding. They are found in the outermost electron shell of an atom. An electronegative atom forms a hydrogen bond with the hydrogen atom by donating its lone electron couple that does not take part in the formation of valent chemical bonds and is suitable to form a hydrogen bond. The free, lone, or unshared electron pair attracts the proton and makes a hydrogen bond with it.

2.2 Nature of the Hydrogen Bond

For many decades the Coulomb attraction (dipole–dipole interaction) between the hydrogen atom and the lone electron pair of an electronegative atom was considered as the only reason for the formation of the hydrogen bond. The hydrogen originality (its tiny size and poor electron density) makes it possible to achieve a high proximity between the hydrogen and the electronegative atom, enhancing the Coulomb interaction, which discriminates both hydrogen from all other atoms and its non-valent interaction (the hydrogen bond) with another atoms.

In the investigation of the hydrogen bond the pure electrostatic point of view was dominating for a long time. What concerns the possibility for the hydrogen atom to interact also with a third atom is that it looks excluded, at the first glance, due to electrostatic repulsion between the electronegative atoms.

According to modern understanding it has become clear that the prescription of the hydrogen bond only to the Coulomb interactions is oversimplified and the important reason for the formation of the hydrogen bond is assumed to be the redistribution of the electron density between atoms X and Y in the chain –X-H···Y-Z (Fig. 2.1) caused due to the proton whereas other electrons of the molecules also take part in the formation of the potential curves for the hydrogen bond although in a less degree.

The quantum calculations performed in [27] showed the change of the electron density distribution around all atoms of the reacting water molecules in the dimer shown in Fig. 2.1. However, the calculations showed that the effective charge of the hydrogen atom which takes part in the formation of the hydrogen bond practically doesn't change as compared with that in the isolated molecule which demonstrates the importance of polarization and redistribution of the electronic density in the space surrounding the molecules. At present time it became clear that the interpretation of the hydrogen bond should be provided with the help of molecular orbital theory.

2.3 The Experimental Crux of the Nature of Hydrogen Bond

Experimenters have confirmed the brilliant idea first proposed by Linus Pauling in the 1930s that the rules of quantum mechanics force the hydrogen bonds between H_2O molecules in ice to get part of their identity from strong covalent bonds within the H_2O molecule. Hydrogen bonds acquire their double identity via minimization of the energy of the electrons forming the hydrogen bond. Electrons strive to get the possible lowest-energy state tending to minimize their total energy including the kinetic energy. However, reduction of an electron's kinetic energy is possible by minimizing its velocity and momentum by spreading out the electron wavefunctions in space, according to the Heisenberg Uncertainty Principle. Thus the overlap of the wavefunctions for the covalent and the hydrogen bonds occurs: the electrons become somewhat indistinguishable and the hydrogen bonds cannot be completely described as electrostatic bonds. Instead, the electrons of the hydrogen bond take on some of the properties of the highly covalent sigma bonds, and vice versa [28].

However, the extent to which hydrogen bonds were being affected by the sigma bonds has never been directly proved by experiment—until the fine experiment described below was carried out. This issue really deserves a detailed description.

A team of scientists used the European Synchrotron Radiation Facility in Grenoble for investigations of the Compton scattering that occurred when the X-ray photons interacted with ordinary ice [28]. This scattering results from interaction of X-ray photon with electrons of a material. The photon transfers a part of its kinetic energy to the electrons being scattered from the material with lower energy and a different direction. Investigation of the properties of many Compton-scattered photons provides comprehensive data concerning the properties of the electrons in a material.

Measuring of the scattered-photon energy losses and its direction, it is possible to evaluate the momentum it transfers to the electrons in a material and to get the information about the original momentum state of the electron previous to the interaction with the X-ray photon, which in turn makes it possible to reconstruct the electron's "ground-state wavefunction" that describes the lowest energy state of an electron [29].

The main idea of the experiment was to observe the overlapping of the electron waves in the sigma and hydrogen bonding sites although it was clear that this effect was very subtle to detect. To facile the detection, solid ice was used instead of liquid water. In ice hydrogen bonds are pointing in only four different directions due to ordering of the H_2O molecules in a regularly repeating structure whereas in liquid water the hydrogen bonds are pointing in all different directions at any given moment.

But even in that case the conditions for the detection of the effect were not very promising, since only a tenth of all electron in the ice take part in formation either the sigma bonds or hydrogen bonds. The Compton scattering yields information on the contributions from all the electrons in ice, which creates an unfavorable situation for the detection of the effect hunted.

However, the use of the technique that produced very intensive beams of X-ray photons which created enough Compton-scattering events to ensure a full statistical analysis made it possible to determine the effect. Furthermore, measuring the differences in the scattering intensity from different angles yielded the possibility to get rid of the "ballast" contributions from electrons nonparticipating in the bonds.

Figure 2.2 shows the plot of the intensity of the scattered x rays against their momentum. One can observe wavelike fringes due to interference between the electrons on neighboring sigma and hydrogen bonding sites.

This fantastic result demonstrates that electrons in the hydrogen bond are quantum mechanically shared (covalent) which was predicted by Linus Pauling. The researchers found also contributions from more distant bonding sites. It was estimated that the hydrogen bond gets about 10% of its behavior from a covalent sigma bond.

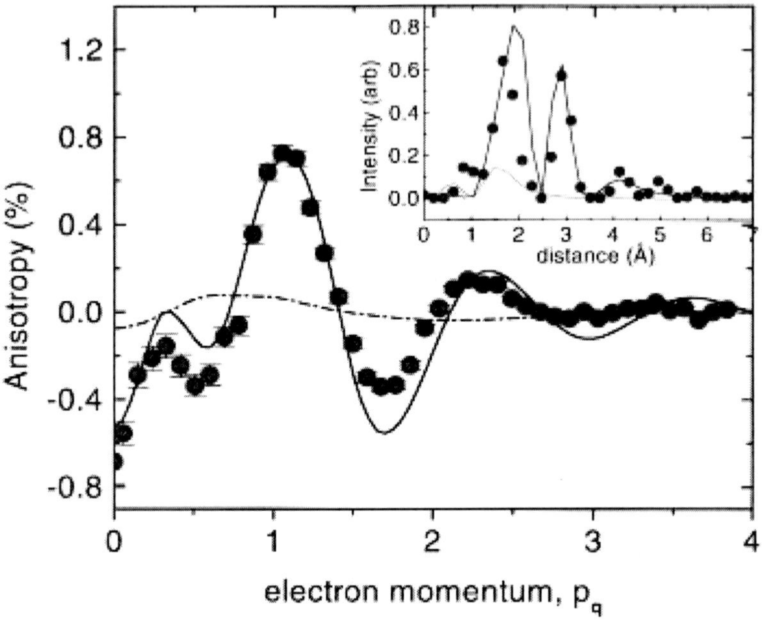

Figure 2.2 Scattering anisotropy against the amount of momentum in the electron scattered in ice; the inset presents the information on the distance between the bonding sites. Reprinted with permission from Ref. [29]. Copyright (1999) by the American Physical Society.

Periodic intensity variations in the anisotropy revealed two distances (see the inset in Fig. 2.2), one of 1.72 Å, near the hydrogen bond length of 1.75 Å, and another at 2.85 Å, close to the nearest-neighbor O–O distance of 2.75 Å [30]. The presence of these two dominant lengths in the Compton profile anisotropy is interpreted as the first direct experimental evidence for the substantial covalent character of the hydrogen bond. Very good quantitative agreement between the data and a fully quantum mechanical bonding model for ice I_h was estimated [31].

The red dots in Fig. 2.2 show the experimental data points along with their error bars, the solid black line shows the fit predicted by an accompanying theory, and the black dots indicate what the data would look like if the electrons on hydrogen bonds were unaffected by the strongly covalent sigma bonds. By performing a mathematical operation known as a Fourier transform on their data, the researchers obtained information on the distances between bonding sites that is presented in the inset of Fig. 2.2.

The intermolecular bonds (hydrogen bonds) are primarily electrostatic in nature, in which the molecules are attracted by means of separated electric charges, but the researchers discovered that the bond is partly covalent in nature, in which electrons are spread out and shared between atoms. The experiment described proved that hydrogen bonds in water posses a significant covalent component. It yielded highly important details concerning water's microscopic properties.

The experiments carried out proved that the hydrogen bond contains both ionic and covalent contributions and there are no arguments to discriminate the hydrogen bond from other intermolecular interactions. For example, the formation of the molecular crystals is accompanied by the intermolecular interactions; the energy of the interactions being calculated for one molecule very often reach the values close to that for the hydrogen bond for molecules in a water dimer. This circumstance could be related to many systems. Also it can no longer be denied that the formation of the –C-H···O-Z hydrogen is possible although this fact was denied by distinguished scientists for many years [32].

There is a special relation of scientists with the hydrogen bond not because of the peculiarities of the interactions but because of the manifestations related mainly to the structural peculiarities of the systems containing the hydrogen bond.

Some of the peculiarities such as small proton mass and low energy of interactions between the molecules forming the hydrogen bond and the changes arising under the deformation in the character of vibrations in the system appear to be very essential. Intent attention to the investigations of the hydrogen bonds is stipulated not by their peculiarities but by their wide abundance, especially in biological systems, and due to the importance of the role they play in biopolymers and vitally important processes.

An enormously important role of the hydrogen bond in biological systems is stipulated first of all by the fact that it determines the secondary protein structure which has importance in all life processes; the base pairs being held in deoxyribonucleic acid (DNA) molecules with the help of the hydrogen bonds, which determines their stable structure in the form of the double spirals, and, finally, the hydrogen bond is responsible for unusual properties of water which is important for the existence of living systems.

These bonds can occur between molecules (intermolecularly), or within different parts of a single molecule (intramolecularly). This type of bond occurs both in inorganic molecules such as water and organic molecules such as DNA.

New knowledge on the nature of the hydrogen bond made it possible to pose the question concerning its redefinition. In the following table there are two definitions of hydrogen bond: the current one adopted by the IUPAC and another proposed by Prof. Desiraju [33], which has good chances to substitute the current one. In the table there are abstracts both from the current and proposed definitions. Considering both definitions it is possible to conclude that the proposed definition is more "liberal" since it does not demand the necessity of an electrostatic interaction.

Current definition	Proposed definition
A form of association between an electronegative atom and a hydrogen atom attached to a second, relatively electronegative atom. It is best considered as an electrostatic interaction, heightened by the small size of hydrogen, which permits proximity of the interacting dipoles or charges.	*The hydrogen bond is an attractive interaction between a hydrogen atom from a molecule or a molecular fragment X–H in which X is more electronegative than H, and an atom or a group of atoms in the same or a different molecule, in which there is evidence of bond formation.*

2.4 Importance of the Hydrogen Bond

In the world scientific literature there is bewildering amount of publications concerning the hydrogen bond, which currently grows at a great rate. Among numerous publications the classical monographs can be recommended [34–38]. Natural scientists from various branches are continuously interested in hydrogen bonding.

The hydrogen bond X-H···Y-Z is an attractive interaction in which an electropositive H atom intercedes between two electronegative species X and Y and brings them closer together. The fundamental importance of the interaction follows from this particular role in molecular association. As it was underlined in [33]:

> "The hydrogen bond is strong enough to hold molecules such as XH and YZ together at normal temperatures and directional enough so that this association is orientationally specific. However, and depending on the nature of X and Y, it can also become weak enough to allow these molecules to come apart and directionless enough to permit loss of orientational specificity in the association of XH and YZ, more like a hydrophobic interaction. This chameleon-like nature of the interaction continues to baffle and excite the researcher".

Accordingly, the hydrogen bond plays a crucial role both in stabilizing static structures and in mediating dynamic processes. This is what accounts for its importance as a structure-defining element in supramolecular chemistry and as a major facilitator of biological reactions. Supramolecular chemistry refers to the area of chemistry beyond the molecules and focuses on the chemical systems made up of a discrete number of assembled molecular subunits or components.

Hydrogen bonding is a fundamental element of chemical structure and reactivity. It is the key to understanding the structure and properties of water, proteins, and the DNA building blocks of life. DNA is a nucleic acid that contains the genetic instructions used in the development and functioning of all known living organisms (with the exception of RNA viruses). Ribonucleic acid, or RNA, is one of the three major macromolecules (along with DNA and proteins) that are essential for all known forms of life.

The DNA segments that carry this genetic information are called genes, but other DNA sequences have structural purposes, or are involved in regulating the use of this genetic information. Along with RNA and proteins, DNA is one of the three major macromolecules that are essential for all known forms of life.

Among the aspects of current interest are its roles in molecular recognition by nucleic acids, proteins, and small molecules. Formation of a hydrogen bond can produce large changes in the kinetics or mechanism of an enzymatic reaction.

The hydrogen bond belongs to the so-called weak interactions. Clear understanding of the exclusive importance of the weak interactions, and especially just of the hydrogen bonds, in the stabilizations of the condensed state for many simple molecular systems (such as water) and what is exclusively important for biopolymers (nucleic acids and proteins) appeared in the last years. However, this weak bond occurred to be extremely important for the vast variety of natural processes. Intermolecular hydrogen bonding is responsible for the high boiling point of water (100°C) compared to other hydrides that have no hydrogen bonds. Intramolecular hydrogen bonding is partly responsible for the secondary, tertiary, and quaternary structures of proteins and nucleic acids. It also plays an important role in the structure of polymers.

The hydrogen bonds let the polymer chains to unite specific three-dimensional structures, which acquire the functional biological activity. On one hand, these structures are stable enough due to a great number of the hydrogen bonds formed, and, on the other hand, they react very sensitively on the changes of the external conditions, for example, on the approach of this or that molecule just due to the fact that these bonds are weak. Rupture of the hydrogen bonds deprives the molecules of their biological functions. The said demonstrates the exclusive importance of the hydrogen bonds in biological process on the molecular level and it is easy to understand the reason behind the keen interest of the researchers who have been investigating the hydrogen bond during the last decades.

2.5 Types of Hydrogen Bond

One of the main geometric features of a hydrogen bond is its symmetry. This symmetry in turn along with the value of an energy barrier determines the predisposition to a proton transfer thus the probability of proton transfer reaction. It is just the time and place to remember a saying of Prof. Bell who said that the formation of the hydrogen bond is the first step in proton transfer [12].

Potential energy curves are used to describe the motion of hydrogen that is attracted to two electronegative atoms [39]. A visual example of the potential energy curves is presented below (Fig. 2.6) where the role of hydrogen bond in proton diffusion will be discussed. It is easy to imagine that formation of an asymmetric double-well potential is possible if the hydrogen is attracted more strongly to one atom than to the other. This often happens, since the two electronegative atoms are not necessarily identical.

In fact the potential energy curve depends strongly on the distance between two atoms. As this distance decreases, the barrier height is reduced. Formation of a low-barrier hydrogen bond (LBHB) becomes possible where the zero-point energy is close to the height of the barrier. The hydrogen is delocalized across a wide region, with "free" motion between the two atoms. Further compression the distance between the two atoms leads the disappearance of the energy barrier, which generally occurs at the distance near 2.5 Å [39]. Then it is possible to form a strong hydrogen bond as in the complex (FHF)–. On creation of this complex the energy equal to a value of 2.17 eV is released [26].

On formation of the complexes with the strong hydrogen bond the configuration of the molecules changes drastically. The proton potential energy has only one flat minimum. For this reason the proton is displaced easily. The easy displacement of the proton under the action of the external field stipulates high polarizability of the complex.

Depending on the nature of X, Y, and Z in the –X-H···Y-Z complex, the energy of a hydrogen bond lies in the wide range 0.03 and 2.8 eV. The strongest hydrogen bonds are stronger than the weakest of covalent bonds, while the weakest hydrogen bonds are practically indistinguishable, in energy terms, from van der Waals interactions [33].

It is the intermediate energy range of many hydrogen bonds that enables them to both associate and dissociate quickly at ambient temperatures—and it is this attribute that accounts for the importance of the hydrogen bond in mechanistic biology. In order to elucidate one more extremely important property of the hydrogen bond it is necessary to consider the hydrogen bonding in water and ice.

2.6 Hydrogen Bond in Water and Ice

In liquid water each molecule is surrounded by four other molecules linked to it via hydrogen bonds forming tetrahedral coordination

(Fig. 2.3). However, the structure of liquid water is disordered. On the contrary, ice has an ordered hexagonal structure, which is shown in Fig. 2.4.

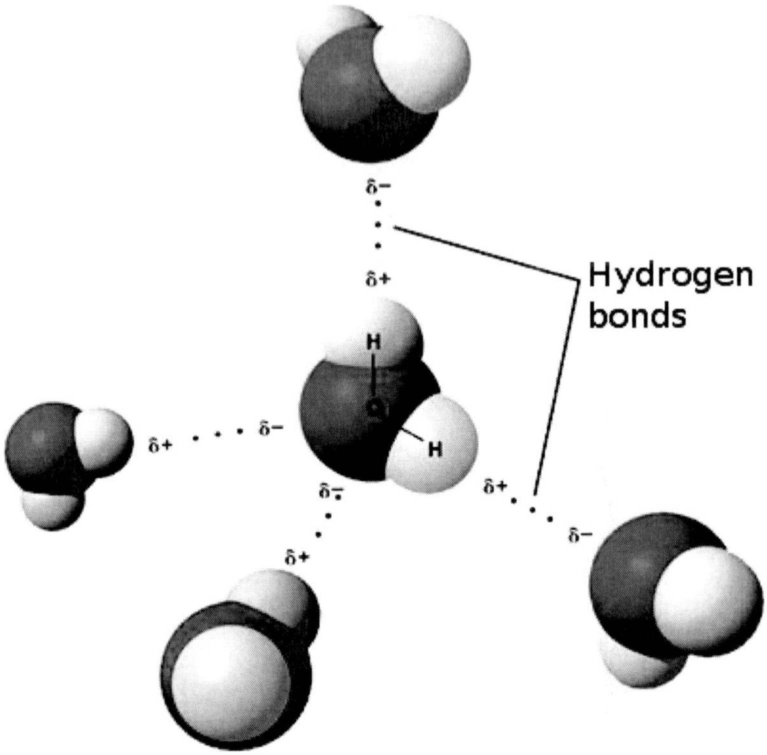

Figure 2.3 Scheme of water structure: formation of tetrahedral coordination (Image: Qwerter) [40].

In all ice structures known, hydrogen is in two-fold coordination with oxygen, and oxygen in four-fold (tetrahedral) coordination with hydrogen (Fig. 2.4). This geometry is dictated by the bond structure of oxygen, not by considerations of ionic radius. In size terms the hydrogen is a mere bump on the oxygen atom. In all known forms of ice, the water molecules retain their identities as distinct molecules and bond to other molecules by hydrogen bonding. Due to the ordered structure, ice has smaller specific mass than liquid water, which is also the reason ice floats in liquid water.

It is ascertained experimentally that ice, alcohols, and some other compounds having hydrogen bonds between the molecules possess enhanced proton conductivity. For example, the proton mobility in ice is only an order of magnitude less as the electron mobility in metals. According to the modern presentations the proton mobility in ice is stipulated by proton transfer across the hydrogen bonds. The ice lattice constitutes of the continuous hydrogen bond net, which unites all the molecules in one giant molecule having the hexagonal structure under the normal conditions. By investigations of the hydrogen transfer in the ice lattice so called the Bernal–Fowler filaments can be isolated that are one-dimensional chains of water molecules bonded by the hydrogen bonds [41–43].

Figure 2.4 Crystal structure of hexagonal ice (I_h) (Image: NIMSoffice) [44].

An ice crystal itself represents a continuous net of hydrogen bonds which links all the water molecules into a single giant molecule (Fig. 2.4). Under normal conditions the latter has a structure in which each oxygen atom is surrounded at a distance of 2.76 Å by four other oxygen atoms generating a regular tetrahedron (Fig. 2.5). Each water molecule in linked with neighboring molecules by four hydrogen bonds. A given proton lies on the line connecting neighboring oxygen atoms at a distance of 0.99 Å from the oxygen atom with which a proton forms a covalent bond and at a distance of 1.77 Å from the other oxygen atom [41].

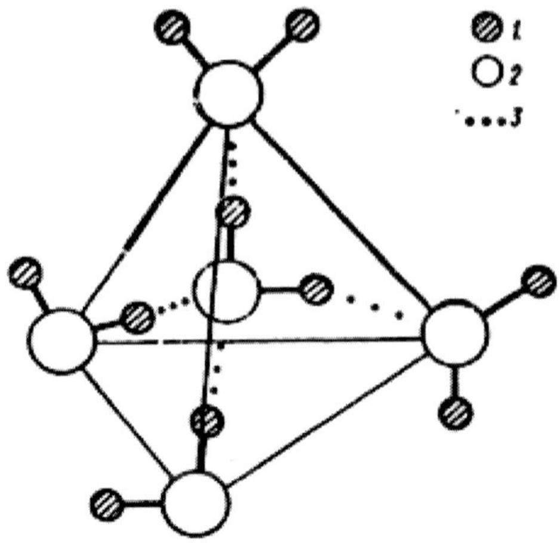

Figure 2.5 The structure of elementary cell of an ice crystal: (1) hydrogen atoms, (2) oxygen atoms, (3) hydrogen bonds. Reproduced from [41] with permission from Naukova Dumka, copyright (1986).

Experiments showed that ice crystals exhibit considerable electrical conductivity. Along the chains it is about 10^3–10^4 times larger than that in the perpendicular direction [see, e.g., 45,46], which makes it reasonable to think that there is a series of hydrogen bonded chains in the ice crystal, the motion of the carriers along these chains causes the electric conductivity of the ice, and the carriers are not the electrons, but the hydrogen ions or protons in these hydrogen bonds due to the very strong electronegativity of oxygen atoms.

To study the process of proton transfer in ice crystals, one can consider the one-dimensional chain of water molecules linked together by hydrogen bonds (Bernal–Fowler filaments). These chains cannot be compared with rigid structures and protons are translocated along them via hydrogen bonds between neighboring water molecules. The difference between disordered liquid water and ordered ice is that in water it is not possible to isolate continuous Bernal–Fowler filaments, which results in comparatively low proton conductivity for water.

Numerical simulations of liquid water in bounded water and thin films by using the Monte Carlo method demonstrated that the water density is inhomogeneous in 10–20 Å thick water layers bounded by solid walls [41–43]. In these layers in turn ~2Å thick layers adjacent to the walls and having a lamellar structure can be isolated [41–43]. In these layers, planes of water molecules are oriented parallel to the walls that bound the water layer. For the difference in the arrangement of water molecules and the hydrogen bonds between them the difference properties are observed for water in narrow pores and in large volumes. The ordered arrangement of hydrogen bonds along the walls of narrow pores prevents the formation of loose hexagonal structure of ice which is confirmed by the fact that in capillary cracks and microscopic pores of rocks water can only freeze at a temperature lower than −45°C and at considerably low temperatures in rather thin capillaries. Due to the layer of near-surface water in pores having a regular distribution of hydrogen bonds between water molecules it is possible to isolate the continuous Bernal–Fowler filaments even at room temperature or higher. Therefore near-surface water must be a good pathway for protons.

An important property of hydrogen bonding is that the proton potential energy curve in the hydrogen bridge in molecules of water and alcohols has a double-minima form whose minima correspond to two possible equilibrium positions of a proton [41]. This curve is shown in Fig. 2.6.

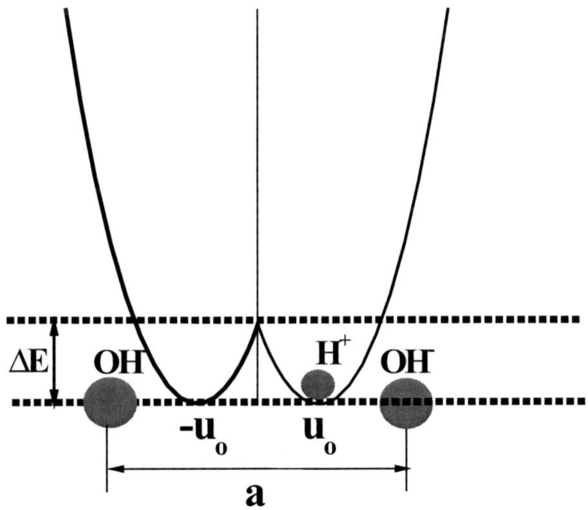

Figure 2.6 Potential wells for a proton in the hydrogen bridge. Adapted with permission from [41].

A study of proton conductivity in chain of water molecules linked by hydrogen bonds is based on the concept that a proton can be translocated (Fig. 2.7) along the chain as ionic defects hydroxonium H_3O^+ and hydroxyl OH^- that are formed under the dissociation of water molecules H_2O due to the transfer of a single proton to the neighboring molecule according with the chemical reaction:

$$2H_2O \leftrightarrow H_3O^+ + OH^- \tag{2.1}$$

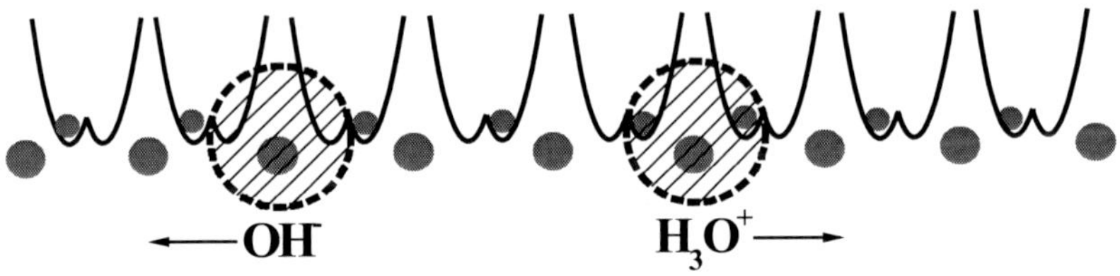

Figure 2.7 The model for the hydroxonium H_3O^+ and hydroxyl OH^- ions transfer along the water molecule chain; the translocations of protons (1) in the potential wells between the hydroxyl ions are denoted by arrows. Adapted with permission from [41].

Along with the migration of ionic defects in the chain one should take into account a possibility of motion of orientational Bjerrum defects, positive (D) and negative (L). For a positive Bjerrum defect, two protons lie between two hydroxyl ions and for negative defects there are no protons. As shown it Fig. 2.8, the orientational Bjerrum defects can arise due to rotation of a water molecule in the neighboring pair of water molecules around the OH bond over 180^0 when one proton comes out of the plane. The second way of generating of the orientational defect is a rotation of one water molecule of the given pair over 120° in the plane of the figure. The processes of the generation of ionic and orientational defects illustrated in Figs. 2.7 and 2.8 are rather simplified. It will be demonstrated further that the process of proton transfer via the hydrogen bonding barrier and the rotation of molecules depicted in Fig. 2.6 actually occur over a certain extent of the chain. In other words, both processes are collective. The aforementioned figures correspond to the limiting case when transition is reduced to the dimension of a single hydrogen bridge. Nevertheless, we preserve further the notations "Bjerrum defects" and "ionic defects" although this implies that in fact both these defects are extended. Inside the chain the ionic and orientational defects are generated only by pairs, whereas at the chain ends the can be formed individually.

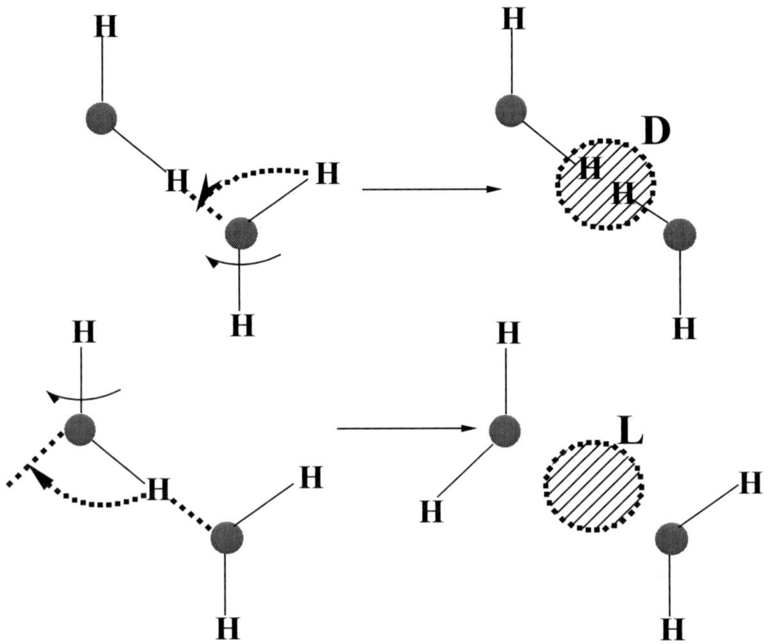

Figure 2.8 Formation of the positive (D) and negative (L) Bjerrum defects under a rotation of one water molecule around the O–H bond. Adapted with permission from [41].

Figure 2.9 illustrates the displacement of a proton (ionic defect) that is coming up to the left end of the chain. The return of the chain into the initial state requires the motion of the Bjerrum orientational defect L from the right end of the chain to the left one. The theory of proton conductivity in a chain of water molecules linked by hydrogen bonds [47,48] use the Bernal–Fowler filaments, represented as infinite chain of water molecules as the model system. In this model only one proton of the each water molecule participates in the formation of the hydrogen bond, whereas the other proton does not take part in the formation of the hydrogen bond being bonded to the oxygen atom by the covalent bond. The proton H^+ together with the oxygen ion O^{2-} was considered as an entity. Thus the chain can be subdivided into two subsystems: the main lattice formed by the hydroxyl ions and the proton subsystem: the main lattice formed by the OH^- groups and the protonic system. If these subsystems interact with each other the excited state of the whole system is two-component. Under the normal conditions of the chain each proton, which takes part in the formation of the hydrogen bond, on one side is bonded to the oxygen atom with the covalent bond and on the other side by the hydrogen bond. These bonds interchange with each other due to proton transition over a potential barrier.

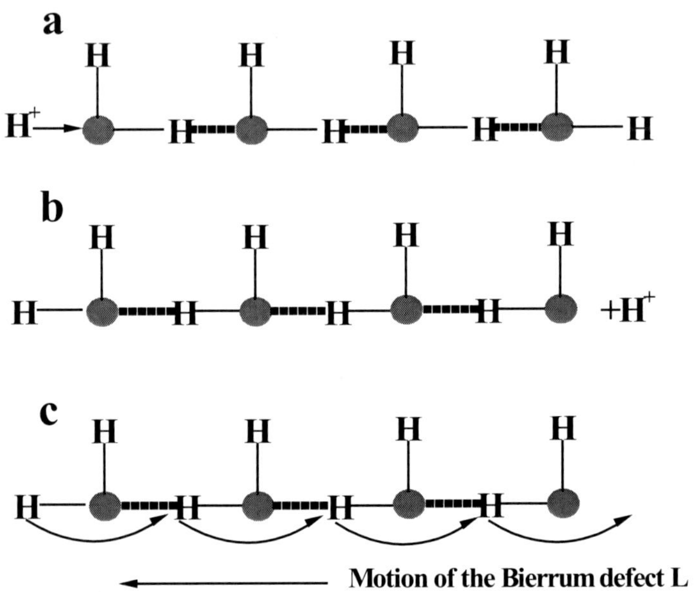

Figure 2.9 The motion of a proton that approaches to the left end of the chain (state a). The final state of the chain (state b). The transition of the chain into the initial state is realized due by the motion of the Bjerrum defect L from the right end of the chain to the left one (state c). Adapted with permission from [41].

It was thus assumed that the potential energy of the proton taking part in the hydrogen bonding can be represented by a symmetric double-minima potential well [41,47,48]:

$$U = \varepsilon_0(1 - u^2/u_0^2)^2 \tag{2.2}$$

where ε_0 is the height of the potential barrier, u is a proton displacement from the equilibrium position; and u_0 is the minimal potential.

In the normal state, that is, where all protons are either in the left or in the right wells the chain is electrically neutral. In this case the molecules form a regular lattice wherein the positive charge is distributed uniformly along the chain. By the local displacements of the protons from the equilibrium positions the uniform distribution of the positive charge is broken. In the compressed domains of the proton density an excessive positive charge arises, whereas in the stretched domains a negative charge occurs. Commonly small part of the water molecules (10^{-7}) dissociates forming the hydroxyl ions and the hydroxonium ions (see Figs. 2.9 and 2.10). The state corresponding to the hydroxonium ion contains the proton near of any OH group in the left potential well on the right hand side and in the right potential well on the left hand side. The state corresponding to the hydroxyl ion in the chain contains a proton in the left potential well on the left hand side of any OH^- group and in the right well on the right hand side. The motion of the hydroxonium ion under either the action of concentration gradient or electric field occurs by the consecutive hopping from the left well to the right one. The chain state corresponding to hydroxyl ion consists of the protons in the left well on the left hand side and in the right well on the right hand side.

Along with this potential energy of a proton, other interactions were also taken into account by theoretical consideration of proton motion along the infinite chain of the hydrogen bonds, namely, interaction of the protons caused by their mutual displacements in the neighboring potential wells; that interaction yielding to collectivization of the displacements of the protons. The third interaction taken into consideration was the interaction between the displacements of the protons and the changes of the equilibrium distances between hydroxyl ions [47].

A system with different types of interactions is usually related to non-linear systems. Taking into account all interactions in the chain of the hydrogen bonds, the displacement of the protons in the chain was described by the system of nonlinear equations. The solution of this system of equations showed that the functions describing the proton motion are related to the specific type of solitons (solitary waves) named as kink [47].

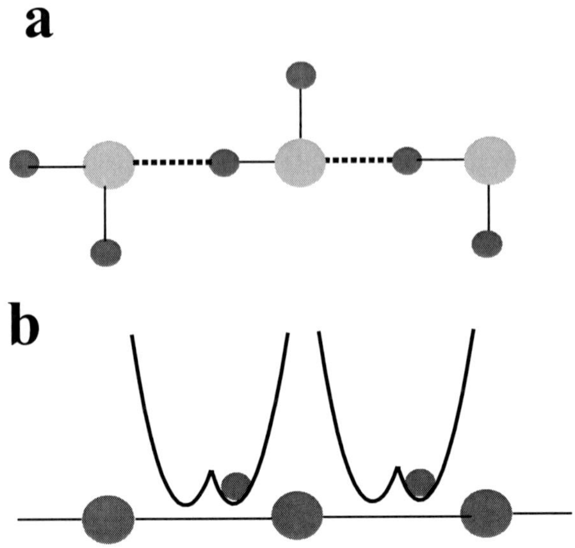

Figure 2.10 (a) Water molecules linked via hydrogen bonds; (b) The model of the one-dimensional system characterizing the position of a proton H^+ in the chain of the hydroxyl ions OH^-. Adapted with permission from [41].

In rigorous sense, the proton displacement in this system can be adequately described as spreading along the chain of the hydrogen bonds of the "blurred" domains that have either enhanced or reduced proton density relatively to its average value. This "blurring" is the result of the cooperative non-linear interactions in the system, which provides exclusive stability of the appropriate proton translocations. It was also shown that the motion of the domains with enhanced or reduced proton density is accompanied by local translocations of the equilibrium positions of the molecules [47]. Consideration of this issue shows that in the latter case an effective reduction of the energy barrier for the proton transport along the chains of the hydrogen bonds takes place, which facilitates the ionic transport of protons along these chains. The same could be said for translocations of the orientational Bjerrum L and D defects since the motion of these defects does not occur abruptly. Along the chain in the definite domains the blurred orientational defects are translocated. Inside of these domains the smooth transition occurs from one orientation of the molecules to another. The "blurring" and translocation of these orientational defects is described by non-linear differential equations analogous to those that describe the translocation of the domains of compression and stretching of the average proton density [47].

2.7 Conclusions

The properties of hydrogen bond were considered in this chapter. The hydrogen bond is related to usual chemical bond containing both ionic and covalent components. Its originality comes out from the tiny size of the hydrogen species and their jealousy for electron density. Small size makes a hydrogen atom ubiquitous. Formation of hydrogen is enormously important for many reactions and processes. By forming the hydrogen bond, hydrogen acts as a "glue" that sticks various molecules together providing the creation of various molecular arrangements and conditions for conformed transformation. But this glue is not very strong and under definite conditions the possibility exists that hydrogen bond can be broken, and this makes it possible to carry many important reactions.

One more important property of the hydrogen bond is its ability to form pathways for hydrogen migration, which occurs due to the creation of specific one-dimensional chains along which hydrogen solitons can migrate. Migration of solitons provides transport of energy. Thus, with the help of the hydrogen bonds, hydrogen can form specific structures that are able to transport energy excitation via solitons. As we see below, this unique property of the hydrogen bond is extremely important for processes of energy transport in biological systems, where hydrogen plays an outstanding role.

Chapter 3

Hydrogen in Nuclear Fusion

3.1 What Is Nuclear Fusion?

Nuclear fusion is the process by which two or more atomic nuclei join together, or "fuse", to form a single heavier nucleus. The fusion is usually accompanied by the release or absorption of large quantities of energy.

In nature fusion is the process that powers active stars, whereas artificial fusion is exploited in the hydrogen bomb and experimental facilities investigating fusion power for electrical generation.

The force that holds protons and neutrons together is attributed to so-called strong forces; this attraction arising from the exchange of quarks between the nucleons. The strong interaction (also called the strong force, strong nuclear force, or color force) is one of the four fundamental interactions of nature [49,50]. At the same time the strong force has a very short range of action which is approximately (1–2) 10^{-15} m that is close to a proton radius.

The density of nuclei is limited by the short range repulsion. The maximum size of nuclei is limited by the fact that the attractive force dies away extremely quickly (exponentially) when nucleons are more than a few fm apart.

In order to get close to each other at the range of the action of the strong forces nuclei must overcome a very serious obstacle that is the Coulomb energy barrier arising due to the electric Coulomb

Hydrogen Energy for Beginners
Alexander Gavrilyuk

ISBN 978-981-4364-77-5 (Hardcover), 978-981-4364-78-2 (eBook)
www.panstanford.com

repulsion between charged nuclei. So, the problem is that the nuclei involved must first overcome the electric repulsion to get close enough for the attractive nuclear strong force to dominate over the Coulomb repulsion to fuse the particles.

Considering the barrier to be the potential energy of two point charges (e.g., protons), the energy U required to reach a separation r is given by

$$U = \frac{ke^2}{r} \tag{3.1}$$

where k is Coulomb's constant and e is the proton charge.

To overcome the Coulomb barrier, extremely high temperatures are required, which is very hard to achieve. Substituting the value of the radius r at which the nuclear attractive force becomes dominant ($r \sim 10^{-15}$ m), the temperature necessary to raise the average thermal energy to that point can be calculated. The temperatures required to overcome the Coulomb barrier for fusion to occur are so high that extraordinary measures are required for their achievement. Such thermally initiated reactions are commonly called thermonuclear fusion; approximate temperatures being for deuterium–deuterium fusion -40×10^7 K and deuterium–tritium fusion -4.5×10^7 K. In the sun, the proton–proton cycle of fusion is presumed to proceed at a much lower because of the extremely high density and high population of particles. Thus, in the interior of the sun, the proton cycle proceeds at $T = 1.5 \times 10^7$ K.

In reality there are two things that facilitate this process and reduce the temperatures which can trigger fusion. First fusion can be initiated by those particles which are out on the high-energy tail of the Maxwell's distribution of particle energies. Second in the case of low-mass particles this barrier can be penetrated by tunneling. Thus, it is clear that protons here again have valuable advantages for fusion against all other nuclei since they have the smallest mass that yields in turn a higher probability for the tunneling under the Coulomb barrier. On the other hand, it is clear from Eq. 3.1 that the Coulomb barrier is smallest for isotopes of hydrogen—they contain only a single positive charge in the nucleus. So, hydrogen species having the smallest values of charge and mass can be considered as the first actors in fusion processes. In the previous chapter the possibility for hydrogen species to penetrate through an energy barrier via tunneling was underlined.

In the case of the proton cycle in stars, this barrier is penetrated by tunneling, allowing the process to proceed at lower temperatures than that which would be required at pressures attainable artificially on the earth.

Generally, when dealing with elements lighter than iron, the lower the ratio of atomic mass to mass number, the heavier the nucleus will be. This is known as mass defect. Fusion of lighter nuclei into heavier nuclei leads to loss of mass when the energy of binding is removed (this energy carries away the lost mass), even though no nucleons are lost. When the nucleons group together to form a nucleus, they lose a small amount of mass, that is, there is mass defect. This lost mass is present in the released energy in accordance with $E = \Delta mc^2$, where Δm is the loss of mass [51]. In nuclear reactions, the energy that must be radiated or otherwise removed as binding energy may be in the form of electromagnetic waves, such as gamma radiation, or as heat. Again, however, no mass deficit can in theory appear until this radiation has been emitted and is no longer part of the system. This energy is a measure of the forces that hold the nucleons together, and it represents energy which must be supplied from the environment if the nucleus is to be broken up. It is known as binding energy, and the mass defect is a measure of the binding energy because it simply represents the mass of the energy which has been lost to the environment after binding.

Most of the excess binding energy is released as kinetic energy of the resulting particles. When these particles are slowed, this energy is available to do work or be converted to electromagnetic radiation or heat.

Fusion reactions power the stars and produce virtually all elements in a process called nucleosynthesis. The fusion of lighter elements in stars releases energy (and the mass that always accompanies it).

The fusion of two nuclei with lower masses than iron (which, along with nickel, has the largest binding energy per nucleon) generally releases energy, while the fusion of nuclei heavier than iron absorbs energy. The opposite is true for the reverse process, nuclear fission. This means that fusion generally occurs for lighter elements only, and likewise, that fission normally occurs only for heavier elements. There are extreme astrophysical events that can lead to short periods of fusion with heavier nuclei. This is the process that gives rise to nucleosynthesis, the creation of the heavy elements during events like supernovas.

3.2 Hydrogen Burning in Stars

In the nebular hypothesis, the majority of the mass of the dust cloud collects at the center. The intense gravitational forces present ultimately lead to nuclear fusion taking place. As most of the matter initially present in the nebula is hydrogen, the process of hydrogen burning takes place. The general scheme of the hydrogen burning is shown in Fig. 3.1.

The main solar nuclear fusion process is so called the proton–proton chain reaction that is the dominant fusion process in the sun. This phenomenon is possible due to tunneling. There are a number of stages through which nuclear fusion in the sun occurs, in a proton–proton chain reaction. Through this process, hydrogen fuel is progressively converted into helium, along with release of energy in the process.

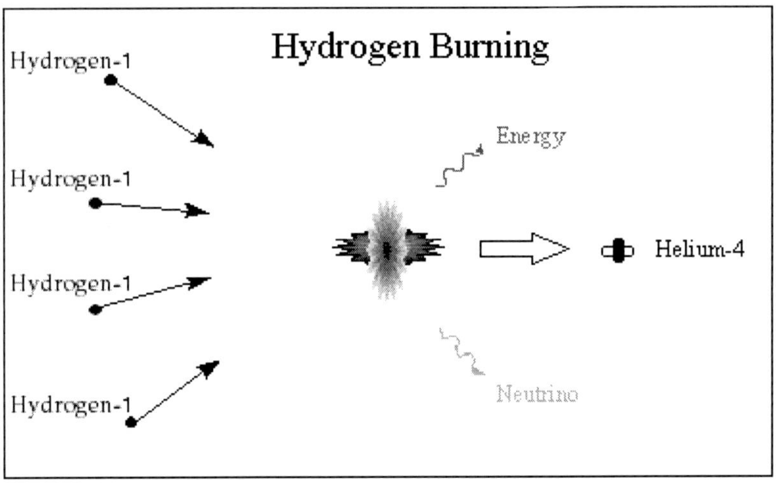

Figure 3.1 General scheme of the proton–proton chain reaction (Image: NASA) [52].

The proton–proton chain dominates in stars the size of the sun or smaller. In general, proton–proton fusion can occur only if the temperature (i.e., kinetic energy) of the protons is high enough to overcome their mutual electrostatic or Coulomb repulsion. The theory that proton–proton reactions were the basic principle by which the sun and other stars burn was advocated by Arthur Stanley Eddington in the 1920s. At the time, the temperature of the sun was considered too low to overcome the Coulomb barrier. After the development of quantum mechanics, it was discovered that tunneling of the wavefunctions of the protons through the repulsive barrier allows for fusion at a lower temperature than the classical prediction.

It was still unclear how proton–proton fusion might proceed, because the most obvious product, helium-2 (diproton), is unstable and immediately dissociates back into a pair of protons. In 1939, Hans Bethe proposed that one of the protons could beta decay into a neutron via the weak interaction during the brief moment of fusion, making deuterium the initial product in the chain [53]. This idea was part of the body of work in stellar nucleosynthesis for which Bethe won the 1967 Nobel Prize in Physics.

In the sun, deuterium-producing events are rare enough (the vast majority of these events produces a diproton instead) that a complete conversion of its hydrogen would take more than 10^{10} (ten billion) years at the prevailing conditions of its core [54]. The fact that the sun is still shining is due to the slow nature of this reaction; if it went faster, the sun would have exhausted its hydrogen long ago. These reactions take place in young stars what happened at the dawn of our solar system. It would be not an overestimation to conclude that just hydrogen triggered the supply of the earth by the sunlight energy released in fusion and just this event predetermined the process of evolution of the earth that eventually yielded the life. The first step involves the fusion of two hydrogen nuclei ^1H (protons) into deuterium, releasing a positron and a neutrino as one proton changes into a neutron [55]. The reaction steps are shown in Fig. 3.2.

$$^1H_1 + {}^1H_1 \rightarrow {}^2D_1 + e^+ + \nu_e + 0.42\ \text{MeV} \tag{3.2}$$

This first step is extremely slow, both because the protons have to tunnel through the Coulomb barrier and because it depends on weak interactions [56,57].

The positron immediately annihilates with an electron, and their mass energy is carried off by two gamma ray photons [55].

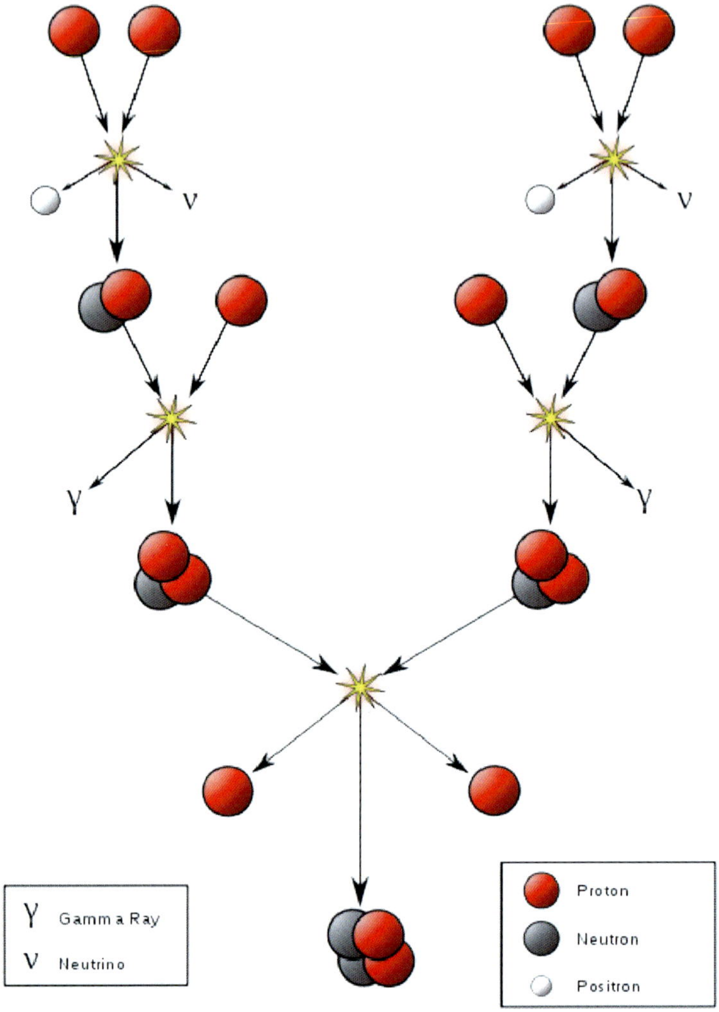

Figure 3.2 Steps of the proton–proton chain reaction (Image: Borb) [58].

$$e^+ + e^- \rightarrow 2\gamma + 1.02 \text{ MeV} \tag{3.3}$$

After this, the deuterium produced in the first stage can fuse with another hydrogen to produce a light isotope of helium, ^{3}He [55].

$$^2D_1 + {}^1H_1 \rightarrow {}^3He_2 + \gamma + 5.49 \text{ MeV} \tag{3.4}$$

From here there are three possible paths to generate helium isotope ^{4}He. In pp I helium-4 comes from fusing two of the helium-3 nuclei produced; the pp II and pp III branches fuse ^3He with a pre-existing ^4He to make beryllium. In the sun, branch pp I takes place with a frequency of 86%, pp II with 14% and pp III with 0.11%. There is also an extremely rare pp IV branch [55].

$$^3He_2 + {}^3He_2 \rightarrow {}^4He_2 + 2\,{}^1H_1 + 12.86\text{ MeV} \quad (3.5)$$

The difference between the "fusing masses" (the four protons) and "fused mass" (helium-4) is 0.7% of the total mass of four protons, which is converted into energy. The total energy produced by the fusion of four protons is 26.73 MeV. The net effect of this is to convert protons (hydrogen nuclei) into ^4He nuclei, along with energetic particles such as photons which reach us here on earth.

3.3 Fusion on the Earth

Creating the required conditions for fusion on the earth is very difficult. For nuclear weapons, some of the energy released by an atomic bomb is used to compress and heat a fusion fuel to the point of "ignition". At this point, the energy released in the fusion reactions is enough to maintain the reaction. Fusion-based nuclear power experiments attempt to create similar conditions using less dramatic means, although to date these experiments have failed to maintain conditions needed for ignition long enough for fusion to be a viable commercial power source.

The fusion of lighter nuclei, which creates a heavier nucleus and often a free neutron or proton, generally releases more energy than it takes to fuse the nuclei together; this is an exothermic process that can produce self-sustaining reactions.

A typical example of a fusion nuclear reaction carried out on the earth using hydrogen species as a fuel material is given in Fig. 3.3, where deuterium and tritium nuclei are fused. This reaction is related to one of the most promising to achieve a controlled fusion synthesis.

If the energy to initiate the reaction comes from accelerating one of the nuclei, the process is called beam–target fusion; if both nuclei are accelerated, it is beam–beam fusion. If the nuclei are part of plasma near thermal equilibrium, the process is called thermonuclear fusion.

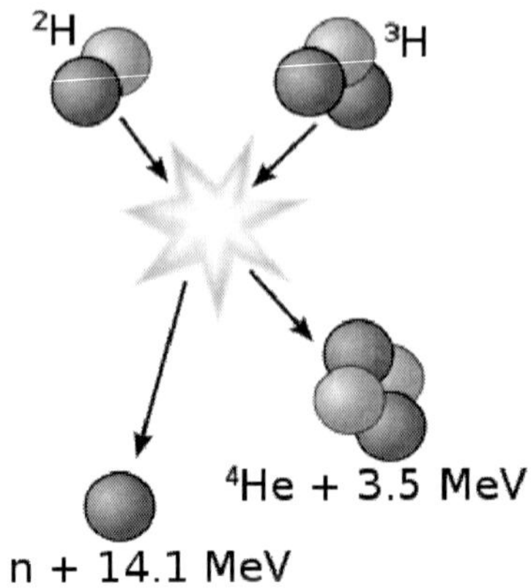

Figure 3.3 Fusion of deuterium with tritium creating helium-4, freeing a neutron, and releasing 17.59 MeV of energy (Image: Wykis) [59].

Energy released in most nuclear reactions is much larger than in chemical reactions, because the binding energy that holds a nucleus together is far greater than the energy that holds electrons to a nucleus. Fusion reactions have an energy density many times greater than nuclear fission; the reactions produce far greater energies per unit of mass even though individual fission reactions are generally much more energetic than individual fusion ones, which are themselves millions of times more energetic than chemical reactions.

Only direct conversion of mass into energy, such as that caused by the collision of matter and antimatter, is more energetic per unit of mass than nuclear fusion.

Using deuterium–tritium fuel, the resulting energy barrier is about 0.01 MeV [60]. In comparison, the energy needed to remove an electron from hydrogen is 13.6 eV, about 750 times less energy. The (intermediate) result of the fusion is an unstable ^{5}He nucleus, which immediately ejects a neutron with 14.1 MeV [60]. The recoil energy of the remaining ^{4}He nucleus is 3.5 MeV, so the total energy liberated is 17.6 MeV [60]. This is many times more than what was needed to overcome the energy barrier. If the energy to initiate the reaction comes from accelerating one of the nuclei, the process is called beam-target fusion; if both nuclei are accelerated, it is beam–beam fusion. If the nuclei are part of a plasma near thermal equilibrium, the process is called thermonuclear fusion. Temperature is a measure of the average kinetic energy of particles, so by heating the nuclei they will gain energy and eventually have enough to overcome this 0.01 MeV. Converting the units between electronvolts and kelvin shows that the barrier can be overcome at a temperature in excess of 120 million kelvin.

The fusion reaction rate increases rapidly with temperature until it maximizes and then gradually drops off. The deuterium–tritium reaction rate peaks at a lower temperature (about 70 keV, or 800 million kelvin) and at a higher value than other reactions commonly considered for fusion energy.

Research into controlled fusion, with the aim of producing fusion power for the production of electricity, has been accompanied by extreme scientific and technological difficulties, but has resulted in progress. At present, controlled fusion reactions have been unable to produce break-even (self-sustaining) controlled fusion reactions. Workable designs for a reactor that theoretically will deliver ten times more fusion energy than the amount needed to heat up plasma up to required temperatures were originally scheduled to be operational in 2018; however this has been delayed and a new date has not been stated.

3.4 Conclusions

In this chapter, we considered the first energy process where hydrogen plays the main role. The energy needed for all vital processes on the earth is produced on the sun via hydrogen burning and delivered to the earth via radiation. Depletion of this hydrogen would mean that the energy supply comes to the end, which would mean that the life on the earth will end. It was shown that the exclusive properties of hydrogen, namely its smallest barion charge and mass, which predetermine predisposition to tunnel through the Coulomb energy barrier provided for its outstanding role in these processes. Hydrogen has tremendous importance in the energy production in the cosmos.

Chapter 4

Photosynthesis

In the previous chapter the functions of hydrogen in the energy processes in the cosmos were covered. It was shown that hydrogen burning during thermo-nuclear reaction on the sun releases great amount of energy, part of which reaches the earth. The paradox is that while hydrogen that plays the main role in processes which supply energy to the earth, it also simultaneously plays a main role in the conversion of this energy. Energy conversion takes place through photosynthesis.

Photosynthesis is a process during which organisms synthesize organic chemical compounds, especially carbohydrates, from carbon dioxide using the energy obtained from light [61]. Photosynthesis is an enormously important natural process which creates the possibility for sustainable existence of living organisms. Investigations on photosynthesis have been carried out by many research groups for centuries. The extreme importance of this scientific trend can be only stressed by the fact that 10 (!) Nobel Prizes have been awarded to researchers working on photosynthesis.

4.1 Introduction

In this chapter the main principles of photosynthesis are elucidated; special attention being paid to the specific role played by hydrogen

Hydrogen Energy for Beginners
Alexander Gavrilyuk

ISBN 978-981-4364-77-5 (Hardcover), 978-981-4364-78-2 (eBook)
www.panstanford.com

which may occur in other systems and materials also and interesting hints and prompts made for implementing the natural design in various man-created devices and technological processes.

4.2 Design of Photosynthetic Systems

In green plants and cyanobacteria, photosynthetic reactions occur in chloroplasts which are autonomous cytoplasmic bodies containing all elements necessary for photosynthesis, namely proteins that are organic molecules consisting of amino acids arranged in a linear chain and folded into a globular form (Fig. 4.1) [62].

The proteins that collect light for photosynthesis are embedded within sub-unit cells (organelles) called chloroplasts. Usually a plant cell contains about 10 to 100 chloroplasts.

Here, membrane design is implemented by nature. The chloroplast is enclosed by a membrane composed of a phospholipid inner membrane, a phospholipid outer membrane, and an intermembrane space between them [62]. Phospholipids are a class of lipids that are a major component of all cell membranes as they can form lipid bilayers. The membrane architecture is an inalienable feature in living organisms; it facilitates the creation of proton gradient across the membrane or, in other words, to charge the biological battery whose energy is employed for the synthesis of adenosine triphosphate (ATP) molecules that will be described below.

Very complicated enzyme–protein complexes, in which all chlorophyll cells are concentrated and the primary act of the light energy conversion occurs, are embedded into three membranes of ~50 Å thickness (Fig. 4.1). The first membrane is a smooth outer membrane which is freely permeable to molecules. The second one is a smooth inner membrane which contains many transporters: integral membrane proteins that regulate the passage in and out of the chloroplast of small molecules like sugars proteins synthesized in the cytoplasm of the cell but used within the chloroplast. Besides the outer and the inner membranes the chloroplast has also a system of thylakoid membranes. A *thylakoid* is a membrane-bound compartment inside chloroplasts and cyanobacteria [62]. The thylakoid membranes enclose a lumen: a system of vesicles (that may all be interconnected).

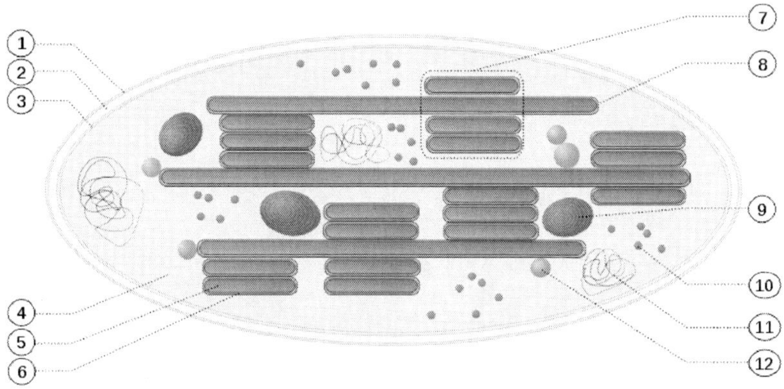

Figure 4.1 Chloroplast structure: (1) outer membrane, (2) intermembrane space, (3) intermembrane space, (4) stroma (aqueous fluid), (5) thylakoid lumen (inside of thylakoid), (6) thylakoid membrane, (7) granum (stack of thylakoids), (8) thylakoid (lamella), (9) starch, (10) ribosome, (11) plastidial DNA, (12) plastoglobule (drop of lipids) (Image: SuperManu) [63].

Within the membrane is an aqueous fluid called the stroma which contains stacks (grana) of thylakoids. The thylakoids are flattened disks, bounded by a membrane with a lumen or thylakoid space within it. The site of photosynthesis is the thylakoid membrane, which contains integral and peripheral membrane protein complexes, including the pigments that absorb light energy, which form the photosystems [62].

The stroma contains all the enzymes needed to carry out the "dark" reactions of photosynthesis; that is, the conversion of CO_2 into organic molecules like glucose, a number of identical molecules of DNA (deoxyribonucleic acid, DNA, is a nucleic acid that contains the genetic instructions used in the development and functioning of all known living organisms), each of which carries the complete chloroplast genome. The genes encode some—but not all—of the molecules needed for chloroplast function. The others are transcribed from genes in the nucleus of the cell translated in the cytoplasm and transported into the chloroplast.

Plants absorb light primarily using the pigment chlorophyll, which is the reason that most plants have a green color. Besides chlorophyll, plants also use pigments such as carotenes and xanthophylls [64]. Algae also use chlorophyll, but various other pigments are present as phycocyanin, carotenes, and xanthophylls in green algae, phycoerythrin in red algae (rhodophytes) and fucoxanthin in brown algae and diatoms resulting in a wide variety of colors.

Four types of protein assemblies are embedded in the thylakoid membranes (Fig. 4.2):

1. Photosystem I (PS I) which includes chlorophyll and carotenoid molecules
2. Photosystem II (PS II) which also contains chlorophyll and carotenoid molecules
3. Cytochromes *b* and *f*
4. ATP synthase that serves for the synthesis of the ATP.

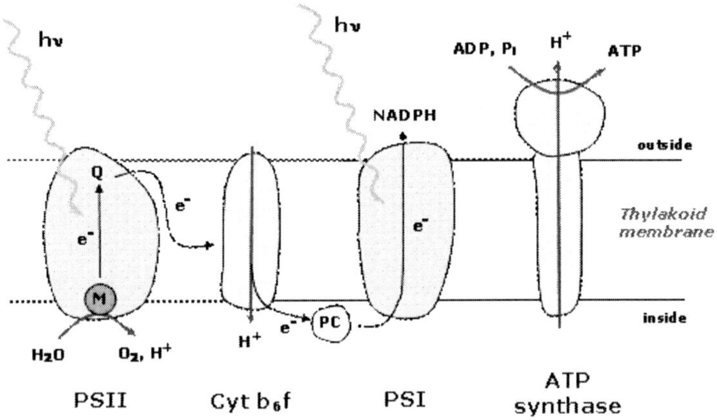

Figure 4.2 Four protein assemblies embedded in the thylakoid membrane. Reproduced with permission from Prof. Alice Haddy, Department of Chemistry and Biochemistry, The University of North Carolina at Greensboro, NC, USA [65].

It should be noted that the objective which Mother Nature put before the part of the photosynthesis that is connected with light reactions is just to generate ATP molecules. For this reason it is noteworthy to cover the ATP facets more attentively.

ATP is a multifunctional nucleotide used in cells as a coenzyme. It is often called the "molecular unit of currency" of intracellular energy transfer [66]. ATP is a unique carrier of chemical energy within cells for metabolism which is the set of chemical reactions that happen in the cells of living organisms to sustain life. It is produced by the processes that we will cover further, namely, photophosphorylation and cellular respiration (substrate-level phosphorylation) and used by enzymes and structural proteins in many cellular processes, including biosynthetic reactions, motility, and cell division [67]. One molecule of ATP contains three phosphate groups, and it is produced by ATP synthase from inorganic phosphate and adenosine diphosphate (ADP) or adenosine monophosphate (AMP). The ATP and ADP structures are shown in Figs. 4.3 and 4.4, respectively.

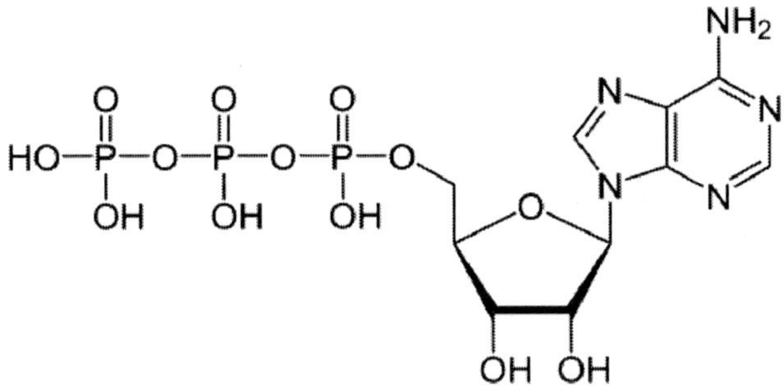

Figure 4.3 Structure of ATP (Image: NEUROtiker) [68].

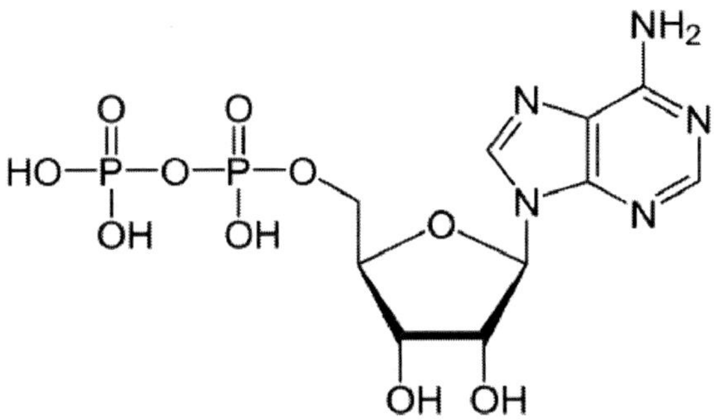

Figure 4.4 Structure of ADP (Image: NEUROtiker) [68].

Metabolic processes that use ATP as an energy source convert it back into its precursors. ATP is therefore continuously recycled in organisms: the daily production of ATP exceeds the weight of a human body [69], which on average contains only 250 grams of ATP [70].

ATP is an unstable molecule in deionized water, in which it hydrolyses to ADP and phosphate since the strength of the bonds between the phosphate groups in ATP are less than the strength of the hydrogen bonds (hydration bonds) between its products (ADP + phosphate, and water).

Thus, if ATP and ADP are in chemical equilibrium in water, the ATP will be continuously converted to ADP. A system that is far from equilibrium is predisposed to energy release, and is capable of doing work. This displacement from equilibrium means that the hydrolysis of ATP in the cell releases a large amount of free energy [71]. The energy is released by cleaving either a phosphate (Pi) or pyrophosphate (PPi) unit from ATP:

$$\text{ATP} + \text{H}_2\text{O} \rightarrow \text{ADP} + \text{P}_\text{i} \qquad (4.1)$$

$$\text{ATP} + \text{H}_2\text{O} \rightarrow \text{AMP} + \text{PP}_\text{i} \qquad (4.2)$$

About 0.5 eV of free energy is released in the reaction shown in Eq. 4.1 [26]. These proteins carry out the so-called light reaction whereas the ATP provides the second essential ingredient for running the so-called Calvin cycle, which leads to fixation of CO_2.

Meanwhile there are other molecules that will be often mentioned concerning the photosynthesis. Nicotinamide adenine dinucleotide (NAD^+) is a coenzyme found in all living cells. The compound is a dinucleotide, since it consists of two nucleotides joined through their phosphate groups, with one nucleotide containing an adenine base and the other containing nicotinamide. The structure of NAD^+ is shown in Fig. 4.5. In metabolism, NAD^+ is involved in redox reactions, carrying electrons from one reaction to another. The coenzyme is, therefore, found in two forms in cells: NAD^+ is an oxidizing agent—it accepts electrons from other molecules and becomes reduced. The scheme of the redox reaction mentioned is presented in Fig. 4.6.

This reaction forms NADH, which can then be used as a reducing agent to donate electrons. These electron transfer reactions are the main function of NAD^+ [72].

When two hydrogen atoms are joined the oxidized form NAD^+ it is transformed into the reduced form according to the following equation:

$$\text{NAD}^+ + 2\text{H}^+ + 2\text{e}^- \rightarrow \text{NAD·H} + \text{H}^+ \qquad (4.3)$$

Another primary acceptor of electrons and protons is nicotinamide adenine dinucleotide phosphate ($NADP^+$) or triphosphopyridine nucleotide (TPN) in older notation, is a coenzyme used in anabolic reactions, such as lipid and nucleic acid synthesis, which require NADPH as a reducing agent [73]. $NADP^+$ has the same active center as NAD⁺. When it joins two protons and two electrons it is turned into the reduced molecule NADP·H:

$$NADP^+ + 2H^+ + 2e^- \rightarrow NADP{\cdot}H + H^+ \quad (4.4)$$

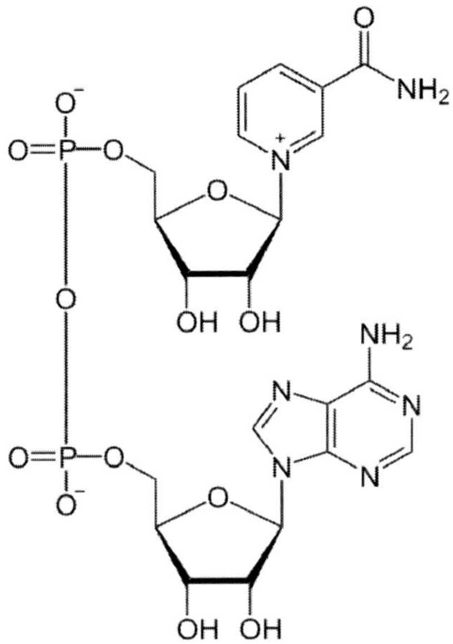

Figure 4.5 Structure of oxidized nicotinamide adenine dinucleotide (NAD^+) (Image: NEUROtiker) [74].

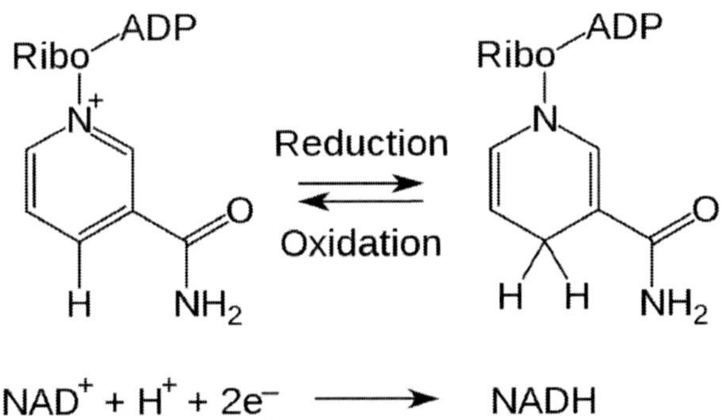

Figure 4.6 Redox reactions of nicotinamide adenine dinucleotide (Image: Fvasconcellos) [75].

NAD^+ and $NADP^+$ molecules act as coenzymes providing the dehydration from the compounds with the groups H–C–OH. In the presence of enzymes and with the help of NAD^+ and $NADP^+$ molecules two protons and two electrons being stripped from H–C–OH group. One electron and two protons join NAD^+ or $NADP^+$ molecules turning them into the reduced forms NAD·H and NADP·H. These reduced forms can in turn serve as electron donors in other reactions.

4.3 How Is Photosynthesis Carried Out?

The spirit of photosynthesis is in two key processes: the photo-triggered cleavage of water molecules with the subsequent removal of hydrogen (H) atoms and the reduction of carbon dioxide (CO_2) along with the fixation of carbon by these hydrogen atoms to form organic molecules. The second process involves a cyclic series of reactions named (after its discoverer) the Calvin cycle. The electrons (e^-) and protons (H^+) are split off separately from water molecules. As the result of the reaction, one gets oxygen and separated electrons and protons.

$$2H_2O \rightarrow 4e^- + 4H^+ + O_2 \tag{4.5}$$

It is interesting and deductive to follow attentively the splendid operation of the sophisticated photosynthetic system designed by Mother Nature. It is constructed to achieve maximally efficient collection of energy delivered by light quanta, to retain it carefully saving against possible recombination, and to transfer to specific reaction centers wherein the products required for the dark reactions are produced.

It is convenient to observe the scheme of the light-dependant reactions is presented in Figs. 4.7 and 4.8 [71] that supplement well each other for the elucidation of the run of the process.

Figure 4.8 shows the so-called Z-scheme of the light-dependant reactions, since the electron pathway includes both uphill and downhill reactions and it presents itself a rotated letter Z on the energy scale.

The process starts in PS II called so just due to the fact that it was discovered later than the PS I. The removal of electrons from water molecules and their transfer to $NADP^+$ requires energy. The electrons are moving from a redox potential of about +0.82 V in water to −0.32 V in NADPH. Reduction potential (also known as redox potential, oxidation/reduction potential, ORP or E_h) is a measure of the tendency of a chemical species to acquire electrons and thereby be reduced. Reduction potential is measured in volts (V), or millivolts (mV). Each species has its own intrinsic reduction potential; the more positive the potential, the greater the species' affinity for electrons and tendency to be reduced.

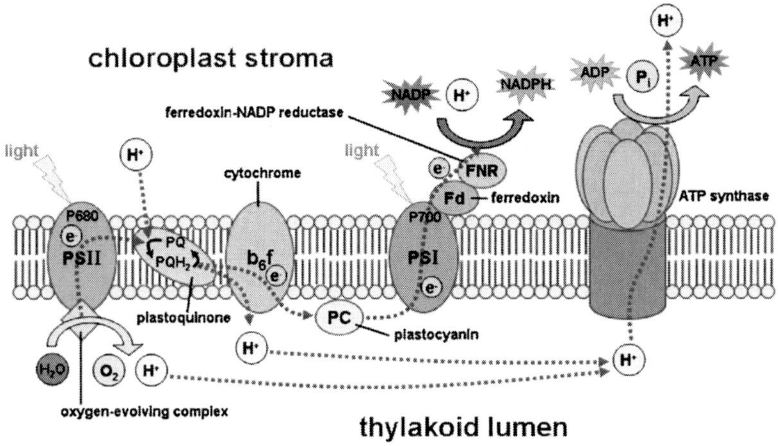

Figure 4.7 Proteins embedded in the thylakoid membrane (Image: Tameeria) [76].

Thus enough energy must be available to move them against a total potential of 1.14 volts. The required energy is delivered by light photons [26].

PS II is bound into the thylakoid membrane (Fig. 4.7). This system contains proteins and chlorophylls subdivided conditionally into three functional units: light harvesting complex, photochemical reaction center where the primary charge separation is achieved, and enzymatic center for the evolution of oxygen.

4.3.1 Light-Harvesting Complex

The first unit must be a collector of the light energy with very high efficiency—the so-called light-harvesting complex (LHC); the energy being collected and transferred further to special reaction centers (Fig. 4.9). This LHC acts as antenna pigments harvesting light and passing its energy on to their respective photosystems.

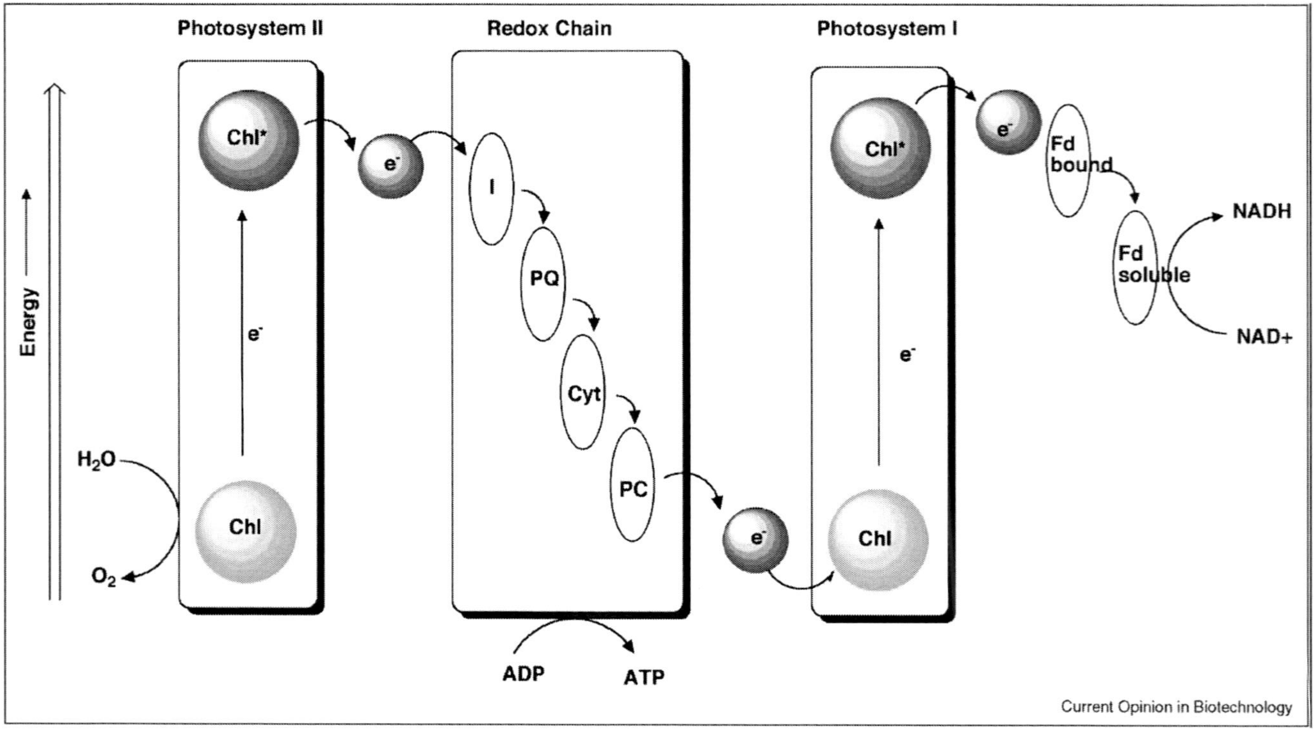

Figure 4.8 Z-scheme of the electron transfer in the electron transport chain. Reprinted from Ref. 71, copyright 2010, with permission from Elsevier.

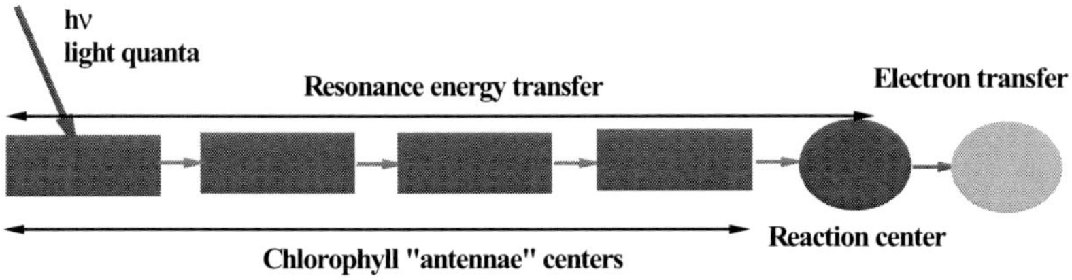

Figure 4.9 Scheme of the transfer of the excitation produced by a light quantum to the reaction center.

A vast majorities of the pigments in a photosynthetic organism are not chemically active, and they primarily function as an antenna [77–80]. The photosynthetic antenna system is organized to collect and deliver excited state energy by transfer of the light excitation to the reaction center complexes where photochemical reaction occurs [77]. The antenna system enhances the effective cross section of photon absorption via enhancement of the number of pigments associated with each photochemical complex. The sunlight intensity is sufficiently dilute so that any given chlorophyll molecule only absorbs at most a few photons per second. By incorporating many pigments into a single unit, the reaction center and electron transport chain can be used to maximum efficiency. A pronounced variety of antenna complexes have been identified from various classes of photosynthetic organisms. Seemingly, there have been multiple evolutionary origins of antenna complexes, as there is no common structural theme evident [77].

The light-collecting ("antenna") unit of green plants contains approximately 200 enzyme molecules of chlorophyll *a*, chlorophyll *b*, and carotenoid positioned on the proteins of the molecular mass from 25000 to 47000 [81].

To absorb the entire visible-light part of the solar radiation (350–700 nm), green plants use chlorophyll *a* as the main light absorber along with a number of accessory pigments such as xanthophylls, carotenoids, and a modified form of chlorophyll (chlorophyll *b*). Chlorophyll *a* absorbs in the blue–violet, orange–red spectral regions while the accessory pigments cover the intermediate yellow–green–orange part. Second important feature is the special arrangement (antenna array) of chlorophyll molecules for efficient collection of light energy and its relay to the reaction center, even when the light flux varies significantly. Photosynthesis occurs at comparable efficiency under bright and diffuse light conditions [81].

In cianobacteria, the pigment–protein units named phycobylins play the role of antenna absorbing also green light which is not sufficiently good collected by the chlorophylls.

If a pigment molecule absorbs a photon, an electron in the molecule becomes excited. For most compounds that absorb light, the electron simply returns to the ground state and the absorbed energy is converted into heat and/or fluorescence. But in an LHC-protein the pigments are so arranged that the excitation energy can be transferred from one molecule to a nearby molecule. The rate of this process, called resonance energy transfer, depends strongly on the distance between the energy donor and energy acceptor molecules. For reasons of conservation of energy, energy transfer must be from a donor in the excited state to an acceptor of equal or lower energy. If the energy of a photon becomes lower, the wavelength also becomes longer. When chlorophyll is isolated from the enzymes it is associated with, the second scenario can be seen to happen. The pigments in an LHC-protein are so arranged that pigments are very close to each other and a pigment is near another pigment that absorbs photons with a longer wavelength. As a consequence, the pigment in the reaction center has to absorb photons with the longest wavelength and cannot transfer this energy to another pigment. The function of LHC-proteins is to create a constant supply of excitation energy to the reaction center pigment. When the energy of light is absorbed by the chlorophyll molecules they occur in the excited states; the excitons being transferred via the resonance interactions, migrating between the pigments, and reaching eventually those chlorophyll molecules which belong to the reaction centers.

4.3.2 The Reaction Center

The second system unit is the mentioned reaction center which transforms the light energy absorbed by the chlorophyll into the potential energy of separated charges which yields the formation of the most powerful biological oxidant, namely, the chlorophyll P680 which had lost one electron ($P680^+$). This chlorophyll is called so because it absorbs light maximally at the wavelength 680 nm.

All main components of the reaction center are embedded into the unit consisting of two proteins with the molecular mass ~32000, the so-called D1 and D2 proteins which span the entire membrane [81,82].

The structure of PS II is shown in Fig. 4.10. The primary photochemical reaction occurs as electron transfer from the excited P680 chlorophyll molecule to a molecule of pheophytin—derivative of the chlorophyll wherein the central magnum atom is substituted by two hydrogen atoms. This process takes only some picoseconds and forms the state $P680^+Ph^-$ with separated charges living 10^{-8} sec due to a possibility of the back transfer. This back transfer is suppressed due to a faster electron transfer ($2\cdot10^{-10}$ sec) from Ph^- to a molecule of an electron acceptor Q_A (a unit of a plastoquinone with a Fe^{2+} atom). Despite approximately one third of the energy delivered by the initial photon is lost by this electron transfer, the more stable state $P680^+PhQ_A^-$ is formed with a life time ~$1.5\cdot10^{-4}$ sec. Thus, at the initial steps of the process the definite stabilization of the excited state is achieved although some energy is lost [81].

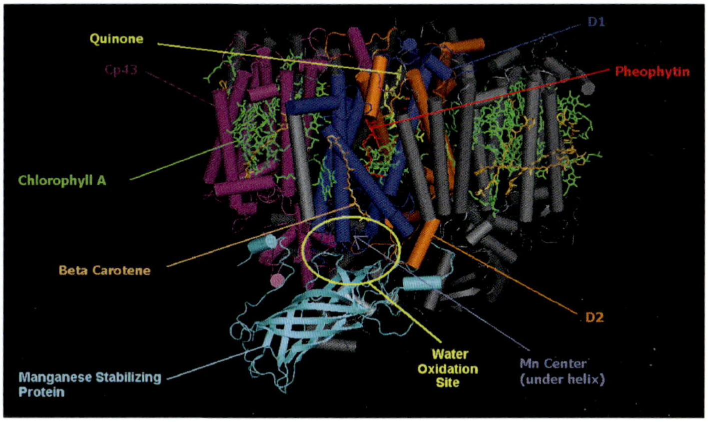

Figure 4.10 PS II (Image: Neveu,Curtis) [82].

It is possible to conclude that as a result of the primary absorption of photon, enormously strong biological oxidant $P680^+$ is formed whose redox potential is +1.1 V relative to the normal hydrogen electrode; it being enough for the oxidation of water [81].

These first stages may be related to the so-called physical charge separation. After the photon excitation the photosynthetic systems makes everything possible for separation of the photogenerated electron and hole, which is achieved with the help of the intermediates; the photogenerated electron hopping along the electron transfer chain consisting of these intermediates via tunneling.

The oxidation of water and evolving of oxygen makes it possible to exchange the photogenerated holes for protons, which yields chemical charge separation that finally stabilizes the excitation.

To evolve oxygen from water it is necessary to split off four electrons and four protons from two water molecules:

$$2H_2O \rightarrow 4e^- + 4H^+ + O_2 \tag{4.6}$$

At the same time, the reaction center of PS II produces one redox equivalent $P680^+$ per one absorbed photon. So, four photochemical acts at the reaction center are required to form an O_2 molecule from two H_2O molecules.

In principle, the water oxidation can be carried out in two ways: (i) either via consecutive splitting off four electrons (one after another) from two water molecules parallel to formation of intermediate products of water oxidation (ii) or with the help of some intermediate system between the reaction center and water that accumulates four redox equivalents in the result of four consecutive photochemical acts at the reaction center and only then carries out simultaneous four-electron oxidation of two water molecules with the formation of one O_2 molecule [81].

It was shown that one-electron path of the water oxidation is barely possible since in that case the redox potential of a value ~2 V is required for the $P680^+$ whereas it is only of value 1.12 V. At the same time the potential of 0.81 eV is required for the water oxidation by the four-electron mechanism. This accumulation of four redox equivalents necessary for the four-electron oxidation of water is carried out at an enzyme center situated in between the photoreaction center and water; the enzyme center consisting of four mixed-valence manganese atoms.

Figure 4.11 shows schematically the manganese center structure where four manganese atoms form two Mn–Mn dimers with the separation of 2.7 Å in the dimer and of 3.3 Å between the dimers. The structure of this center, which includes along with four manganese ions, a calcium ion, a chloride ion, and a tyrosine residue that forms a radical, has not been fully established. In the process of the accumulation of the redox equivalents in the Mn complex due to its oxidation by the photoreaction center, the Mn valence at first is changing from +2 up to +4, and then the electrons transferred at the Mn atoms due to the water oxidation reduce it back [81] (Fig. 4.12).

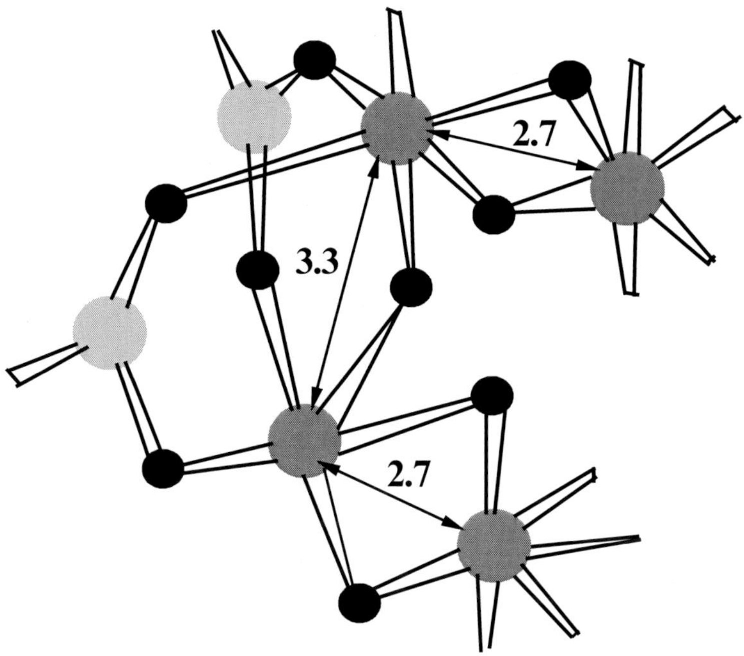

Figure 4.11 Scheme of a four-manganese center for photosynthetic water oxidation. Manganese ions are depicted by dark gray, oxygen atoms by black, whereas carbon atoms by light gray. The distances between the manganese centers are given in angstroms. Adapted from [81].

The recurrence of the water oxidation process and of oxygen release was confirmed experimentally: under illumination of greens by a series of short light pulses the maximum evolution of oxygen is observed at each fourth pulse. This is the evidence of the four-tact accumulation of the redox equivalents at the Mn-containing system of water oxidation [81].

Manganese was apparently evolutionarily selected for this role because of its ability to exist in multiple oxidation states (Mn^{2+}, Mn^{3+}, Mn^{4+}, Mn^{5+}) and to form strong bonds with oxygen-containing species. The center is oxidized, one electron at a time, until two bound H_2O molecules are linked to form a molecule of O_2, which is then released from the center. A tyrosine residue (not shown) also participates in the coupled proton–electron transfer steps [81].

In other words, the accumulation of the photogenerated holes is being carried out at the manganese center reaching the value of

4 unless these holes are scavenged by four electrons from water; two water molecules blow up simultaneously, releasing one oxygen molecule and four protons.

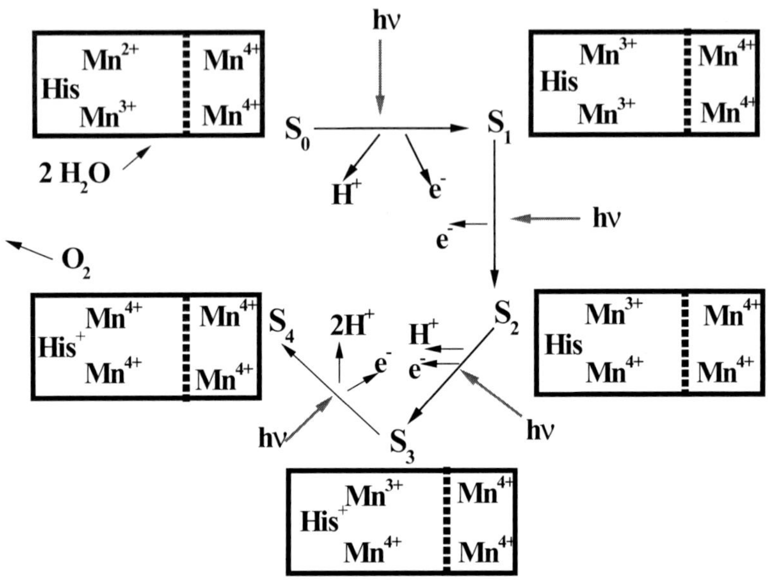

Figure 4.12 Scheme of light-induced cyclic transitions of the water oxidation system from the initial state (S_0) up to the state with four electrons lost (S_4).The system is oxidized by the reaction center of PS II and subsequently returns to the initial state S_0 via the oxidation of two water molecules accompanied by formation of one molecule of oxygen. His is the residue of histidine acid. Adapted from [81].

This is the moment when the protons come to the scene. And what for? What are their functions? Nature constructed the photosynthetic system to maintain photon excitation by all means. For this reason the photoexcited electron is removed swiftly from its "birth place" to avoid the recombination being passed along the chain of the transmitters; the electron pays for this by losing some energy but nevertheless the excitation remains preserved. And then the final stabilization of the excitation is achieved when the water molecule is split, or, in other words, when the photochemical reaction occurs and the photogenerated holes are exchanged for the protons and the light energy is transformed into the energy of chemical charge separation. Thus, instead of the electron–hole pair generated under the action of light, the electron–proton pair arises that is a universal energy transmitter. Below we shall consider as these energy transmitters both being elements comprising a hydrogen atom transfer, transform energy in many vitally important process.

Under the action of light, the electrons and the photochemical protons appear in the photosynthetic system. So, photosynthesis is connected with hydrogen production under the action of light via cleavage of water. This idea may occur very promising and later we can see how this is used in technological research that mimics the natural photosynthesis. It makes sense to remember always the performance of the water splitting center to reproduce its action artificially for technological purposes. The protons detached from water are released into the lumen, enhancing the concentration of protons inside of the thylakoid, whereas the electrons start their long way along the electron transfer chain (Figs. 4.7 and 4.8).

4.3.3 Electron Transfer Chain and Proton Chemiosmosis

Water splitting under the action of light is not the only source of the protons in the interior of thylakoid. Protons are also always present in the aqueous media of a cell. On their way along the electron transfer chain (see Figs. 4.7 and 4.8), the photogenerated electrons set up an electrochemical charge that provides the energy for pumping protons from the stroma of the chloroplast into the interior of the thylakoid (the lumen).

Leaving the PS II, the electron first passes to plastoquinone that is a mobile carrier; being reduced by the photogenerated electron the plastoquinone consumes two protons from the stroma and transports the photogenerated electron to the cytochrome b_6f protein complex (Figs. 4.7 and 4.8). The photogenerated electrons lose the energy on their way along the electron transfer chain; thus exciting the membrane and facilitating the passage of the protons across the membrane from the stroma to the lumen. Then these protons are released in the lumen when the reduced plastoquinone is oxidized by the cytochrome b_6f protein complex on the lumen side of the thylakoid membrane. The appearance of the photochemical protons and the protons consumed from the stroma in the interior of the thylakoid leads to the establishment of a proton gradient between the stroma and the lumen with an access of the protons in the interior of the thylakoid.

As electrons pass down the gradient between PS II and PS I, the energy released is harnessed by the cytochrome b_6f complex to pump protons (H^+) against their concentration gradient from the stroma of the chloroplast into the interior of the thylakoid (an example of active transport). As their concentration increases inside (which is the same as saying that the pH of the interior decreases), a strong diffusion gradient is set up. The only exit for these protons is through the ATP synthase complex (Figs. 4.7 and 4.8). As they flow back out from the interior of the thylakoid (by facilitated diffusion), passing down their concentration gradient), the energy they give up is harnessed to the conversion of ADP to ATP. Because it is driven by light, this process is called photophosphorylation.

$$ADP + P_i \rightarrow ATP \tag{4.7}$$

The process providing the proton gradient is called chemiosmosis and is an example of facilitated diffusion. The electrons are transferred along the electron transport chain where they provide the energy for chemiosmosis. The hypothesis of the chemiosmosis was first elaborated by Peter Mitchell and was confirmed further numerous times experimentally. For this Peter Mitchell received a Nobel Prize, and in his Nobel lecture he presented the essence of the phenomenon [83]: "At this conceptual level one makes use of the general principle of coupling by proticity, the protonic analogue of electricity. Separate proton motive redox (or photoredox) and reversible proton motive ATPase complexes are conceived as being plugged through a topologically closed insulating membrane between two proton-conducting aqueous media at different protonic potential. Thus, coupling may occur, not by direct chemical or physical contact between the redox and reversible ATPase systems, but by the flow of proticity around an aqueous circuit connecting them. I use the word proticity for the force and flow of the proton current by analogy with the word electricity, which describes the force and flow of an electron current."

To provide the energy for the chemiosmosis the electrons are transferred along the transport chain. The chemiosmosis links the photochemical reaction with the production of ATP.

The energy of Mitchell's proticity is used to convert ADP to ATP when it is dissipated on passing ATP synthase under the action of the proton motive force arising due to the proton gradient between the stroma and the lumen.

One can observe an elegant work of the couple converting energy: electrons passing along the membrane are losing its energy, the energy being used to excite the membrane and to pump the protons from the stroma into the lumen creating the chemiosmosis, the proton gradient, and the proton motive force, in other words, Mitchell's proticity, which brings the energy to ATP synthase to create ATP molecules; it is possible to observe vectorial transfer of energy directly to ATP synthase which provides the formation of ATPs.

Then the electrons are transported further from the cytochrome b_6f complex to PS I, called so because this PS I was discovered earlier than PS II (Figs. 4.7 and 4.8). PS I operates in a way similar to PS II. Its maximum absorption of light occurs at 700 nm; for this reason it is also named as P_{700}. Activation of P_{700} in PS I enables it to pick up electrons arrived from the cytochrome b_6f complex [by way of another mobile carrier plastocyanin (PC) in Fig. 4.8] and push them uphill to a sufficiently high redox potential that, after passing through ferredoxin (Fd in Fig. 4.8), they can reduce $NADP^+$ to NADPH.

4.4 Role of Hydrogen in Photosynthesis

It is now time to distinguish and emphasize the exclusive role of hydrogen in the light-dependent reactions of photosynthesis.

(i) First, hydrogen atoms released from water make a decisive income in the stabilization of the photo-induced excitation: the photogenerated holes are exchanged for the protons, which stabilize reliably the charge separation making the recombination of the electron–hole pairs impossible.

(ii) Then the photogenerated protons together with their "colleagues" consumed from the stroma take part in the creation of the chemiosmosis and move under the action of the concentration gradient from the lumen to the stroma through the ATP synthase delivering the energy for the ATP production. So, the protons take part in conversion of the photon energy in other energy forms and transport this energy to the reaction center wherein it is used for the production of ATP.

(iii) According to the Davydov's hypothesis, ATP synthase has narrow inner channels filled with water [41–43]. This gives possibility to form the quasi one-dimensional chain of the water molecules bonded by the hydrogen bonds. As it was considered above the proton transport in such filaments is provided by the hydrogen solitons. So, the important properties of the hydrogen bonds are exploited for the facilitation of the proton transport.

(iv) Since the oxygen atoms bonded covalently to the hydrogen atoms within water molecules get the electrons back on evolving during photosynthesis, namely the protons released from water donate the electrons to scavenge the holes accumulated into the reaction center; thus the hydrogen atoms in water are electron donors for the whole process of the photosynthesis; they supply the energy being used to produce ATP and NADPH this energy being necessary for the CO_2 fixation in the light-independent reactions.

(v) It is important to understand that water is not an obligatory reactant in the photosynthesis. Photosynthetic bacteria don't evolve molecular oxygen. Instead of water they use either non-organic substances (H_2S, H_2, etc.) or organics (milky acid, isopropanol) as the electron donors. For example, green bacteria consume hydrosulfate [26]:

$$12H_2S + CO_2 + \text{light} \rightarrow C_6H_{12}O_6 + 6H_2O + 12S \tag{4.8}$$

For this reason the general equation of the photosynthesis may be written in the generalized form [26]:

$$12D(H_2) + CO_2 + \text{light} \rightarrow C_6H_{12}O_6 + 6H_2O + 12D \tag{4.9}$$

Where $D(H_2)$ is a donor of hydrogen atoms, and D is its oxidized form. Water, hydrosulfate, isopropanol can play the role of the hydrogen donor. The photosynthetic organisms do not use only CO_2 as the hydrogen acceptor. Sometimes for nitrogen-fixing organism along with CO_2 this role is played by molecular nitrogen [26]:

$$3D(H_2) + N_2 + \text{light} \rightarrow 2NH_3 + 3D \tag{4.10}$$

So, the circles of the reactants can be different in different photosynthetic systems but only hydrogen atoms are absolutely irreplaceable.

(vi) Meanwhile, photosynthesis and photophosphorylation can also occur without participation of the chlorophyll. In March

1976 it was reported first that halobacteria utilize light energy without chlorophyll which was unknown that time [84–86]. Halobacteria live only in some very salty aqueous reservoirs with high concentration of NaCl and Mg^{2+} salts and in which salt is extracted by evaporation [84–86]. Halobacteria posses a cylindrical rod-shaped form with a diameter of 0.5 μm and a length of 5 μm and are enclosed by a glycoprotein wall. Tights are attached to its both ends to provide mobility of a bacterium.

The main color of the bacteria is determined by its violet fraction that consists of single-type proteins and lipids in the ratio 3:1. This fraction is a part of the halo bacteria membrane. In undamaged bacteria it forms islets (violet and purplish patches) on the membrane surface. These islets have a two-dimensional hexagonal crystalline structure. Each unit cell contains three protein molecules of the same type with a molecular weight of 26000. Triples of molecules, triads are combined into the regular hexagons. The greater part of the surface of these molecules is hydrophobic and is immersed in the lipid layer of the membrane [84–86].

A colored grouping, the residue of a retinal molecule (aldehyde of vitamin A), is involved in the protein chain of each molecule. By their structure, the molecules of the violet patches are close to those of rhodopsin that is a vision pigment of the retina in animal eyes which are formed by a complex of the retinal molecules with the protein opsin. For this reason, the protein revealed in violet patches of halobacteria is called bacteriorhodopsin [84–86].

In the bacteriorhodopsin molecule, the aldehyde end of the retinal molecule binds to the amino group of the lysine amino acid residue of the protein molecule with the release of a water molecule. After this, a covalent bond forms between the nitrogen of the lysine residue and the carbon of the aldehyde end of the retinal molecule, forming a compound called the Schiff base. The absorption of light in the 570 nm range by the bacteriorhodopsin molecule is due to the presence of a bond of the Schiff-base type.

When light is absorbed, a proton is split off from the nitrogen atom belonging to the Schiff-base bond. Since the final state of the proton is related to the continuous spectrum, the absorption band is very wide. When the proton is stripped and deprotonation occurs the absorption band of the molecule is shifted to higher energies and the molecules become lighter.

After returning back of the proton the absorption band is again shifted in the range with the maximum at 570 nm. By investigations of the absorption spectra of the suspensions of the undamaged bacteria, it was inferred that the medium becomes more acidic during the expelling of the protons out of the membrane and less acidic when those protons return back into the membrane [84–86].

On the ground of the experiments it was possible to make a conclusion that in the undamaged cell the red phase of the membrane acts as the proton pump. It was supposed that the release of the protons occurs at the external side of the membrane whereas their capture occurs at the inner side.

Thus the gradient of the concentration of the protons is formed along with the potential difference across the membrane. The free energy of such electrochemical gradient is used for the synthesis of ATP molecules from molecules of ADP and inorganic phosphate P by the participation of the ATP synthase enzymes that, apparently, are included into the red phase of the membrane.

It is astonishing that halo bacteria are capable of synthesizing ATP molecules both either using the energy released by oxidative respiration, or using the energy of light absorbed by the red islets of the membrane. The latter statement is confirmed by the fact that the bacteria synthesize the islets on the membrane only when they grow under light and at low oxygen concentration. The ability of the halobacteria to synthesize ATPs by both ways was established by direct experiments [86] that discovered a steep decrease in concentration of ATPs down to one-third of the initial value in the halo bacteria in salty solution when nitrogen was being passed through it.

Under illumination of the molecules by a light of 570 nm wavelength in the same nitrogen atmosphere the ATP concentration rapidly enhances and drops down again when the illumination is stopped. If oxygen is blown through this solution in dark, the ATP concentration is enhanced abruptly due to oxidative phosphorylation. So, here again hydrogen is an irreplaceable element to provide the photosynthesis and performs its functions even in the photosynthetic systems without chlorophyll. Only this element can provide the Mitchell's proticity conjugating the photochemical reactions with the ATP production via photophosphorylation.

4.5 Light-Independent Reactions in Photosynthesis

It was shown that under the action of light the ATP and the NADP·H molecules are formed in thylakoids. These molecules are consumed in the stroma of the chloroplasts to provide the final phase of the photosynthesis, the "dark phase" transformation of CO_2 into hydrocarbons. This phase of photosynthesis will be considered only briefly here since the main objective of the present book is hydrogen and its functions and in the dark stage of photosynthesis hydrogen does not reveal new functions compared with other metabolic processes that will be considered below.

It was shown in [87] that the energy of two NADP·H and three ATP molecules is required to transform one CO_2 molecule. In turn the formation of two NADP·H molecules requires four light quanta, since according to the reaction:

$$2(\text{Fer})^- + \text{NADP}^+ \rightarrow \text{NADP·H} + \text{H}^+ + 2(\text{Fer}) \qquad (4.11)$$

where Fer is ferrodoxin, two electrons are required which can be released in the reactive center of PC I only if two light quanta are absorbed (see Figs. 4.7 and 4.8). If to suppose that the energy of the translocation of one electron is consumed by each phosphorylation act, then the synthesis of three ATPs requires the absorption by both PC I and PC II not less than of three quanta. Thus, not less than seven light quanta are required to transform (fix) one CO_2 molecules in the photosynthesis. Hydrogen plays here the role that will be considered in detail in the next chapter when the oxidative-reduction reactions in metabolism will be considered.

The dark reactions of CO_2 transformation into hydrocarbon proceed in a similar way for all photosynthesis systems (see Fig. 4.13).

The fixation or reduction of carbon dioxide is a process in which carbon dioxide with the help of the special enzyme (Ribulose-1,5-bisphosphate carboxylase oxygenase, most commonly known by the shorter name RuBisCO) combines with a five-carbon sugar, ribulose 1,5-bisphosphate (RuBP), to yield two molecules of a three-carbon compound, glycerate 3-phosphate (GP), also known as 3-phosphoglycerate (PGA). GP, in the presence of ATP and NADPH from the light-dependent stages, is reduced to glyceraldehyde 3-phosphate (G3P). This product is also referred to as 3-phosphoglyceraldehyde

(PGAL) or even as triose phosphate. Triose is a three-carbon sugar (see carbohydrates). Most (five out of six molecules) of the G3P produced is used to regenerate RuBP so the process can continue (Calvin–Benson cycle). One out of six molecules of triose phosphates, which are not "recycled" often condense to form hexose phosphates, which ultimately yield sucrose, starch, and cellulose. The sugars produced during carbon metabolism yield carbon skeletons that can be used for other metabolic reactions like the production of amino acids and lipids. The scheme of the carbon fixation is presented in Fig. 4.13.

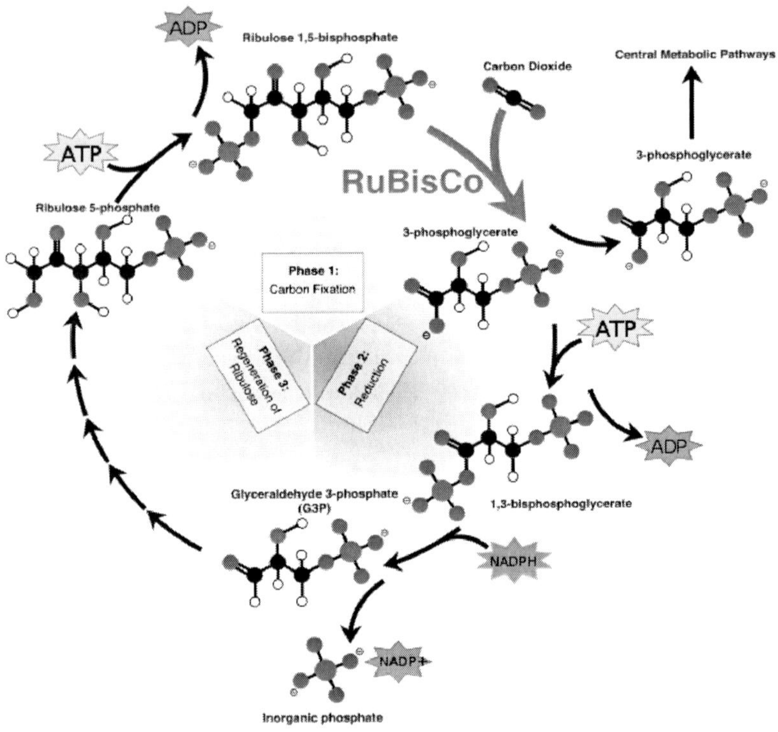

Figure 4.13 Schematic diagram of the Calvin cycle and carbon fixation (Image: Mike Jones) [88].

The overall equation for the light-independent reactions in green plants is as follows [89]:

$$3CO_2 + 9ATP + 6NADPH + 6H^+ \rightarrow C_3H_6O_3\text{-phosphate} + 9ADP + 8P_i + 6NADP^+ + 3H_2O \qquad (4.12)$$

As it is seen from Eq. 4.12, the hydrogen atoms (protons) participate in the light independent reactions; their functions are similar to those which will be considered below in the oxidation–reduction reactions. For the interested, the details of the carbon fixation process and the Calvin cycle can be easily found in the literature.

4.6 Conclusions

New facets of hydrogen were elucidated in this chapter. It was shown that hydrogen species play an extremely important role in photosynthesis. Under the action of solar light, hydrogen atoms from water donate their electrons to facilitate the reduction process in the electron transport chain.

At the same time hydrogen atoms bound oxygen atoms in water easily turn into protons and electrons which is typical for many reactions in which hydrogen atoms participate. This yields charge separation that is important to trigger other processes for energy conversion.

The charged protons serve as a tool to carry out chemiosmosis and vectorial energy transfer to ATP synthase.

Hydrogen atoms actively participate in electron transfer reactions, which will be highlighted in the next chapter.

Chapter 5

Hydrogen in Metabolism: Oxidative Respiration

5.1 Hydrogen and Metabolism

This chapter is devoted to the activity of hydrogen in biological processes and the functions of hydrogen in metabolism.

While metabolism is the set of chemical reactions that proceeds in living organisms to maintain life, light that arises due to the hydrogen burning in the sun is the energy source for all existing living organisms on the earth. As it was in shown in Chapter 3, hydrogen is the main fuel to produce the energy suitable for use on the earth. Hydrogen helps to convert energy from the sun, which is then delivered to the earth as photons to power all biological processes.

However, the pattern in which various living organism consume the energy of light is different. What is really striking and deserves special attention is that hydrogen—this uncanny cosmic hydrogen which powers the sun energy converted with the help of at the same time plays an outstanding role in the processes of energy conversion, accumulation, and transport in metabolism on the earth. Indeed, hydrogen is a ubiquitous matter and enormously functional; new functions of hydrogen in energy processes will be now elucidated.

Green plants and photosynthetic algae use the energy of the sun directly. As discussed in the previous chapter, during the process

Hydrogen Energy for Beginners
Alexander Gavrilyuk

ISBN 978-981-4364-77-5 (Hardcover), 978-981-4364-78-2 (eBook)
www.panstanford.com

of the photosynthesis molecules of glucose are formed from water and carbon dioxide and other more complicated organic compounds are formed if some other inorganic compounds participate in the process. By participation of the enzymes (proteins–catalysts) the reaction of glucose formation takes place under the action of light [26]:

$$6H_2O + 6CO_2 \rightarrow C_6H_{12}O_6 + 6O_2 \tag{5.1}$$

The green plants use the photosynthesis for their growth and development evolving, at the same time, oxygen which is required for the entire living nature. However, animals cannot transform the energy of light into chemical energy directly. They consume the energy by eating green plants or other animals, getting proteins, fats, and hydrocarbons or, in other words, high-energy products produced with the help of the photosynthesis; these being passed via oxidation through a series of consecutive stages with the help of great number of enzymes into simpler lower-energy substances, eventually into water, carbon dioxide, and nitrogen compounds.

Meanwhile, there are also anaerobic organisms which do not require oxygen getting the energy from chemical reactions. For example, nitrite bacteria oxidize nitrogen to nitrite; nitrate bacteria oxidize nitrite to nitrates. These bacteria play a leading role in the nitrogen cycle: they convert ammonia into nitrates , making it easily accessible for green plants.

The energy released by chemical reactions in the cells of animals is transformed, accumulated, and used for the synthesis of new compounds, for creation of non-equilibrium distributions of substances and ions inside the cells, for displacement of substances, organ motion, etc.

The reaction opposite to photosynthesis (Eq. 5.1) is called oxidation. This is a downhill reaction and it is accompanied by 30 eV energy release; about 40% of this energy is used by the cell for useful work while the remaining part of the energy is turned into heat [26].

This oxidation is achieved via a chain of consecutive series of hydrogen atom transfers between various molecules. It proceeds in a way that electron are consecutively passed along a specific chain from an oxidizing molecule to another molecule which is considered to be reduced.

The first point to be stressed in this chapter is that the entire chain of the oxidation–reduction reactions always requires hydrogen atom transfer. Although some steps in this reaction chain can occur as a result of only electron transfer, generally, along the entire chain, it always happens that reaction steps can be passed only by hydrogen atom transfer or proton-coupled electron transfer. A proton is required as an inalienable partner for an electron passing along the electron transfer chain to fulfill the electroneutrality principle and to facilitate the conformation transformations in enzymes, for which proton transfer could also appear as a useful tool.

Enzymes are proteins that catalyze (i.e., increase the rates of chemical reactions). In enzymatic reactions, the molecules at the beginning of the process, called substrates, are converted into different molecules, called products. Almost all chemical reactions in a biological cell need enzymes in order to occur at rates sufficient for life. Since enzymes are selective for their substrates and speed up only a few reactions from among many possibilities, the set of enzymes made in a cell determines which metabolic pathways occur in that cell.

However, the outstanding role of hydrogen atom very often is not properly underlined. Since protons are always present in an aqueous media of the cell, it is usually assumed that the required number of the protons could be consumed from the aqueous media. Because of this, often only electron transfer is considered in the oxidation process, which sometimes hides the role of the protons and incorrectly presents the process.

The oxidation reaction proceeds inside the cell in several stages. Under the action of special enzymes two electrons are transferred from the food product to any primary acceptor. Some other enzymes transfer the electrons further along the chain of the electron transfer to the secondary acceptor and so on. The process finishes with the formation of a water molecule for which each oxygen atom requires two electrons and two protons.

As in photosynthesis, the oxidized forms, NAD^+ and $NADP^+$, of the molecules NAD and NADP play a very important roles here as primary electron acceptors. Similarly, FAD and FMN molecules play an important role in the oxidation–reduction process.

A primary acceptor of electrons flavin adenine dinucleotide (FAD) is a redox cofactor involved in several important reactions in metabolism. FAD can exist in two different redox states, to which it converts by accepting or donating electrons. The molecule consists of a riboflavin moiety (vitamin B_2) bound to the phosphate group of an ADP molecule. The flavin group is bound to ribitol, a sugar alcohol, by a carbon–nitrogen bond, not a glycosidic bond. Thus, riboflavin is not technically a nucleotide; the name flavin adenine dinucleotide is a misnomer [90–92].

FAD can be reduced to $FADH_2$ by accepting two hydrogen atoms (a net gain of two electrons). Via participation of the enzymes, FAD performs dehydrogenation of substances that contain –C–C–groups.

Flavin mononucleotide (FMN) is a biomolecule that, during catalytic cycle, undergoes the reversible interconversion of oxidized (FMN), semiquinone (FMNH) and reduced ($FMNH_2$) forms in the various oxidoreductases. FMN is a stronger oxidizing agent than NAD and is particularly useful because it can take part in both one- and two-electron transfer [93].

In all those reactions the electrons and the protons act together providing the redox process. Their partnership looks natural since they have small sizes, are mobile and with the same value of electric charge of opposite sign.

Studying the oxidation process in a mitochondria helps in deducing the functions of hydrogen in an oxidation process.

5.2 Mitochondrion and Its Functions

5.2.1 Mitochondrion Design

The main function of a mitochondria is to provide the cells with energy via the reaction called oxidative phosphorylation of food substrates; a mitochondrion functions as specific energy power station in a cell.

Each mitochondria is surrounded by two membranes (Fig. 5.1). The membrane architecture is an inalienable feature in living organisms; it makes it possible to separate electric charges and to create a proton gradient across the membrane or, in other words, to charge the biological battery whose energy is employed for the synthesis of ATP molecules.

And is it a function of the mitochondria to generate ATP molecules? How do they perform this function?

Mitochondria are specially constructed supramolecular complexes that span the cell cytoplasm and have a width of 0.2–5 μm. The shape of mitochondria varies from ellipsoidal to that of elongated stick (Fig. 5.1). In some cells the mitochondria span 15–20% of the total cytoplasm volume. The mitochondria in the cells can translocate, and change their size and shape by merging together or decomposing for the smaller ones. Usually they are situated in that cell domain where the metabolism is the most intensive.

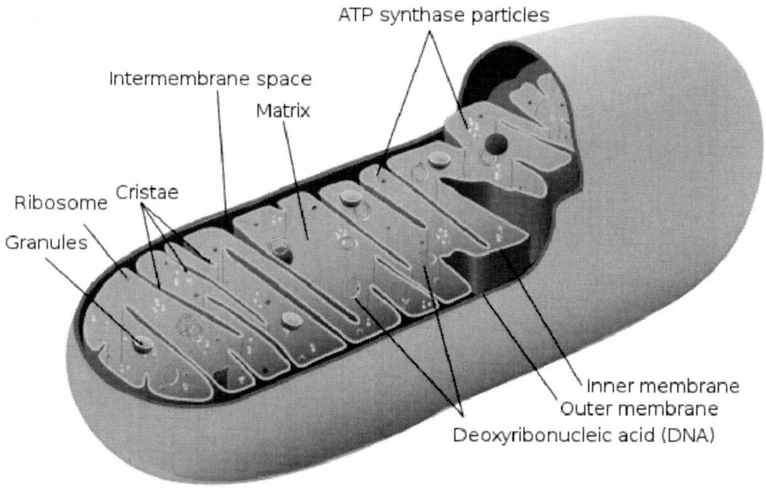

Figure 5.1 Schematic diagram of a mitochondrion structure (Image: Mariana Ruiz Villarreal) [94].

The outer mitochondrial membrane, which encloses the entire organelle (sub-unit), has a protein-to-phospholipid ratio similar to that of an eukaryotic plasma membrane (about 1:1 by weight). It contains large numbers of integral proteins called porins. These porins form channels that allow molecules 5000 Daltons or less in molecular weight to freely diffuse from one side of the membrane to the other [95]. Larger proteins can enter the mitochondrion if a signaling sequence at their N-terminus binds to a large multisubunit protein called translocase of the outer membrane, which then actively moves them across the membrane [96]. The N-terminus is the first part of the protein that exits the ribosome during protein biosynthesis. It often contains signal peptide sequences, "intracellular postal codes" that direct delivery of the protein to the proper organelle. The outer relatively smooth membrane has a structure similar to that of the external membrane of the cell itself.

The intermembrane space is the space between the membranes. Because the outer membrane is freely permeable to small molecules, the concentrations of small molecules such as ions and sugars in the intermembrane space is the same as the cytosol (intracellular fluid) [96]. However, large proteins must have a specific signaling sequence to be transported across the outer membrane, so the protein composition of this space is different from the protein composition of the cytosol. One protein that is localized to the intermembrane space in this way is cytochrome *c* [97].

The inner membrane presents itself a closed sack with numerous folders and prominences directed perpendicular to the mitochondrial (Fig. 5.1). These folders called cristae enhance drastically the inner membrane surface of the membrane. A semi-liquid inner containment restricted by the inner membrane is called a matrix (Fig. 5.1). The matrix is the space enclosed by the inner membrane. It contains about two-third of the total protein in a mitochondrion [96]. The matrix is important in the production of ATP with the aid of the ATP synthase contained in the inner membrane. The matrix contains a highly concentrated mixture of hundreds of enzymes, special mitochondrial ribosomes, tRNA (transfer RNA), and several copies of the mitochondrial DNA genome. The major functions of the enzymes include oxidation of pyruvate and fatty acids, and the citric acid (CA) cycle.

The matrix has no communication with the inter-membrane space. The thickness of the inner membrane is ~70–90 Å. The number of proteins included in the membrane composition exceeds approximately by a factor of two the number of lipids. About one-third of the proteins perform the function of the electron transfer (cytochromes, ferroproteids, and flavoproteins). It is much easier to achieve the necessary concentration of ADP and inorganic substrate necessary for the synthesis of ATP molecules in the space between the external and internal mitochondrial membranes than in the large cell volume. The inner mitochondrial membrane contains proteins with five types of functions [95]:

(i) Those that perform the redox reactions of oxidative phosphorylation

(ii) ATP synthase, which generates ATP in the matrix
(iii) Specific transport proteins that regulate metabolite passage into and out of the matrix
(iv) Protein import machinery
(v) Mitochondria fusion and fission protein

Contrary to the outer membrane, the inner membrane does not contain porins and is highly impermeable to all molecules. Almost all ions and molecules require special membrane transporters to enter or exit the matrix.

Mitochondria possess an autonomous mechanism for the synthesis of the proteins and nucleic acids which gives the possibilities to synthesize their own proteins.

5.2.2 Glycolysis

Prior to carrying out the set of oxidation–reduction reactions along the electron transport chain to produce ATPs, mitochondria prepare necessary substrates.

First stage of preparation of these substrates is called glycolysis. Glycolysis is the process in which glucose is broken down into two molecules of pyruvate (pyruvic acid). It occurs in the cytoplasm of the cell and is an anaerobic (without oxygen) process. Glucose is broken down to extract as much stored energy as possible. The energy that is captured is stored in molecules of ATP.

Glycolysis, as a process may be divided into two steps. In the first step two ATP molecules are required to push one molecule of glucose up the energy hill, and two molecules of G3P result. Glyceraldehyde-3-phosphate, also known as triose phosphate or 3-phosphoglyceraldehyde, and abbreviated as G3P, GADP, GAP, TP, GALP, or PGAL, is a chemical compound that occurs as an intermediate in several central metabolic pathways of all organisms.

In the second stage of glycolysis the two G3P molecules are broken down into two molecules of pyruvate. Concerning this stage several note should be made that are important. First, 2NAD are reduced to 2NADH. Second, 2ADP are phosphorylated to 2ATP. This step pays back the two ATP that were used in the first stage to take glucose up to 2G3P. Third, another 2ADP are phosphorylated to 2ATP. Finally, 2 pyruvate molecules are synthesized. There is a net gain of 2 ATP from glycolysis: 2 ATP used, then 4 ATP synthesized = net gain of 2 ATP.

5.2.3 Transition Stage between Glycolysis and Citric Acid Cycle

So far, the breakdown of glucose into two molecules of pyruvate has resulted in a gain of 2NADH and only 2ATP, which is rather poor result. It is necessary to extract more energy from glucose than that, and mitochondria and oxygen perform this function. The pyruvate molecule is passed into the mitochondrion where it will be completely broken down into carbon dioxide. For simplicity, only one molecule of pyruvate will be followed, although for each glucose molecule that enters glycolysis two pyruvate molecules are produced.

The first step in the next process is to convert the pyruvate into a form that can enter the so-called citric acid (CA) cycle. This step may be characterized as transition stage between glycolysis and the CA cycle. The pyruvate enters the mitochondrion and one carbon is removed as carbon dioxide. Some of the energy released by this reaction is used to reduce NAD to NADH. A molecule of coenzyme A (CoA) attaches to the two-carbon molecule that remains and activates it. The result is acetyl-CoA, an activated form of acetic acid (vinegar). It is acetyl-CoA that enters the CA cycle. So, of the three carbons that entered the mitochondrion in each pyruvate, only two are left to enter the CA cycle.

5.2.4 Citric Acid Cycle

The goal of the oxidation is to generate ATP by breaking the bonds in glucose and capturing as much as possible of the energy stored in that molecule. The CA cycle will produce very little ATP directly, but will generate many molecules of reduced coenzymes NAD and FAD as NADH and $FADH_2$. These will be of critical importance.

The first thing that happens is that the two carbons from acetyl-CoA combine with a four-carbon compound present in the cycle. CA is formed using the six carbons. Now we will follow the cycle until we get back to the four-carbon compound with which we started. (It is named oxaloacetate.)

Two NAD are reduced to two NADH and two carbons are removed that leave as two CO_2. A carbon count will make the situation visual. There were six carbons in glucose. Glycolysis results in two pyruvates, each with three carbons. Each pyruvate that enters the mitochondrion loses one carbon in the "bridge" and two more in the CA cycle. Thus, glucose completely oxidizes to carbon dioxide.

There is a substrate-level phosphorylation of ADP to ATP (phosphorylation without chemiosmosis).

The cycle continues with reactions that resynthesize the original four-carbon oxaloacetate. FAD is reduced to $FADH_2$. Another NAD is reduced to NADH.

Since there are two pyruvates from each glucose molecule, so the complete oxidation of glucose requires two turns of the CA cycle.

Each cycle takes only a fraction of a second. It is important to realize that there are hundreds of CA cycles operating simultaneously in a single mitochondrion and there can be more than a hundred mitochondria in an active cell. Since an organism like you contains billions of cells. So far each glucose molecule has provided us with two ATP from glycolysis and two more from the CA cycle (one ATP from each cycle), thus each pyruvate entering the mitochondrion yields four NADH, one FAD_2, one ATP and three CO_2 (as waste). The bulk of the ATP obtained from aerobic respiration comes from the reduced coenzymes, NADH and $FADH_2$.

5.3 The Behavior of Hydrogen in the Respiration Process

We consider the process highly schematically following only its backbone since the main objective of the consideration is to highlight the role, importance, and behavior of hydrogen. It is clear that at this stage hydrogen atoms participate in the reactions of dehydration preparing coenzymes such as NADH, FAD_2, $FADH_2$, and NADPH.

Several examples which demonstrate this follow below.

The NAD^+ and $NADP^+$ molecules are coenzymes that provide dehydration of compounds containing the group [26]:

$$
\begin{array}{c} | \\ \mathrm{H{-}C{-}OH} \\ | \end{array}
$$

With the help of the enzymes and by the participation of the NAD^+ and $NADP^+$ two hydrogen atoms (two protons and two electrons) are detached from that atomic group. One proton and two electrons combine with the NAD^+ or $NADP^+$ turning them into the reduced forms NADH and NADPH; the second proton is released. For example, the oxidation of the lactate is carried out by this mechanism with formation of pyruvate [26]:

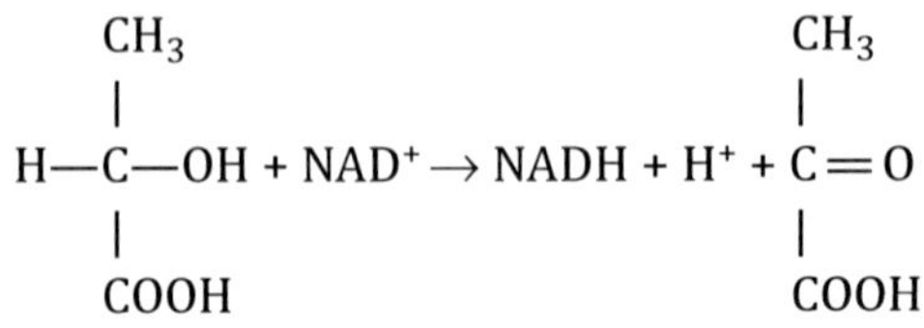

(24)

Lactate Pyruvate

The reduced molecules NADH and NADPH can serve as electron donors in other reactions. They take part in numerous processes of biosynthesis.

Other similar reaction can be also carried out. For example, the primary electron acceptor, namely coenzyme FAD, by participation of enzymes provides dehydration of the substances having the group [26]:

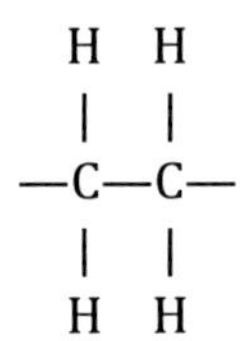

One more example is the oxidation of succinate with the formation of fumarate [26]:

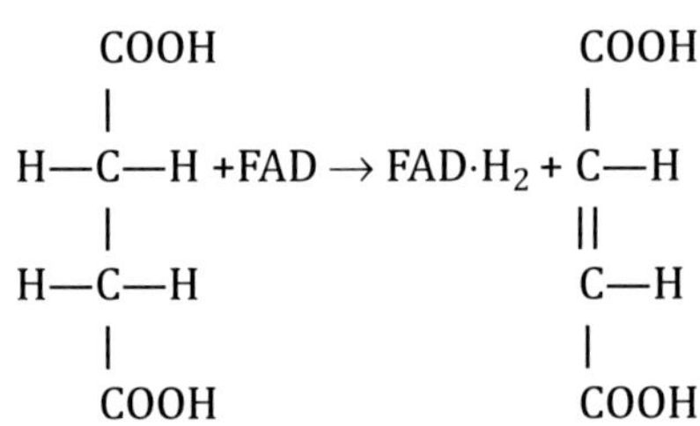

Succinate Fumarate

The electron transfer in the mitochondrial membranes is consecutively provided by the system of the intermediates that are united into highly organized complexes of protein molecules embedded into the membranes. At every stage electron is transferred from one molecule to another lowering its free energy. A molecule donating an electron is called donor whereas a molecules accepting an electron is called an acceptor. The energy transfer is accompanied by oxidation of the donor and reduction of the acceptor, and because of that it is called oxidation–reduction reaction. Often this oxidation–reduction reaction (electron transfer) is proton-coupled. In the latter case a donor looses a hydrogen atom whereas the acceptor acquires the hydrogen atoms. The split off a hydrogen atom from a molecule is called dehydration. In definite sense dehydration is equivalent to oxidation. Sometimes it happens that an electron and a hydrogen atom are transferred simultaneously. The donor and the acceptor form a conjugated donor–acceptor couple. The ability of a donor to donate an electron to a conjugated acceptor is characterized by a value of the redox potential φ.

The electron (hydrogen atom) transfer in the conjugative oxidation–reduction couple is provided by special enzymes [97–99].

The appropriate scheme of the electron transfer along the respiratory chain of the inner membranes looks as: from the food products (pyruvates, succinates, malates), substrates (S·2H), enzymes, and so on; two hydrogen atoms are transferred to NAD^+ and FAD^+ molecules which act as the primary acceptors of electrons.

5.4 Electron Transport Chain

The reduced coenzymes, NADH and $FADH_2$, which were formed in the CA cycle, are used here. These go to the electron transport system in the mitochondrion where they pass their electrons to the carriers in the system. The coenzymes become oxidized FAD and NAD and return to a CA cycle to be reduced again. The electron transport chain is embedded in the inner membrane of the mitochondria. It consists of four large protein complexes, and two smaller mobile carrier proteins. Electrons passing along the electron transport chain lose their energy, some of which is used to energize the membrane. The latter being energized opens its channels for protons what makes it possible to pump the protons out from the matrix into the outer membrane, thus, creating the proton gradient and the potential difference.

NADH is the initial donor in the electron transfer chain. It passes two electrons and two protons to NADH dehydrogenase (complex I, Fig. 5.2). The electrons are transferred to FeS proteins whereas the protons are pumped out from the matrix space of the mitochondria into the intermembrane space.

The electrons are then transferred to the mobile carrier protein known as ubiquinone (Q, Fig. 5.2). Ubiquinone transports the electrons, two at a time, to the next complex in the chain namely to cytochrome b–c_1 (complex III, Fig. 5.2).

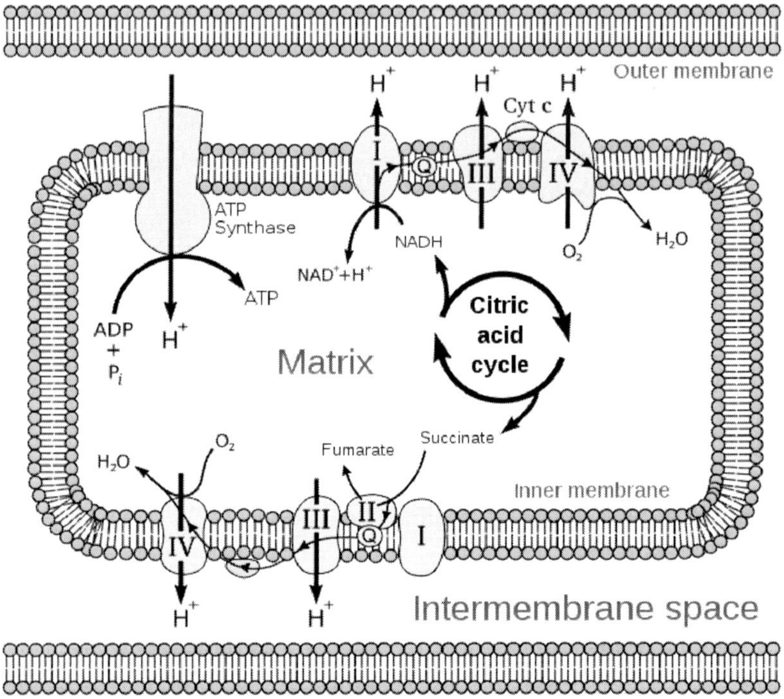

Figure 5.2 The electron transport chain in the mitochondrion. The NADH and succinate generated in the CA cycle are oxidized, providing energy to power ATP synthase (Image: Fvasconcellos) [100].

As each electron makes its way through the complex, a hydrogen ion, or proton, is pumped from the matrix space of the mitochondria into the intermembrane space, helping to maintain the proton gradient.

After affecting the pumping of a proton across the membrane, the electron leaves cytochrome b–c_1 and enters the mobile carrier protein, cytochrome c (cyt c).

The mobile carrier protein cytochrome c transfers electrons, one at a time, to cytochrome oxidase (complex IV, Fig. 5.2). Four electrons must be transferred to the oxidation complex for the next reaction to occur. The next major event is the reaction of the four electrons, a molecule of O_2 (oxygen), and eight protons. The reaction results in the pumping of four hydrogen ions across the inner membrane into the intermembrane space, and the release of two H_2O (water) molecules into the matrix space.

Similar process originates also by succinate which is reduced with the help of the complex II (the enzyme succinate dehydrogenase). The following transport is identical to that originated from NADH.

When the proton gradient and the potential difference arise, the system strives to diminish both but the inner membrane is impermeable to the protons. The protons then have the only option to return in the matrix by passing through the ATP synthase that they normally do (see Fig. 5.2).

ATP synthase accepts one proton from the intermembrane space and releases a different proton into the matrix space to create the energy it needs to synthesize ATP. It does this three times to synthesize one ATP from the substrates ADP and Pi (inorganic phosphate).

The energy of the proticity is used to convert ADP to ATP when it is dissipated on passing ATP synthase under the action of the proton-motive force arising due to the proton gradient between the intermembrane space and the matrix.

It is possible to observe again how elegantly the energy pair (electron and proton that in essence presents a hydrogen atom) acts: the electron passing along the membrane loses its energy; the energy is used to excite the membrane and to pump the protons from the matrix into the intermembrane space, creating the chemiosmosis, the proton gradient, and the proton motive force, or in other words, Mitchell's proticity, which delivers the energy to ATP synthase to create ATP molecules. This behavior is repeated in other biological processes and will be considered later.

One can see that in the oxidation the protons exhibit sophisticated and paradoxical behavior. First they are irreplaceable electron assistants in passing the electron transport chain: without them the electrons could not make this pathway along a chain of downhill reactions. All reactions in the chain run with the release of energy. The protons assist the electrons in loosing this energy being very natural assistants in that case: they are small and mobile. But at the same time, the paradox is just this energy is going exactly to the protons: the electrons energize the inner membrane that makes it possible to pump out the protons from the matrix into the inter membrane space; the proton gradient being installed, or according to Mitchell, the proticity. This is a vectorial transfer of energy; the energy being not dissipated but vectorially transferred to the ATP synthase; the protons being the transmitters that deliver energy to the ATP synthase. The energy produced due to chemical reaction is stored then as a difference of potentials across the membrane. This process is analogous to the charging of a battery, and this biological battery triggers the ATP synthesis.

Furthermore, hydrogen atoms pave the way for protons to pass through the ATP synthase. According to the Davydov's hypothesis [26], the ATP synthase enzyme has narrow inner channels filled with water. This gives possibility to form the quasi one-dimensional chain of the water molecules bonded by the hydrogen bonds. As it was considered in Chapter 2, the proton transport in such filaments is provided by the hydrogen solitons [41,43]. So, the important properties of the hydrogen bonds are exploited for the facilitation of the proton transport.

5.5 Conclusions

Resuming what has been said above, it is possible to state that the hydrogen atoms perform various important functions in the oxidation process (cellular respiration); they provide the oxidation–reduction reaction, participate in creation of the hydrogen osmosis (proticity), and deliver the energy required for the ATP synthesis with the help of hydrogen solitons.

So, step by step we discover the great range of various functions of hydrogen as an element for energy. It can power the stars and, at the same time, participates in fine processes of the energy conversion in living organisms on the earth. Hydrogen is the main actor in the energy cycle of the earth, which is in its essence the electron cycle pumped by solar irradiation (Fig. 5.3).

It is important to stress that the electron source is a hydrogen compound, namely water; dioxygen O_2 is liberated via abstraction of electrons in a four-electron process ($2H_2O \rightarrow O_2 + 4e^- + 4H^+$). The electrons from water are pumped by solar visible light through a series of reactions to higher potentials and then CO_2 is reduced to produce carbohydrates ($C_6H_{12}O_6$). These high-energy electrons stored in the products are consumed by various living organisms as foods and they combine again with O_2 to liberate free energy that is used for biological activities (respiration) reproducing H_2O and CO_2. Free energy is also liberated and used for human needs when the high-energy electrons stored as fossil fuels combine with O_2 by combustion reproducing again H_2O and CO_2. In such energy cycles, solar energy and water (hydrogen product) are the most important since CO_2 is just used to produce materials that can be utilized by biological existence [101].

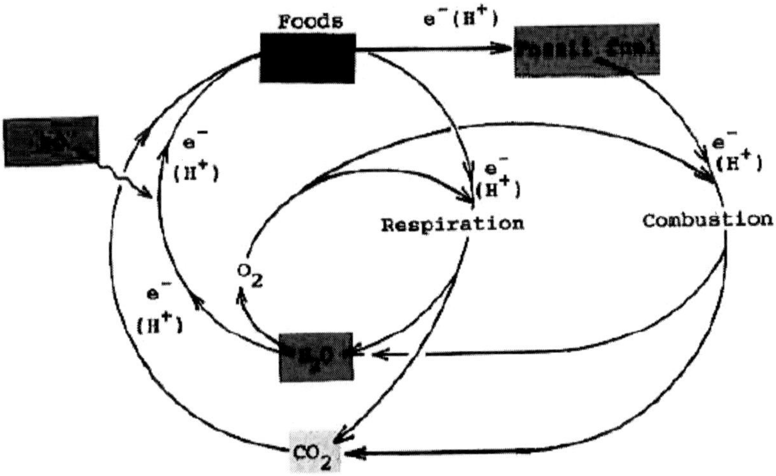

Figure 5.3 Energy cycle on the earth represented by the electron cycle pumped up by solar radiation. Reproduced with permission from Prof. Masao Kaneko, Ibaraki University and The Institute of Biophotochemonics Co. Ltd., Japan [101].

But along with the electron cycle one can also observe the hydrogen cycle, without which it would be impossible to drive the electron cycle. The protons are released from water under the action of solar light and the pumping of the electrons up would not be possible without the chemiosmosis or, in other words, without the Mitchell's proticity that triggers photophosphorylation. The protons are returned back to water when carbohydrates are oxidized either by respiration or combustion. Thus both the electron and proton cycles serve to provide the biological energy conversion, and the role of hydrogen here is irreplaceable.

Chapter 6

Biohydrogen Production

6.1 Introduction

In previous chapters a series of biological processes was being considered where hydrogen plays the main role in energy conversion processes. It makes sense also to elucidate here natural processes which provide production of molecular hydrogen (H_2) that is a dream fuel for future energetics and to make a bridge between the natural processes with participation of hydrogen and the development of new technologies of hydrogen production which often implement the elements of the natural processes and even mimic them.

Biological production of hydrogen (biohydrogen), using microrganisms, is an exciting new area of technology development that offers the potential production of usable hydrogen from a variety of renewable resources. Biological systems provide a wide range of approaches to generate hydrogen that will be highlighted in this chapter. These diverse methods require to be developed thoroughly to make biohydrogen an economically serious alternative to other methods of hydrogen production. To date all these biomethods are not free from limitations that retard their use for hydrogen production. This lecture elucidates the main hydrogen metabolic processes utilized by these different production methods and explains each of their limiting factors. To overcome these limitations, new, exciting, and sophisticated strategies of metabolic engineering are employed

Hydrogen Energy for Beginners
Alexander Gavrilyuk

ISBN 978-981-4364-77-5 (Hardcover), 978-981-4364-78-2 (eBook)
www.panstanford.com

that provide a way to develop the technologies on a higher level. We follow here the review [102,103] that clearly presents the state of art in this swiftly developing field.

6.2 Metabolic Pathway Engineering as a Strategy

Metabolic engineering deals with the redirection of metabolic pathways for enhanced production of existing natural products, production of unnatural products, or degradation of undesired molecules (e.g., environmental contaminants) [102].

Metabolic engineering can be employed for hydrogen production. There are various pathways for the hydrogen production, and each one is dependent on the hydrogen-producing enzymes acting inside of the individual microorganism.

Although these enzymes catalyze a very simple redox reaction,

$$2H^+ + 2e^- \leftrightarrow H_2 \quad (6.1)$$

in order to realize it, most hydrogen-evolving enzymes must have complex metalloclusters as active sites and require special proteins [102,103].

While biohydrogen systems can produce H_2, no commercial systems are yet available, and problems concerning the practical application of biohydrogen still remain unsolved. However, biohydrogen production has a huge variety of technologies that can be potentially developed. At the same time, the success in this field is also dependent on creation of commercial systems of hydrogen storage. Nevertheless, for us the scientific volubility of this method is the most interesting since the ideas can excite fantasy which in turn can yield the technological success.

6.3 Nitrogenase and Nitrogen Fixation

There are three main hydrogen-evolving enzymes operating in most biohydrogen-evolving systems: nitrogenases, [NiFe]-hydrogenases, and [FeFe]-hydrogenases [102,103].

Nitrogenases catalyze a reaction that fixes nitrogen into ammonia and is accompanied by the obligatory reduction of protons (H^+) to hydrogen [104].

The reaction that this enzyme performs is:

$$N_2 + 8H^+ + 8e^- + 16ATP \rightarrow 2NH_3 + H_2 + 16\ ADP + 16P_i \quad (6.2)$$

Here it is worthwhile to underline the exclusive importance of hydrogen (hydrogen protons) in the nitrogen fixation that is a vitally important process. This process is exclusively essential for life since it fixes nitrogen required to biosynthesize the basic building blocks of life, for example, nucleotides for DNA and RNA and amino acids for proteins. However, to fix nitrogen is not an easy task since an N_2 molecule is very stable and it is difficult to break it apart into individual nitrogen atoms.

Nitrogen-fixing bacteria have the ability to convert nitrogen gas into ammonia with the help of hydrogen protons, which is easily combined with other raw materials to form the building blocks of proteins and nucleic acids. In nitrogen-fixing bacteria, the enzyme nitrogenase drives the reaction with a large quantity of ATP, and uses a collection of metal ions, including an unusual molybdenum ion, to drive the reaction.

Being a strong catalyst, nitrogenase performs this function using hydrogen protons, and is powered by ATP molecules.

The nitrogenase complex consists of two proteins encoded by three so-called structural nif genes [105]. The MoFe protein, shown in blue and purple (Fig. 6.1), contains all the necessary machinery to perform the reaction, but requires addition of six electrons for each nitrogen molecule that is split into two ammonia molecules.

The Fe protein, shown in green, uses the breakage of ATP to pump these electrons into the MoFe protein. In the typical reaction, two molecules of ATP are consumed for each electron transferred. Nitrogenase also converts hydrogen ions to hydrogen gas at the same time, thus consuming even more ATP in the process. This is a large investment in energy, but necessary if nitrogen is not available in the environment. Fortunately, nitrogen-fixing bacteria are widely spread through the world, and are often found in the vicinity of plants. For instance, legumes have special nodules in their roots that provide an accommodation to the bacteria. The plants provide shelter and even few essential nutrients to make the accommodation of bacteria comfortable so that they can supply nitrogen steadily.

At the heart of nitrogenase is an unusual complex of iron, sulfur and molybdenum ions that apparently performs the nitrogen-fixing reaction. A series of cofactors supply this MoFe-cluster with electrons

[105]. As seen in Fig. 6.2 on the left, electrons start at a pair of ATP molecules (two at each end of the dimeric complex), move inwards into the iron–sulfur cluster, then to the P-cluster, and finally to the MoFe-cluster. The three metal clusters are shown in the right side of Fig. 6.2. The MoFe-cluster is at the bottom, with the molybdenum atom in bright red. A homocitrate molecule, shown with white carbon atoms and pink oxygen atoms, helps to stabilize this unusual metal ion [105]. The P-cluster is in the middle and the iron–sulfur cluster of the Fe protein is at the top.

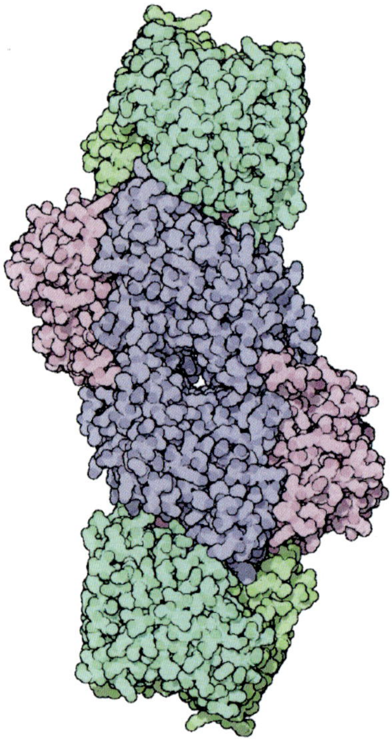

Figure 6.1 Schematic diagram of nitrogenase complex. Image reproduced with permission from the RCSB PDB February 2002 Molecule of the Month by David Goodsell [105].

The metal clusters are the centerpiece of nitrogenase that contains both the MoFe protein (in blue and purple at the center) and two copies of the Fe protein dimer bound on either end (shown in green, see Fig. 6.3). The metal ions are easily displayed using a space filling representation, which reveals the iron–sulfur cluster, the P-cluster, and the FeMo-cluster arranged in a row. The ATP binding site is revealed in this structure by using an unusual analogue of ATP: an ADP molecule with an aluminum fluoride ion. Two of these molecules bind at each end, forming a stable but inactive complex with the Fe protein, essentially gluing the Fe protein to the FeMo protein so its structure can be solved [105].

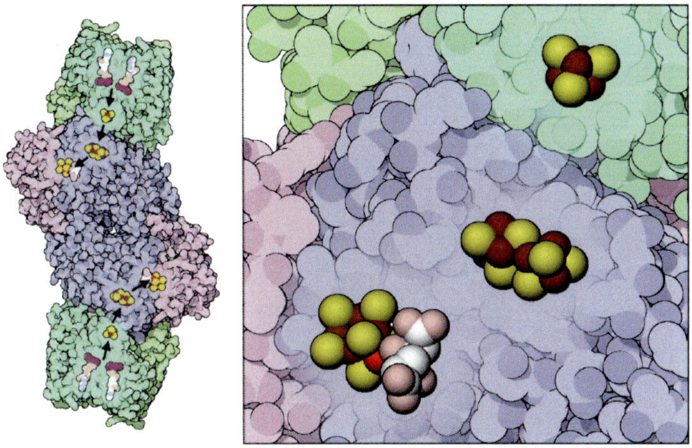

Figure 6.2 Schematic diagram of reaction center in the nitrogenase. Image reproduced with permission from the RCSB PDB February 2002 Molecule of the Month by David Goodsell [105].

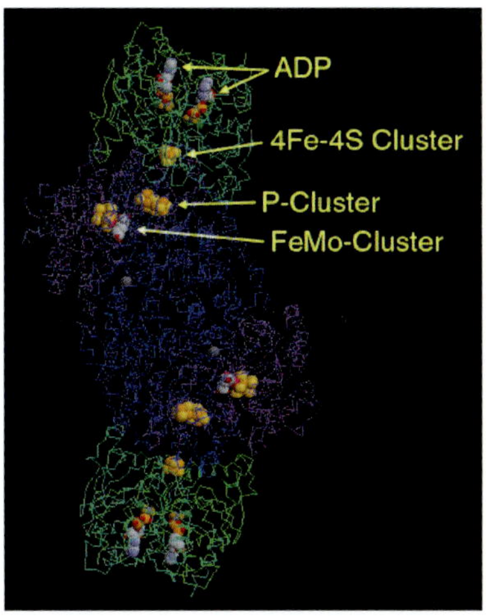

Figure 6.3 Structure of metal clusters of the nitrogenase. Image reproduced with permission from the RCSB PDB February 2002 Molecule of the Month by David Goodsell [105].

6.4 Hydrogenases

However, there are other enzymes named hydrogenases that work more efficiently for hydrogen production. For instance, [NiFe]-hydrogenases (Fig. 6.4) work about 15 times more productive than nitrogenases for hydrogen and do not require ATP [102,103]. Furthermore, [NiFe]-hydrogenases can perform two functions: they produce hydrogen, but can also work as uptake hydrogenases, utilizing the electrons from hydrogen to reduce NADP [102,103].

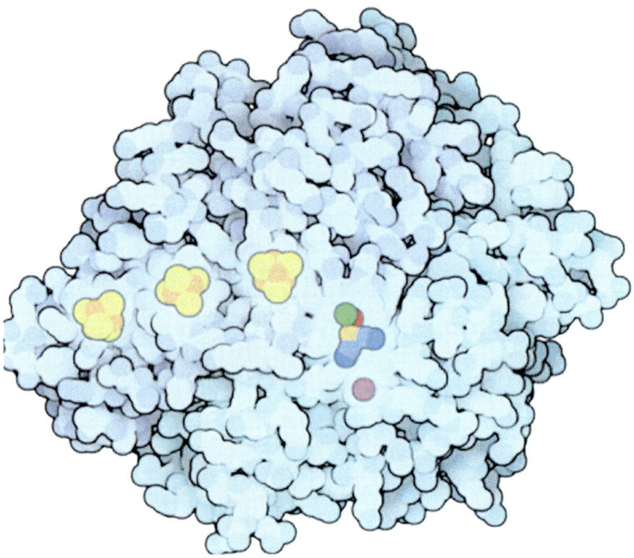

Figure 6.4 The structure of Ni–Fe hydrogenase. Image reproduced with permission from the RCSB PDB March 2009 Molecule of the Month by David Goodsell [106].

The most-efficient hydrogen-producing enzymes are [FeFe]-hydrogenases (Fig. 6.5), which can achieve an efficiency 1000 times higher than nitrogenases and ~10–100 times higher than [NiFe]-hydrogenases [107]. These hydrogenases consist of one protein containing a Fe–Fe catalytic core and can have a variety of electron donors and acceptors [106]. [FeFe]-hydrogenases can either produce or consume hydrogen depending on environment.

Hydrogenases constitute a family of enzymes found in certain microorganisms, which catalyze the reversible oxidation of hydrogen gas. The interest in the enzyme is heated up since these cultures of hydrogenase-containing microorganisms have the ability to produce a constant output of hydrogen gas (H_2) from using only sunlight and water. Electrons pumped into the catalytic core combine with hydrogen ions from the cell environment at the hydrogenase's buried H-cluster active site to produce H_2 (Fig. 6.5). If harnessed properly, hydrogenase and/or hydrogenase-containing organisms could be used to supply affordable and renewable H_2 to be used as an energy fuel, and thus solve the "supply" aspect of the future hydrogen economy.

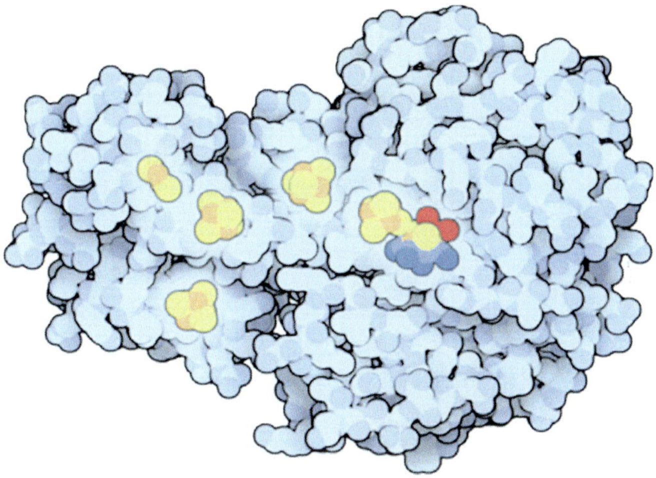

Figure 6.5 A structure of Fe–Fe hydrogenase. Image reproduced with permission from the RCSB PDB March 2009 Molecule of the Month by David Goodsell [106].

Now it is time to consider various methods of biohydrogen production.

6.5 Direct Photobiological Production

Photosynthetic production of hydrogen from water is a biological process that can convert sunlight into useful, stored chemical energy by the following general reaction:

$$2H_2O \rightarrow \text{(light energy) } 2H_2 + O_2 \qquad (6.3)$$

Direct photolysis, the most extensively studied method of biohydrogen production, is carried out using photosynthetic microalgae or cyanobacteria (also known as blue-green algae, *cyanophyceae, or cyanophytes)*. The examples of green algae and cyanobacteria are shown in Figs. 6.6 and 6.7, correspondingly.

Figure 6.6 An example of green algae (Image: Kristian Peters) [108].

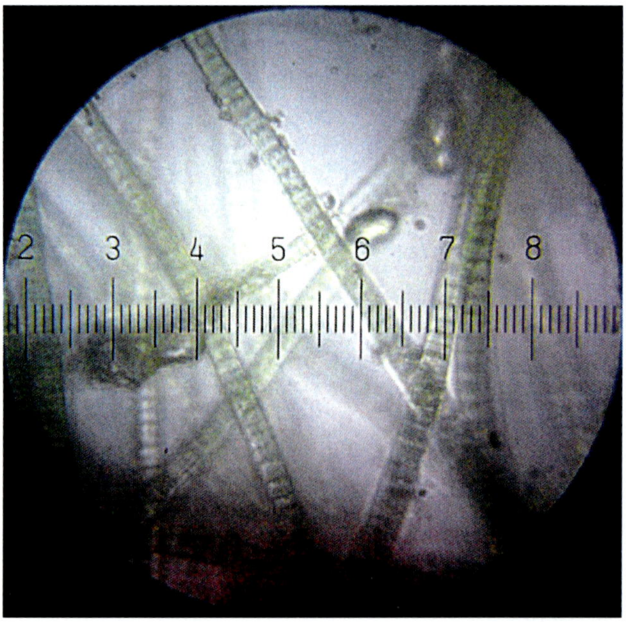

Figure 6.7 An example of cyanobacteria. (Image: Matthewjparker) [109].

This method utilizes the [FeFe]-hydrogenases and [NiFe]-hydrogenases discovered in these photosynthetic organisms to convert solar energy and water into hydrogen.

Figure 6.8 shows the already familiar to us process of photosynthesis which is redirected to the production of hydrogen. Green algae, under anaerobic conditions, can either use H_2 as an electron donor in the CO_2-fixation process or evolve H_2. Hydrogen production by green microalgae requires several minutes to a few hours of anaerobic incubation in the dark to induce the synthesis and/or activation of enzymes involved in H_2 metabolism, including a reversible hydrogenase enzyme. The hydrogenase combines protons (H^+) in the medium with electrons (donated by reduced ferredoxin) to form and release H_2.

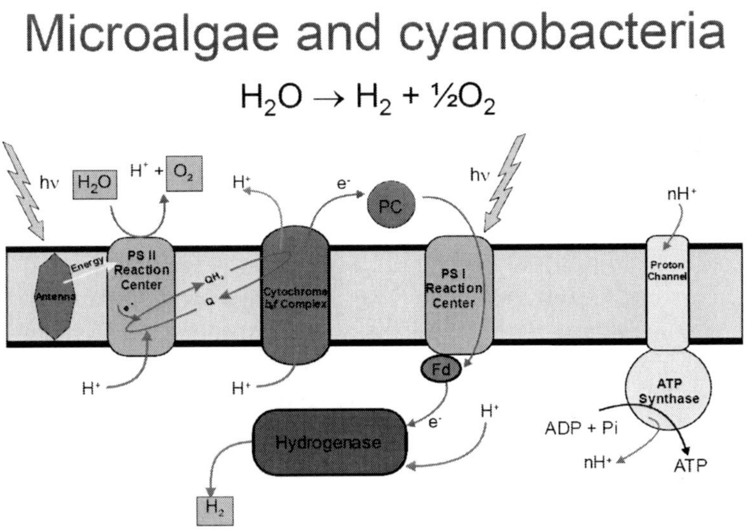

Figure 6.8 Schematic diagram of direct photobiological production of hydrogen by microalgae and cyanobacteria. Reproduced with permission from Prof. Devens Gust, ASU Center for Bio-Inspired Solar Fuel Production and Department of Chemistry and Biochemistry, Arizona State University, AZ, USA [110].

Thus, green microalgae possess the genetic, enzymatic, metabolic, and electron-transport machinery to photoproduce H_2 gas. The synthesis of H_2 permits sustained electron flow through the electron-transport chain, which supports synthesis of ATP [111]. Here, ferrodoxin (Fd), instead of supplying electrons to form NADP, transfers them to the hydrogenase that consumes the protons also from the lumen and converts both into molecular hydrogen.

The process of algal photosynthesis oxidizes H_2O and evolves O_2. Light energy absorbed by photosystem II (PS II) generates electrons which are transferred to ferredoxin, using light energy absorbed by photosystem I (PS I). A reversible hydrogenase accepts electrons directly from the reduced ferredoxin to generate H_2 [112]. Unfortunately, the weak point of the process is that the hydrogenase enzyme responsible for evolution of molecular H_2 is highly sensitive to O_2; thus, photosynthetic production of H_2 and O_2 must be temporally and/or spatially separated.

In a two-phase process, CO_2 is first fixed into hydrogen-rich substrates during normal photosynthesis (Phase 1), followed by light-mediated generation of molecular H_2 when the microalgae are incubated under anaerobic conditions (Phase 2). To achieve Phase 2 of the two-stage process the microalgae must be incubated in medium that does not contain sulfur-containing nutrients [111].

When cultures of the green alga *Chlamydomonas reinhardtii* are deprived of inorganic S, the rates of O_2 synthesis and CO_2 fixation decline significantly within 24 h (in the light). The reason for this loss of activity is due to the need for frequent replacement of the H_2O-oxidizing protein D1 in the PS II reaction center [112,113]. Depletion of sulfur blocks the synthesis of the D1 polypeptide chain which contains many sulfur containing amino acids, such as cysteine and methionine. Despite the greatly reduced, photosynthetic capability respiration continues, and after ~22 h of sulfur deprivation, *C. reinhardtii* cultures maintained in the light become anaerobic, and begin to synthesize H_2 [111]. Based on this phenomenon, systems for sustained H_2 gas production in *C. reinhardtii* have been developed [114,115].

Cyanobacteria can also perform direct biophotolysis, but do so primarily through nitrogenases and low-activity bidirectional [NiFe]-hydrogenases (Fig. 6.9) [116]. These enzymes do not produce hydrogen at the high rates inherent to [FeFe]-hydrogenases.

The nitrogenases in cyanobacteria are coupled to uptake [NiFe]-hydrogenases and serve to provide possibilities for the cell to reutilize and regain the hydrogen formed by the nitrogenases [116]. The bidirectional hydrogenase, which is present in both nitrogen and non-nitrogen fixing strains, can evolve hydrogen under anaerobic conditions but only at low hydrogen partial pressures [117].

Therefore, utilizing this bidirectional hydrogenase also requires constant sparging of an inert gas. In order to combat this problem, the bidirectional, oxygen-tolerant [NiFe]-hydrogenase genes, hydS and hydL have been recently expressed from *Thiocapsa roseopersicina* in the cyanobacteria *Synechococcus PCC7942* [118]. Figure 6.9 shows the scheme of hydrogen production in cyanobacteria.

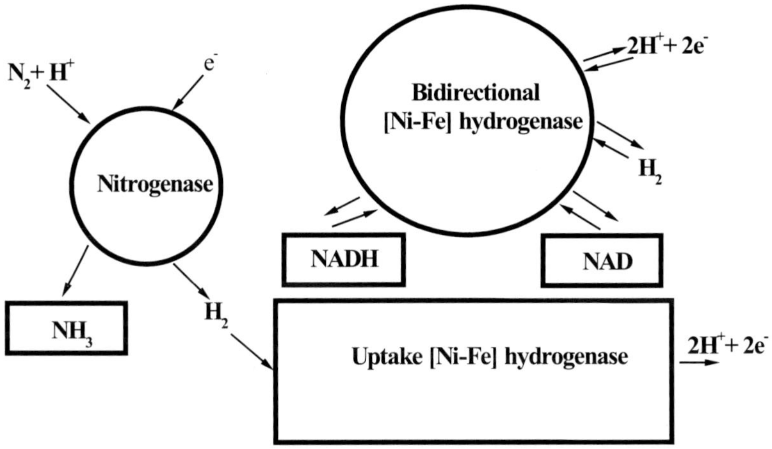

Figure 6.9 Scheme of hydrogen production in cyanobacteria. Ammonia production by the nitrogenase yields hydrogen formation, some of which is consumed by the uptake hydrogenase. The bidirectional hydrogenase catalyzes the formation of hydrogen. It must be protected from oxygen inactivation. Adapted from [119].

The bidirectional hydrogenase from cyanobacteria also does not require ATP to function and can suffer from a buildup of ATP which then inhibits electron flow [117]. Fortunately, a variety of genetic tools exist to metabolically engineered cyanobacteria [120]. Improvement of cyanobacterial hydrogen production can be achieved by expressing a clostridial [FeFe]-hydrogenase in the cyanobacteria *Synechococcus* elongates [121,122]. It must be noted, however, that the mechanism by which cyanobacteria are able to express an active [FeFe]-hydrogenase, without the co-expression of the appropriate maturation proteins, is still unknown [123,124]. The uptake of hydrogen by the [NiFe]-hydrogenase in cyanobacteria can also be eliminated to increase hydrogen yield [125]. For further information a recent review [117] can be advised.

6.6 Indirect Photobiological Production

Cyanobacteria can also synthesize and evolve H_2 through photosynthesis via the following processes [103]:

$$12H_2O + 6CO_2 \rightarrow \text{(light energy) } C_6H_{12}O_6 + 6O_2 \quad (6.4)$$

$$C_6H_{12}O_6 + 12H_2O \rightarrow \text{(light energy) } 12H_2 + 6CO_2 \quad (6.5)$$

Cyanobacteria are a large and diverse group of photoautotrophic (an autotroph, or producer, is an organism that produces complex organic compounds from simple inorganic molecules) microorganisms, which evolved and diversified early in the earth's history [125]. Cyanobacteria contain photosynthetic pigments, such as *chl a*, *carotenoids*, and *phycobiliproteins* and can perform oxygenic photosynthesis. They have different morphologies that include unicellular, filamentous, and colonial species. Within the filamentous cyanobacteria, vegetative cells may develop into structurally modified and functionally specialized cells, such as the akinetes (resting cells) or heterocysts (specialized cells that perform nitrogen-fixation) [119]. The nutritional requirements of cyanobacteria are simple: air (N_2 and O_2), water, mineral salts, and light [126].

Species of cyanobacteria may possess several enzymes directly involved in hydrogen metabolism and synthesis of molecular H_2. These include nitrogenases which catalyze the production of H_2 as a by-product of nitrogen reduction to ammonia, uptake hydrogenases which catalyze the oxidation of H_2 synthesized by the nitrogenase, and bi-directional hydrogenases which have the ability to both oxidize and synthesize H_2 [119].

Hydrogen production by cyanobacteria has been studied for over three decades and has inferred that efficient photoconversion of H_2O to H_2 is influenced by many factors [127]. Hydrogen production has been assessed in a very wide variety of species and strains under a vast range of culture conditions [127]. The rates of H_2 evolution are as wide-ranging as the species and conditions used in these studies. Rates of H_2 production by non-nitrogen-fixing cyanobacteria range from 0.02 μmol H_2/mg *chl a*/h (*Synechococcus PCC 6307*) to 0.40 μmol H_2/mg *chl a*/h (*Aphanocapsa montana*) [128].

These rates are very low compared with those of *heterocystous cyanobacteria*, which range from 0.17 μmol H_2/mg *chl a*/h (*Nostoc linckia* IAM M-14) to 4.2 μmol H_2/mg *chl a*/h (*Anabaena variablilis* IAM M-58) [129].

Because of the higher rates of H_2 production by *Anabaena* species and strains, these have been subject of intensive investigations for the past several years. Mutant strains of *A. variabilis* have demonstrated significantly higher rates of H_2 production compared with wild-type strains. *A. variabilis PK84*, for example, produced H_2 at a rate of 6.91 nmol/μg of protein/h (in 350 ml cultures). When *A. variablis PK84* was cultured under conditions of nitrogen starvation, the rate of H_2 synthesis was 12.6 nmol/μg of protein/h (in 350 ml cultures). The concentration of total protein in the culture was 28.2 μg/ml of culture [130]. Assuming no change as the culture volume is scaled up, 1 l of culture would contain 28,200 μg protein and would produce 355,320 nmol H_2/h or approximately 0.355 mmol $H_2/(l \times h)$.

6.7 Photofermentation

Purple non-sulfur bacteria evolve molecular H_2 catalyzed by nitrogenase under nitrogen-deficient conditions using light energy and reduced compounds (organic acids):

$$C_6H_{12}O_6 + 6H_2O \rightarrow \text{(light energy)}\ 12H_2 + 6CO_2 \qquad (6.6)$$

These photoheterotrophic bacteria have been investigated for their potential to convert the energy of light into H_2 using waste organic compounds as substrate [131–135] in batch processes [136], continuous cultures [137,138], or cultures of bacteria immobilized in carrageenan (linear sulfated polysaccharides) [139], in agar gel [140], on porous glass [134], on activated glass [141], or on polyurethane foam [133].

In general, rates of hydrogen production by photoheterotrophic bacteria are higher when the cells are immobilized in or on a solid matrix, than when the cells are free-living. Continuous cultures of *Rhodopseudomonas capsulata* and *Rhodobacter spheroides* were reported to produce H_2 at rates that range from 40 to 50 ml H_2/l of culture/h [131,142,143], 80 ml to 100 ml H_2/l of culture/h [135]. Continuous cultures of *Rhodospirillum rubrum* were reported to produce H_2 at a rate of 180 ml H_2/l of culture/h [138]. Cultures of *Rb. spheroides* immobilized on porous glass, on the other hand, were reported to produce H_2 at a rate of 1:3 ml H_2/ml/h (1:3 l H_2/l of immobilized culture/h) [134]. Rates of H_2 production by *Rb. spheroides* GL1 immobilized on activated glass were 3.6–4.0 ml H_2/ml/h [141,144]. If the system of culturing *Rb. spheroides* on porous glass could be scaled-up without compromising the rate of H_2 synthesis, this would result in rates of 3.6–4.0 l H_2/l of immobilized culture/h, which would correspond to 0.145 mmol $H_2/(l \times h)$ to 0.161 mmol $H_2/(l \times h)$.

6.8 Hydrogen Synthesis via the Water–Gas Shift Reaction of Photoheterotrophic Bacteria

Certain photoheterotrophic (A heterotroph is an organism that cannot fix carbon and uses organic carbon for growth) bacteria within the superfamily *Rhodospirillaceae* can grow in the dark using CO as the sole carbon source to generate ATP with the concomitant release of H_2 and CO_2 [145–147]. The oxidation of CO to CO_2 with the release of H_2 occurs via a water–gas shift reaction:

$$CO(g) + H_2O(l) \rightarrow CO_2(g) + H_2(g) \; \Delta G^\circ = -20 \text{ (kJ/mol)} \qquad (6.7)$$

In these organisms, however, the reaction is mediated by proteins coordinated in an enzymatic pathway. The reaction takes place at low (ambient) temperature and pressure.

Thermodynamics of the reaction are very favorable to CO-oxidation and H_2 synthesis since the equilibrium is strongly to the right of this reaction.

Stoichiometric amounts of CO_2 and H_2 are produced during CO-oxidation [147]. The enzyme that binds and oxidizes CO, carbon monoxide: acceptor oxidoreductase is part of a membrane bound enzyme complex [147,148]. *Rubrivivax gelatinosus* CBS is a purple non-sulfur bacterium that not only performs the CO–water–gas shift reaction in darkness, converting 100% CO in the atmosphere into near stoichiometric amounts of H_2, it also assimilates CO into new cell mass in the light (via CO_2 fixation) when CO is the sole source of carbon [149,150]. Even when an organic substrate is available with CO, *R. gelatinosus* CBS will utilize both substrates simultaneously, indicating that the CO-oxidation pathway is fully functional even when a more favorable substrate is included [151].

6.9 Dark Fermentation

Hydrogen can be produced by anaerobic bacteria, grown in the dark on carbohydrate-rich substrates. Fermentation reactions can be operated at mesophilic (25–40°C), thermophilic (40–65°C), extreme thermophilic (65–80°C), or hyperthermophilic (>80°C) temperatures. While direct and indirect photolysis systems produce pure H_2, dark-fermentation processes produce a mixed biogas containing primarily H_2 and carbon dioxide (CO_2), but which may also contain lesser amounts of methane (CH_4), CO, and/or hydrogen sulfide (H_2S).

Bacteria known to produce hydrogen include species of *Enterobacter, Bacillus, and Clostridium.* Carbohydrates are the preferred substrate for hydrogen-producing fermentations.

Glucose, isomers of hexoses, or polymers in the form of starch or cellulose, yield different amounts of H_2 per mole of glucose, depending on the fermentation pathway and end-products. When acetic acid is the end-product, a theoretical maximum of 4 mole H_2 per mole of glucose is obtained:

$$C_6H_{12}O_6 + 2H_2O \rightarrow 2CH_3COOH + 3H_2 + 2CO_2 \quad (6.8)$$

Hydrogen production by these bacteria is highly dependent on the process conditions such as pH, hydraulic retention time (HRT), and gas partial pressure, which affect metabolic balance. Thus, fermentation end-products produced by a bacterium depend on the environmental conditions in which it grows. Reduced fermentation end-products like ethanol, butanol, and lactate, contain hydrogen that has not been liberated as gas. To maximize the yield of H_2, the metabolism of the bacterium must be directed away from alcohols (ethanol, butanol) and reduced acids (lactate) towards volatile fatty acids (VFA). *C. pasteurianum* is a classic H_2 and VFA producer, but its metabolism can be directed away from H_2 production and towards solvent production by high glucose concentrations (12.5% w/v), by CO (which inhibits Fe-hydrogenase), and by limiting Fe concentrations [152].

The partial pressure of H_2 (p_{H2}) is an extremely important factor for continuous H_2 synthesis. Hydrogen synthesis pathways are sensitive to H_2 concentrations and are subject to end-product inhibition. As H_2 concentrations increase, H_2 synthesis decreases and metabolic pathways shift to production of more reduced substrates such as lactate, ethanol, acetone, butanol, or alanine. As the temperature increases, however, conditions that favor reaction (6.8), shown above, are less affected by H_2 concentration.

6.10 Perspectives of Biohydrogen Production

Both light-dependent (direct photolysis, indirect photolysis, and photofermentation) and dark-fermentation biohydrogen systems are under intensive investigation to find ways to improve both the rates of H_2 production and the ultimate yield of H_2.

Hydrogen production by direct photolysis using green algae is currently limited by three parameters [153]: (i) solar conversion efficiency of the photosynthetic apparatus; (ii) H_2 synthesis processes (i.e., the need to separate the processes of H_2O oxidation from H_2 synthesis); and (iii) bioreactor design and cost. A number of approaches to improve H_2 production by green algae are currently under investigation.

These include genetic engineering of light-gathering antennae [154], optimization of light input into photobioreactors [155], and improvements to the two-phase H_2 production systems used with green algae [156,157].

Hydrogen production via indirect photolysis using cyanobacteria can be improved by screening for wild-type strains possessing highly active hydrogen evolving enzymes (nitogenases and/or hydrogenases), in combination with high heterocyst formation [127]. Genetic modification of strains to eliminate uptake hydrogenases and increase levels of bidirectional hydrogenase activity may yield significant increases in H_2 production. For example, a mutant strain of *Anabaena* (AMC 414), in which the large subunit of the uptake hydrogenase (hupL) was inactivated by a deletion event [158], produced H_2 at a rate that was more than twice that of the parent wild-type strain, *Anabaena* PCC 7120 [159]. Finally, optimization of cultivation conditions such as light intensity, pH, temperature, and nutrient content, as well as maintaining low partial pressures of H_2 and CO_2 will contribute to increased H_2 production.

Many of the parameters that limit H_2 production by green algae and cyanobacteria also apply to photoheterotrophic bacteria used in photofermentation systems.

Efforts to improve H_2 production in these bacteria also include elimination of competing microorganisms, such as microalgae, using light filters [160], co-cultures of photoheterotrophic bacteria with different light-utilization characteristics [161], two-phase fermentation systems in which photoheterotrophic bacteria utilize substrates produced by anaerobic bacteria during dark-fermentation processes [162], novel photobioreactor designs [161,163], and use of specific waste streams as substrate for photofermentation [164].

Dark-fermentation systems also appear to have the great potential to be developed as practical biohydrogen systems. Substantial improvements, however, can be made through rapid gas removal and separation, bioreactor design, and genetic modifications in the microorganisms.

Improvements in gas separation will contribute to significant increases in H_2 production. As discussed above, the p_{H2} is an extremely important factor for continuous H_2 synthesis. As H_2 concentrations increase, H_2 synthesis decreases and metabolic activity shifts to pathways that synthesize more reduced substrates.

The concentration of CO_2 also affects the rate of synthesis and final yield of H_2. Cells synthesize succinate and formate using CO_2, pyruvate, and reduced nicotinamide adenine dinucleotide (NADH) via the hexose monophosphate pathway [165]. This pathway competes with reactions in which H_2 is synthesized by NADH-dependent hydrogenases (which oxidize NADH to NAD+). Efficient removal of CO_2 from the fermentation system would reduce competition for NADH and thus result in increased H_2 synthesis.

In dark-fermentation processes, this problem is compounded by the fact that the gas produced is primarily a mixture of H_2 and CO_2, but may also contain other gases such as CH_4, H_2S, or ammonia (NH_4). Moreover, the H_2 content of the gas mixture may be low (~50%). PEMFCs require high-purity H_2 (~99%) and cannot tolerate CO at concentrations ~10 ppm [127]. To both maintain continuous H_2 synthesis and remove diluting (CO_2; CH_4) and/or contaminating (CO) gases, rapid removal of the gases and purification of the H_2 are essential.

Methods to enhance H_2 production by removal of H_2 and CO_2 include sparging with N_2 or argon (Ar) gas and applying a vacuum to the head space to reduce the H_2 partial pressure [166]. An increase in H_2 production of over 50% was obtained after periodic sparging with N_2 [167]. Too much sparging, however, dilutes the H_2 and creates a serious problem with respect to separation of the H_2 from the sparging gas.

Removal and selective purification of H_2 has been demonstrated using membrane technologies. A hollow fiber/silicone rubber membrane effectively reduced biogas partial pressure in a dark-fermentation system, resulting in a 10% improvement in the rate of H_2 production and a 15% increase in H_2 yield [168], and a non-porous, synthetic polyvinyl trimethylsilane membrane was used for production of high-purity H_2 from three different H_2 producing bioreactor systems [169].

Substantial gains in H_2 production can also be achieved through optimization of bioreactor designs. Using fixed-bed bioreactors containing an undefined consortium of mesophilic bacteria, Chang et al. [170] observed rates of H_2 synthesis far greater than other studies [121 mmol H_2 = (l × h)]. This remarkable rate of H_2 production was achieved using activated carbon as a support matrix that allowed retention of the H_2 producing bacteria within the bioreactor, plus a membrane filter system to remove the biogas and maintain low gas partial pressures.

A theoretical study of H_2 production by (hyper) thermophilic bacteria in a high rate bioreactor was conducted [102]. In this system, (hyper) thermophilic bacteria would be growing as a biofilm within an anaerobic trickling filter containing packing material with a very high surface area. The liquid-suspended biomass substrate would be passed continuously through the filter, so that the biomass substrate, the H_2 producing bacteria, and the resulting gas phase would be in close proximity. Low H_2 and CO_2 partial pressures would be maintained by stripping the H_2 gas from the bioreactor using steam within the filter. Finally, gains in H_2 production may be achieved through genetic modification of H_2 producing bacteria. Hydrogen producing strains of bacteria can be genetically modified in several ways to increase H_2 synthesis, including (1) over-expression of cellulases, hemi-cellulases, and lignases that can maximize substrate (glucose) availability; (2) elimination of uptake hydrogenases; (3) over-expression of H_2 evolving hydrogenases that have themselves been modified to be oxygen tolerant; and (4) elimination of metabolic pathways that compete for reducing equivalents required for H_2 synthesis.

Biohydrogen technologies are still in their infancy. Existing technologies offer potential for practical application, but if biohydrogen systems are to become commercially competitive they must be able to synthesize H_2 at rates that are sufficient to power fuel cells of sufficient size to do practical work. Further research and development aimed at increasing rates of synthesis and final yields of H_2 are essential.

Optimization of bioreactor designs, rapid removal and purification of gases, and genetic modification of enzyme pathways that compete with hydrogen producing enzyme systems offer exciting prospects for biohydrogen systems.

Even a 10-fold increase in the rate of H_2 synthesis by some dark-fermentation systems would reduce bioreactor size dramatically. This would greatly facilitate overcoming the engineering challenges of scale-up, and create new opportunities for practical applications.

Other interesting applications for the biohydrogen production may unexpectedly appear.

The team at the Harbin Institute of Technology has optimized hydrogen production from organic matter between 4°C and 9°C by using a microbial electrolysis cell (MEC) (see Fig. 6.10) [171].

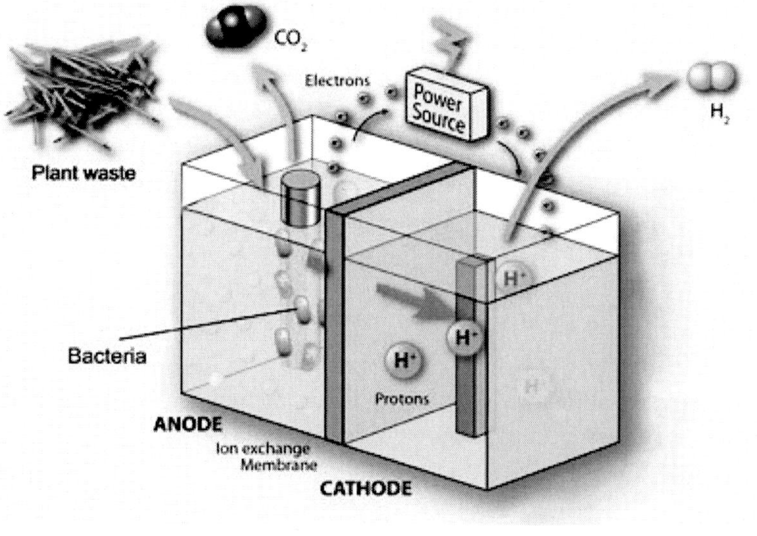

Figure 6.10 A microbial electrolysis cell allows hydrogen production from organic materials at low temperatures. Reproduced from Ref. [172] by permission of The Royal Society of Chemistry, copyright (2011).

Normally, hydrogen production by bacterial metabolism is reduced at lower temperatures because it slows down the enzymes that catalyze the reactions. This eliminates the cost of heating and could enable hydrogen production to be carried out at high latitudes and mountainous regions where the air temperature is below 10°C.

MECs generate hydrogen directly upon applying an electric current to bacteria. Bacteria consume acetic acid, which is produced from fermenting plant matter and release protons, electrons, and CO_2. Addition of an electric current enables the protons and electrons to join together to make hydrogen gas, and the higher the current, the more is the hydrogen produced.

In order to achieve a high yield of hydrogen in MECs, it is important to make both the electron transfer and hydrogen recovery highly efficient.

Traditionally, methanogenesis, or methane formation, is a common problem in MECs, which occurs at higher temperatures as a result of bacterial anaerobic respiration. This can reduce the efficiency of electron transfer to the cathode, reducing the overall output of hydrogen. However, at temperatures below 10°C, no methane was produced since the growth of the methane-producing organisms was inhibited and the yield of hydrogen produced was comparable to that at temperatures above 25°C.

The next aim is to enhance hydrogen recovery and to explore new electrode materials [171]. In the future the MEC technology could be considered for biohydrogen production in cold environments.

6.11 Conclusions

We have considered a non-usual method of hydrogen production provided by living organisms (bacteria). This natural process, carried out with the help of specific enzymes (hydrogenases), allows numerous modifications carried out already with the help of various and different technologies, including even genetic engineering. This chapter represents the transitional stage of hydrogen production between the natural biological processes and those provided by human technologies.

Chapter 7

Hydrogen Production via Water Splitting under the Action of Light

7.1 Introduction

In the previous chapter the methods of biohydrogen production were highlighted. These methods occur in nature but can be technologically modified and improved to get enhanced uptakes of dihydrogen. This is a good example of mimicking natural processes in high technologies.

Photosynthesis makes a useful hint concerning production of hydrogen. As it was shown in Chapter 4, during photosynthesis photosystem II releases protons under the action of sunlight along with a simultaneous release of molecular oxygen. Meanwhile the demands of ecologically friendly and simple production of dihydrogen (H_2) can be achieved via water splitting. Since our aim is production of hydrogen for energy, this chapter considers the mechanisms of splitting of water under the action of sunlight and evolution of hydrogen.

Decomposition of water to its elements as molecular gases H_2 and O_2 has been a big challenge for chemists.

$$2H_2O \rightarrow 2H_2 + O_2 \; E^\circ = 1.23 \text{ V versus NHE} \quad (7.1)$$

The photon energy is converted to chemical energy accompanied by a largely positive change in the Gibbs free energy through water

Hydrogen Energy for Beginners
Alexander Gavrilyuk

ISBN 978-981-4364-77-5 (Hardcover), 978-981-4364-78-2 (eBook)
www.panstanford.com

splitting. This reaction is similar to photosynthesis by green plants because both are uphill reactions. Therefore, photocatalytic water splitting is regarded as artificial photosynthesis and is an attractive and challenging topic in science. Specially selected semiconductors play the role similar to that of chlorophyll [173].

To oxidize water to O_2 one needs first a strong oxidant (with $E° \geq 1.23$ V at pH = 0). A hole in the valence band of a semiconductor generated by a light quantum is used for this purpose (see Fig. 7.1).

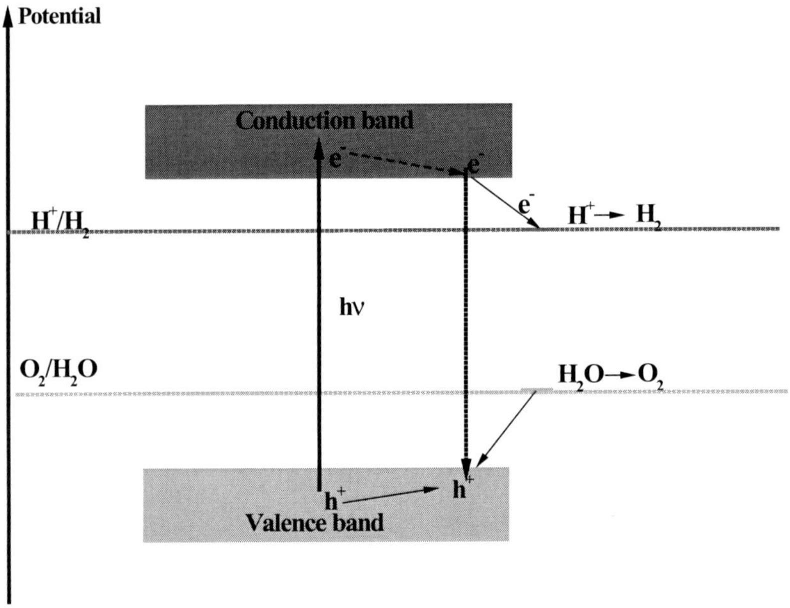

Figure 7.1 Scheme of water splitting with the help of a solid catalyst.

But this semiconductor must perform another important function for the photosystem: namely, it must be a suitable redox catalyst to assist the formation of O_2 without going through one-electron intermediates. Then the reaction of water splitting runs in a way as

$$2H_2O_{(liquid)} \rightarrow O_{2(gas)} + 4H^+_{(aqua)} + 4e^- \ (E^\circ_{ox} = 1.23 \text{ V}) \tag{7.2}$$

The same applies for water reduction to H_2—one needs a strong reductant and redox catalysts to reduce protons to H_2. A photogenerated electron loses its energy via thermolization and eventually occurs at the energy level as the bottom of the conduction band of the semiconductor playing the role of the reductant and the semiconductor again performs its catalytic function. The reaction then runs as

$$2H^+_{(aq)} + 2e^- \rightarrow H_{2(gas)}\ (E^\circ_{(red)} = 0.00\ V) \quad (7.3)$$

Figure 7.1 shows a schematic diagram of water splitting into H_2 and O_2 under the action of light with the use of a semiconductor photocatalyst.

Photocatalysis on semiconductor particles passes through three main steps: (i) absorption of photons with energies exceeding the semiconductor bandgap width, which results in the generation of electron (e^-)–hole (h^+) pairs in the semiconductor particles; (ii) charge separation followed by migration of these photogenerated carriers in the semiconductor particles; (iii) surface chemical reactions between these carriers with various compounds (e.g., H_2O) [173]. Obviously, some electrons and holes recombine with each other without participating in chemical reactions.

7.2 Requirements for Semiconductor Catalysts

From simple thermodynamic consideration it is obvious that a semiconductor photocatalyst used for water splitting must have the bottom of the conduction band more negative than the reduction potential of water to produce H_2, and the top of the valence band more positive than the oxidation potential of water to produce O_2, as it is shown in Fig. 7.1. This requirement makes serious restrictions in the list of semiconductors suitable for water splitting. For this reason, the position of the valence and the conduction bands in energy scale, relative to reduction potential of water to produce O_2 and H_2 is a very important parameter for being an effective catalyst for water splitting. The positions of electron bands in energy scale for a series of semiconductors is shown in Fig. 7.2.

However it is difficult to develop an oxide semiconductor photocatalyst that possesses both a sufficiently negative conduction band for H_2 production and a sufficiently narrow band gap (i.e., <3.0 eV) for visible light absorption, hence the O 2p orbital forms the highly positive valence band (at ca. +3.0 V *vs.* NHE) [174]. From Fig. 7.2 one can easily understand that many visible light-absorbing oxide photocatalysts, such as WO_3, SnO_2, Fe_2O_3, or $BiVO_4$ [175–181], cannot produce H_2 from water due to their conduction bands being too low for water reduction.

Meanwhile, there is another important restriction for this function: the photocatalyst must be stable in aqueous solutions against photocorrosion under photoirradiation. Oxide semiconductors are generally highly stable against photocorrosion and have thus been extensively used as heterogeneous photocatalysts. Some non-oxide (e.g., sulfides [182–187] and nitrides [188–190]) semiconductors possess appropriate band levels for water splitting under visible light, but they are generally unstable and readily become deactivated through photocorrosion or self-oxidation.

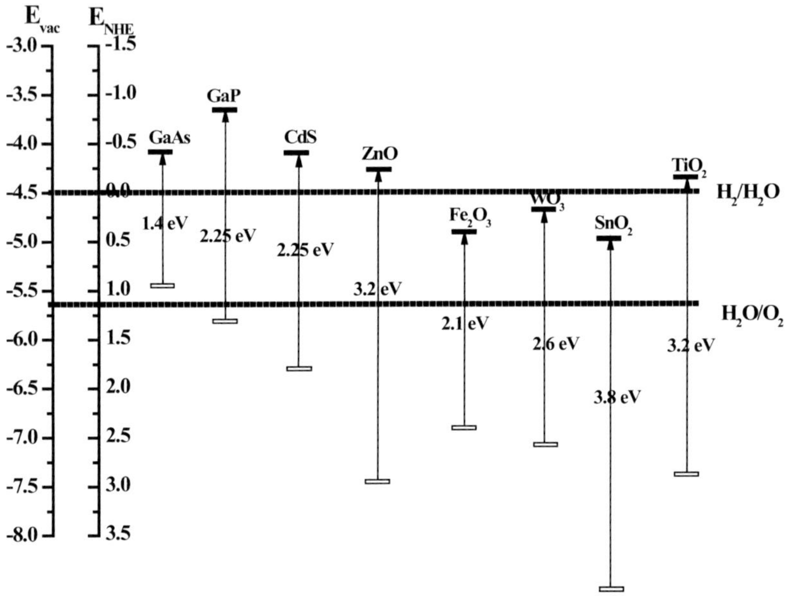

Figure 7.2 Band positions of several semiconductors in contact with aqueous electrolyte at pH = 0. The lower edge of the conduction band (black solid bars) and upper edge of the valence band (white hollow bars) are presented along with the band gap in electron volts. The energy scale is shown in electron volts using either the normal hydrogen electrode (NHE) or the vacuum level as a reference. On the right side the standard potentials of water splitting are presented against the standard hydrogen electrode potential.

The most important step in the photocatalytic water splitting involves the surface chemical reactions. The important issues for this step are surface quality (active sites) and quantity (surface area). Even if the photogenerated electrons and holes possess thermodynamically sufficient potentials for water splitting, they will have to recombine with each other if the active sites for redox reactions do not exist on the surface. Co-catalysts such as Pt, NiO, and RuO_2 are usually loaded to introduce active sites for H_2 evolution because the conduction band levels of many oxide photocatalysts are not high enough to reduce water to produce H_2 without catalytic assistance. Back reactions to form water by reactions between evolved H_2, O_2, and intermediates easily proceed because of an uphill reaction. Therefore, poor properties for back reactions are required for the surface of co-catalyst and photocatalyst.

When engineering the photocatalytic systems for water splitting one should also pay attention to the processes of charge separation and migration of photogenerated carriers. Crystal structure, crystallinity and particle size strongly affect these issues. The higher the crystalline quality is, the smaller the amount of defects is. The defects operate as trapping and recombination centers between photogenerated electrons and holes, resulting in a decrease in the photocatalytic activity. If the particle size becomes small, the distance that photogenerated electrons and holes have to migrate to reaction sites on the surface becomes short and this results in a decrease in the recombination probability [173].

Small particle size sometimes gives a quantum size effect as seen in colloidal particles resulting in widening of band gap and blue shift in the absorption spectrum [190]. The resultant photocatalytic activity is dominated by the balance among these factors. A high degree of crystallinity is often required rather than a high surface area for water splitting because recombination between photogenerated electrons and holes is especially a serious problem for uphill reactions.

These examples show that the semiconductor must combine several important qualities for being an effective photocatalyst in the process of water splitting. Sometimes these requirements sound contradictory as a narrow bandgap and the stability against photocorrosion. So, it is understandable that photocatalysts should be highly functional materials. For this reason the search of new classes of materials suitable to perform the catalytic function was being launched.

Another class of non-oxide semiconductor photocatalyst is on the agenda; that is, homogeneous sensitizer molecules, namely, organic dyes and metal complexes. A new strategy was implemented to achieve effective visible light harvesting, spectral sensitization of wide-band-gap semiconductors (e.g., TiO_2) by using such sensitizer molecules for photocatalytic H_2 production from water [191–197]. The adsorption of molecular sensitizers, such as the dye $Ru(bpy)_3^{2+}$ on platinized titanium oxide (Pt/TiO_2) particles resulted in efficient H_2 evolution from water under visible light irradiation in the presence of a sacrificial electron donor such as triethanolamine or ethylenediaminetetraacetic acid [191–206].

A sacrificial donor is a molecular entity that acts as the electron donor in a photoinduced electron-transfer process and is not restored in a subsequent reduction process but is destroyed by irreversible chemical conversion (IUPAC definition).

Sacrificial reagents are often employed to evaluate the photocatalytic activity for water splitting as shown in Fig. 7.3, because overall water splitting is a tough reaction. When the photocatalytic reaction is carried out in an aqueous solution including a reducing reagent, in other words, electron donors or hole scavengers, such as alcohol or a sulfide ion, photogenerated holes irreversibly oxidize the reducing reagent instead of water [207]. It enriches electrons in a photocatalyst and an H_2 evolution reaction is enhanced as shown in Fig. 7.3a. This reaction will be meaningful for realistic hydrogen production if biomass and abundant compounds in nature and industries are used as the reducing reagents [207–213]. On the other hand, photogenerated electrons in the conduction band are consumed by oxidizing reagents (electron acceptors or electron scavengers), such as Ag+ and Fe^{3+}, resulting in an enhanced O_2-evolution reaction as shown in Fig. 7.3b [207]. These reactions using sacrificial reagents are studied to evaluate whether a certain photocatalyst satisfies the thermodynamic and kinetic potentials for H_2 and O_2 evolution. These reactions are regarded as half reactions of water splitting and are often employed as test reactions for photocatalytic H_2 or O_2 evolution. Even if a photocatalyst is active for these half reactions the results do not guarantee a photocatalyst to be active for overall water splitting into H_2 and O_2 in the absence of sacrificial reagents. From this point, the term "water splitting" should be distinguishably used for H_2 or O_2 evolution from aqueous solutions in the presence of sacrificial reagents. Water splitting means to split water into H_2 and O_2 in a stoichiometric amount in the absence of sacrificial reagents [207].

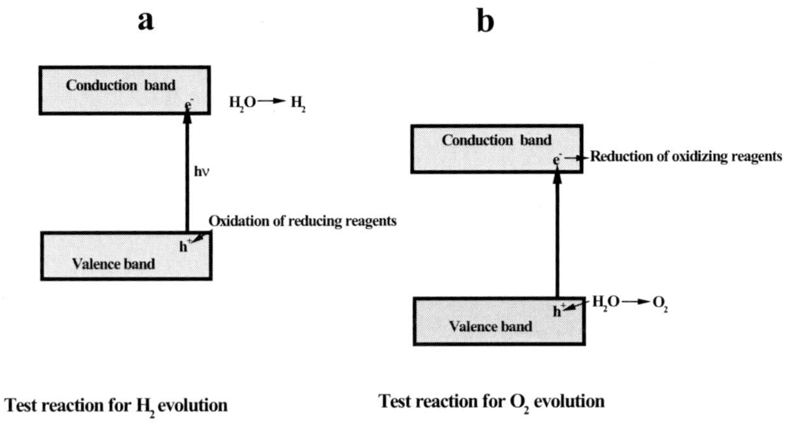

Figure 7.3 Scheme of H_2 or O_2 evolution in the presence of sacrificial reagents and half reactions of water splitting.

Another example of a system employing a sacrificial reagent is that containing Ru and Rh complexes whose photoexcitation leads to the photoinduced electron transfer [214]:

$$[Ru(bpy)_3]^{2+} + h\nu \rightarrow {}^*[Ru(bpy)_3]^{2+}, \quad (7.4)$$

$$^*[Ru(bpy)_3]^{2+} + [Rh(bpy)_3]^{3+} \rightarrow [Ru(bpy)_3]^{3+} + [Rh(bpy)_3]^{2+} \quad (7.5)$$

Then follows the catalytic detachment of molecular hydrogen over a Pt catalyst.

$$[Rh(bpy)_3]^{2+} + H_2O \rightarrow (Pt)\ [Rh(bpy)_3]^{3+} + 0.5H_2 + OH^- \quad (7.6)$$

A sacrificial donor (triethanolamine) provides the regeneration of the catalyst [207]:

$$[Ru(bpy)_3]^{3+} + N(C_2H_4OH)_3 \rightarrow [Ru(bpy)_3]^{2+} + [N(C_2H_4OH)_3]^+ \quad (7.7)$$

However, the sensitizer molecules themselves cannot oxidize H_2O to O_2 and they generally decompose through self-oxidation under photoirradiation in the absence of a sacrificial electron donor. Because the evolution of O_2 from water requires the abstraction of four electrons, it is quite difficult for it to occur on simple sensitizer molecules, in which only one positive or negative charge can be stored. Only a few studies have achieved appreciable O_2 generation from water using a Ru complex combined with a colloidal hydrated iridium oxide ($IrO_2 \cdot nH_2O$) [215,216] or a molecular manganese complex [217], which can scavenge the positive charge in the Ru complex and act as a catalyst for O_2 evolution.

7.3 Two Approaches for Achieving Water Splitting

At present, it makes sense to distinguish two main approaches for achieving water splitting using visible light. One approach is to use a two-step photoexcitation in two different photocatalysts (Fig.7.4a), whereas the other approach is the conventional one-step system (Fig.7.4b), [218–232]. The second process was inspired by natural photosynthesis in green plants and is known as the Z-scheme. In two-step systems, the water-splitting reaction is broken up into two stages: one for H_2 evolution and the other for O_2 evolution; these are combined by using a shuttle redox couple (Red/Ox) in the solution. Over a H_2 evolution photocatalyst (referred to as a H_2 photocatalyst), the photoexcited electrons reduce water to H_2, and the holes in the valence band oxidize the reductant (Red) to an oxidant (Ox). The oxidant is reduced back to the reductant by photoexcited electrons generated over an O_2 evolution photocatalyst (referred to as an O_2 photocatalyst), where the holes oxidize water to O_2. This system lowers the energy required for photocatalysis, allowing visible light to be utilized more efficiently than in conventional water-splitting systems. The Z-scheme enables a semiconductor to be used that has either a water-reduction or -oxidation potential on one side of the system. For example, visible-light responsive oxides such as WO_3 can be used as O_2 photocatalysts if they can reduce the oxidant to a reductant. Similarly, non-oxide photocatalysts (e.g., sulfides, oxynitrides, and dyes) can be used as H_2 photocatalysts if they can oxidize the reductant to an oxidant. Consequently, various semiconductor materials can be used in the Z-scheme even if they do not satisfy all the strict requirements for a one-step system. Another advantage of Z-scheme systems is the ability to separate production of H_2 and O_2 by employing a separator, such as a porous glass filter, that permits only redox mediators to be transferred. It is impossible in principle to achieve separate production of H_2 and O_2 in a conventional one-step water-splitting system since H_2 and O_2 are evolved simultaneously on small semiconductor particles. However, it should be also mentioned that the Z-scheme systems have a significant disadvantage that the two-step photoexcitation system requires the number of photons two-fold larger than the one-step system to achieve water splitting [173].

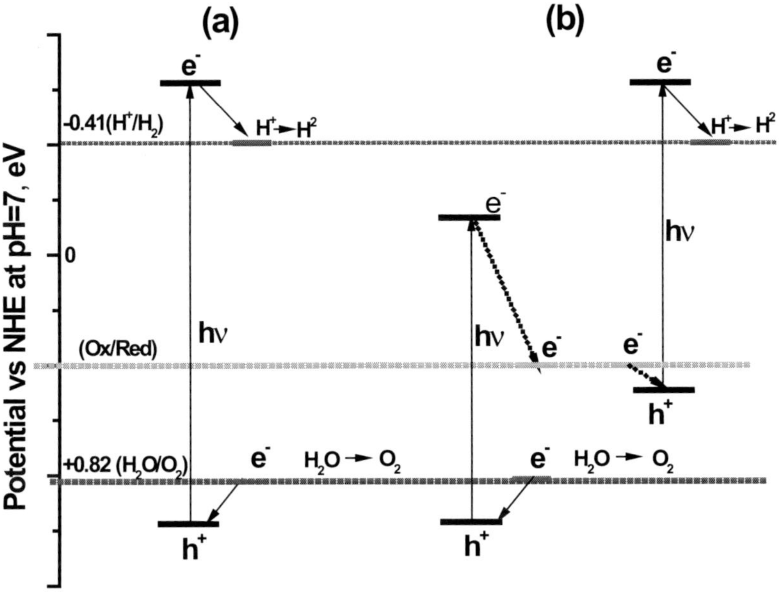

Figure 7.4 Schematic energy diagrams of photocatalytic water-splitting systems: (a) conventional one-step system and (b) two-step photoexciation system.

A good example of a two-step photoexcitation system was presented in [190] based on two photosystems connected in series as shown in the electron flow diagram of Fig. 7.5. A thin film of nanocrystalline tungsten trioxide, WO_3 [233], serves as the top semiconductor absorbing the blue part of the solar spectrum. The valence band holes (h^+) created by band-gap excitation of the film oxidize water to oxygen:

$$4h^+ + H_2O \rightarrow O_2{}^+ + 4H^+ \tag{7.8}$$

and the conduction-band electrons are fed into the second photosystem consisting of the dye-sensitized nanocrystalline TiO_2

cell. The latter is placed directly under the WO_3 film, absorbing the green and red part of the solar spectrum that is transmitted through the top semiconductor. The photovoltage generated by the second photosystem enables hydrogen to be generated by the conduction-band electrons.

$$4H^+ + 4e^- \rightarrow 2H_2 \quad (7.9)$$

The overall reaction corresponds to the splitting of water by visible light. There is close analogy to the Z-scheme (named for the shape of the flow diagram) that operates in photosynthesis. In green plants, there are also two photosystems connected in series, one that oxidizes water to oxygen and the other generating the compound NADPH used in fixation of carbon dioxide. As discussed above, the advantage of the tandem approach is that higher efficiencies can be reached than with single junction cells if the two photosystems absorb complementary parts of the solar spectrum. At present, the overall conversion efficiency from standard solar light to chemical energy achieved with this device stands at 4.5%, and further improvements are eventually possible.

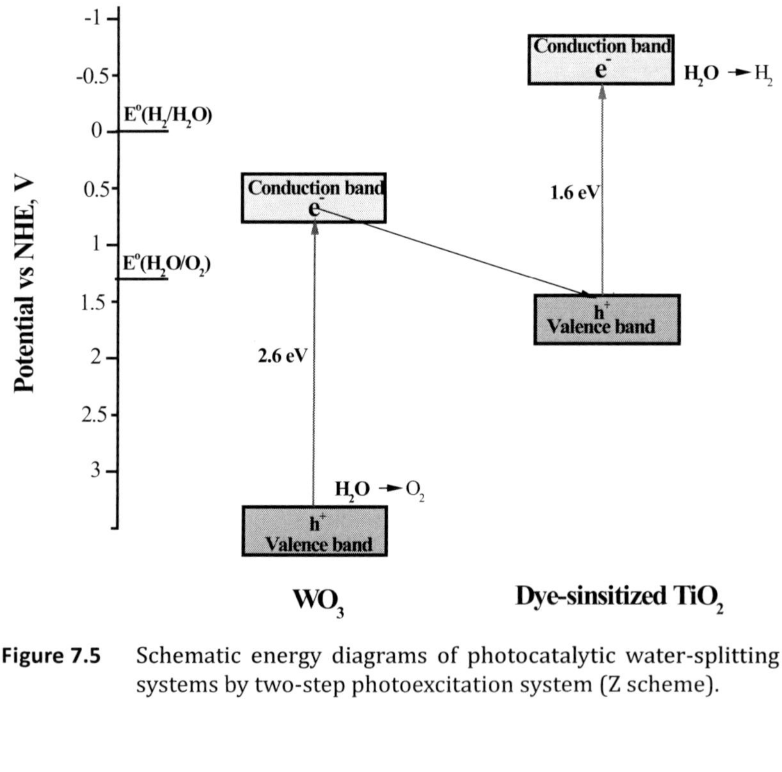

Figure 7.5 Schematic energy diagrams of photocatalytic water-splitting systems by two-step photoexcitation system (Z scheme).

The other approach for achieving water splitting using visible light is to split water into H_2 and O_2 using a single visible-light responsive photocatalyst (a one-step system), as illustrated in Fig. 7.4a. Because few stable semiconductors can absorb visible light and have a sufficiently high potential for water splitting, band engineering of semiconductors is required to artificially develop new photocatalyst materials that satisfy the following requirements for water splitting under visible light: (i) have a narrow band gap, (ii) are stable under photoirradiation, and (iii) have suitable conduction and valence band levels for H_2 and O_2 production.

Various band modification methods have been studied, as formation of a midgap electron donor level (doped materials) [234–243], hybridization of the O 2p orbital with other orbitals (oxides [179,244–248], oxynitrides [249–255] and oxysulfides [256–259]), and the formation of solid solutions [260–265]. However, very few reliable photocatalysts for one-step water splitting under visible light [260–262] are appropriate due to the strict requirements. An effective one-step water-splitting system that uses a solid solution photocatalyst such as $(Ga_{1-x}Zn_x)(N_{1-x}O_x)$ or $(Zn_{1+x}Ge)(N_2O_x)$, may be mentioned as a rare case [260–262].

As mentioned above, two-step water-splitting (Z-scheme) systems have significant advantages over conventional one-step systems and they look to be the most promising to achieve efficient water splitting under visible light. A two-step water-splitting system was proposed at first in 1979 [266], and since then, a lot of work has been done to construct such a Z-scheme system trying to develop new semiconductor materials and effective shuttle redox mediators [267–277].

Many workers have investigated half reactions (i.e., either H_2 or O_2 evolution) by combining various redox mediators with conventional photocatalysts such as TiO_2. Some combinations resulted in gas evolution (either H_2 or O_2) under light irradiation; however, in most cases, gas evolution was interrupted during the reaction.

In photocatalytic reactions with reversible redox mediators, backward reactions proceed readily over the photocatalyst and suppress gas (H_2 or O_2) evolution (Fig. 7.6). Over the H_2 photocatalyst, reduction of Ox to Red (indicated by the broken lines in Fig. 7.6), a thermodynamically advantageous backward reaction proceeds preferentially over reduction of water to H_2. At the same time, over the O_2 photocatalyst, oxidation of Red to Ox proceeds preferentially over oxidation of water to O_2 [173].

Furthermore, it is much more difficult to achieve simultaneous evolution of H_2 and O_2 in two-step water-splitting systems in which the two different photocatalysts coexist in an aqueous solution containing both Red and Ox, because the two reactions have different favorable redox concentrations. For example, a high concentration of Red is favorable for the H_2 evolution photocatalyst, but it causes a significant backward reaction (i.e., reoxidation of Red to Ox) over O_2 photocatalyst, which suppresses O_2 evolution.

On the other hand, a high concentration of Ox, which is desirable for the O_2 photocatalyst, reduces the H_2 evolution rate. To realize a two-step water-splitting system, it is thus necessary to develop a photocatalytic system that has a high selectivity for the forward reactions (indicated by the solid lines in Fig. 7.6). Indeed, there had been no report of simultaneous and stoichiometric evolution of H_2 and O_2 even under UV irradiation, prior to the demonstration using a combination of Pt-loaded anatase TiO_2 and bare rutile TiO_2 photocatalysts in the presence of an iodate/iodide (IO_3^-/I^-) shuttle redox mediator in 2001 [278].

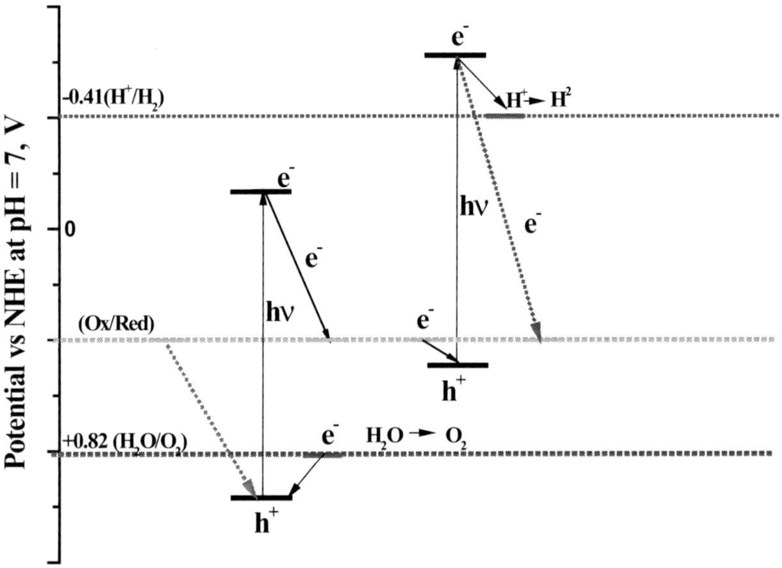

Figure 7.6 Forward and backward reactions in the two-step photoexcitation system. The solid arrows indicate electron transfer in forward reactions and dotted arrows indicate electron transfer in backward reactions.

7.4 Photoelectrochemical Water Splitting Using Semiconductor Photoelectrodes under Visible Light

Since the first report of photoelectrochemical (PEC) water splitting using TiO_2 electrodes under UV light [279], extensive efforts have been made to construct an efficient PEC water-splitting system and to develop new semiconductor materials for efficient photoelectrodes [280–324]. Figure 7.7a shows PEC water splitting based on *n*-type-semiconductor (e.g., TiO_2) electrodes. Electrons and holes are generated when an *n*-type semiconductor absorbs photons with higher energies than its band gap. The photoexcited electrons transfer to a counter electrode (e.g., Pt) through the outer circuit and then reduce water to H_2, whereas the holes transfer to the semiconductor surface and then oxidize water to O_2. In most cases, an external bias supplied by a power supply or a pH difference (a chemical bias) between the electrodes is necessary for efficient charge separation, even when the *n*-type semiconductor possesses a suitable conduction band energy level for reducing water to H_2, possibly due to the resistance produced between the two electrodes in the solution and to that produced at the interfaces between semiconductor particles within the electrode. In the case of *p*-type semiconductor materials, water is reduced on the semiconductor surface, while water is oxidized on the counter electrode. The efficient use of visible light is also essential for efficient production of H_2 by PEC water splitting under solar light.

Therefore, the development of a stable photoelectrode material that functions efficiently under visible light has been extensively investigated. Oxide semiconductors generally possess high stability against photocorrosion. They have thus been employed as photoanodes for water oxidation in combination with a counter electrode such as Pt for H_2 generation. Although the conduction band levels of visible light–responsive oxides (e.g., WO_3, Fe_2O_3, and $BiVO_4$) are generally too positive for H_2 evolution, applying an external bias between the photoanode and counter electrode enables oxide semiconductors to be used for PEC water splitting.

However, applying a large external bias using a power supply is undesirable in terms of the energy consumption of the whole system. Thus, it is desirable to develop a semiconductor material that has both a sufficiently high band level for H_2 (or O_2) production and a narrow band gap (i.e., <3.0 eV) for visible light absorption, just as for heterogeneous photocatalytic systems. New semiconductor materials such as oxynitrides, oxysulfides, sulfides, selenides, and *p*-type oxide $CaFe_2O_4$ have been developed.

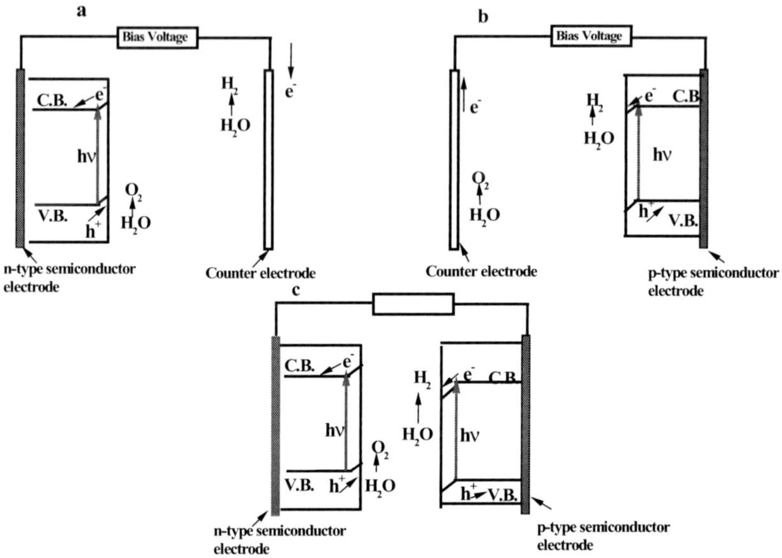

Figure 7.7 PEC water-splitting systems using *n*-type semiconductor photoanode (a), *p*-type semiconductor photocathode (b), and tandem system.

One of the most attractive ways for harvesting a wide wavelength range of visible light in a PEC system without applying an external bias is to introduce "tandem" systems. One system in tandem systems consists of a photoelectrode with a solar cell that supplies the external bias required between the photoelectrode and the counter electrode. The other system consists of an *n*-type semiconductor electrode with an appropriate *p*-type semiconductor electrode (see Fig. 7.7c). Such PEC systems have recently been demonstrated.

7.5 Water Splitting and Nanotechnologies

The progress in the development of the technology of the hydrogen production via various methods of water splitting under the action of visible light is dependent upon the creation of the appropriate catalysts. And the concentration of the efforts of many workers is just directed to this approach; nanotechnologies being the main clue to solve this problem. In order to drive the surface reactions efficiently, one should strive to create a great surface area. To achieve that, new technologies to prepare highly porous mesoscopic materials are being currently developed, which possess an enormous internal surface area. Mesoporous films of various materials are made up of arrays of tiny crystals measuring a few nanometers across. Oxides such as TiO_2, ZnO, SnO_2, and Nb_2O_5, or chalcogenides such as CdSe are the preferred compounds. These are interconnected to allow electronic conduction to take place.

Polymerization complex method is often used to prepare the photocatalysts. This preparation method gives fine and well-crystalline powders with a high surface area at relatively low calcination temperature and short calcination time compared with a conventional solid state method [207 and references therein].

Between the particles are mesoscopic pores filled with a semiconducting or a conducting medium, such as a *p*-type semiconductor, a polymer, a hole transmitter or an electrolyte. The net result is a junction of extremely large contact area between two interpenetrating, individually continuous networks.

The semiconductor structure, typically 10 μm thick and with a porosity of 50%, has a surface area available for dye chemisorption over a thousand times that of a flat, unstructured electrode of the same size. If the dye is chemisorbed as a monomolecular layer, enough can be retained on a given area of electrode to provide absorption of essentially all the incident light. Figure 7.8 shows an electron micrograph of a nanocrystalline TiO_2 film with a grain size in the range of 10–80 nm [325].

An interesting feature specific to nanocrystalline materials is the appearance of quantum confinement effects. These appear when the films are made up of small quantum dots, such as 8 nm-sized CdTe particles [326]. Such layers have a larger band gap than the bulk material, the band edge position being shifted with respect to the positions indicated in Fig. 7.2 for macroscopic materials. The conduction band redox potential is lowered and that of the valence band is increased. As a consequence, electrons and holes can perform reduction and oxidation reactions that cannot proceed on bulk semiconductors.

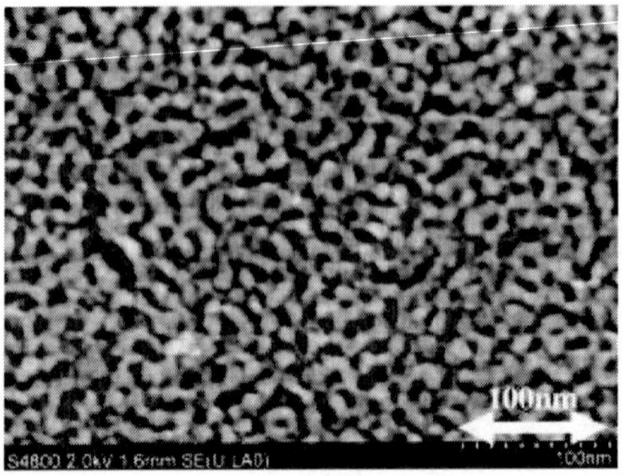

Figure 7.8 Scanning electron micrograph of the surface of a mesoporous TiO_2 film. Reprinted from Ref. [325], copyright (2008), with permission from Elsevier.

In the recent years various methods to develop catalysts appropriate for water splitting under the action of light are being more intensively investigated because producing hydrogen from water is the most attractive challenge. In this way new and interesting ideas are being developed. Although water splitting was first inferred using an externally biased TiO_2 photoelectrode, a powdered TiO_2 photocatalyst cannot split water without any modifications such as loading co-catalysts. At the initial stage of the research, it was not quite clear that a platinized TiO_2 photocatalyst could split water because the activity was usually low and no O_2 evolution was often observed. After that, coating with NaOH [327] and addition of alkali carbonates [328] have been found to be effective for water splitting with the Pt/TiO_2 photocatalyst.

Different than TiO_2, $SrTiO_3$ [329,330] and $KTaO_3$ [329,331] photoelectrodes with perovskite structure can split water without an external bias due to of their high-conduction band levels which makes it possible to use them as powdered photocatalysts. It was reported that NiO-loaded $SrTiO_3$ powder can split pure water into H_2 and O_2 [332–337]. NiO-loaded K4Nb6O17 is also appropriate for this purpose [338,339]. The NiO co-catalyst for H_2 evolution is usually activated by H_2 reduction and subsequent O_2 oxidation to form a NiO/Ni double-layer structure which facilitates electron migration from a photocatalyst substrate to a co-catalyst [335]. The pretreated NiO co-catalyst is often denoted as NiOx in literature. It is important that the NiO co-catalyst does not cause the back reaction between H_2 and O_2, being different from Pt. The excellent NiO co-catalyst has often been employed for many photocatalysts for water splitting [207].

Since it is hard to determine the real amount of photons absorbed by a photocatalyst in a dispersed system because of scattering, the apparent quantum yield (AQY) is usually determined to value the photocatalyst efficiency. AQY is estimated to be smaller than the real quantum yield because the number of absorbed photons is usually smaller than that of incident light [207].

$$\text{AQY} = \left(\frac{\text{number of reacted electrons}}{\text{number of incident phonons}}\right) \times 100\%$$

It was also reported that various titanate, niobate, and tantalate photocatalysts with layered perovskite structure give high quantum yield in the water-splitting reaction [207].

As an example, Fig. 7.9 clarifies the functioning of the layer photocatalysts on $K_4Nb_6O_{17}$ which possesses two kinds of interlayers in which ion-exchangeable potassium cations exist. H_2 evolution proceeds in one interlayer with a nickel co-catalyst while O_2 evolution occurs in another interlayer. It is the characteristic of the $K_4Nb_6O_{17}$ photocatalyst that H_2 evolution sites are separated from O_2 evolution sites by the photoactive niobate sheet [207].

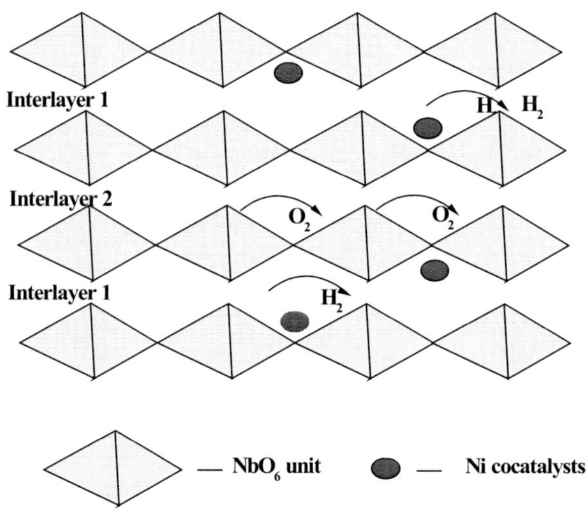

Figure 7.9 Water splitting over $K_4Nb_6O_{17}$ photocatalyst with layered structure.

It is noteworthy to isolate here $NiO/NaTaO_3$ catalysts which demonstrate high activity that highly increases with doping of lanthanide ions [340,341]. An optimized NiO (0.2 wt%)/$NaTaO_3$:La (2%) photocatalyst shows high activity with an AQY 56% for water splitting. The activity is stable for more than 400 h under irradiation of light from a 400 W high pressure mercury lamp. Bubbles of H_2 and O_2 evolved can be observed when the photocatalyst is irradiated with UV from a 200 W Xe–Hg lamp.

The mechanism of high efficiency of this photocatalyst is clarified by Fig. 7.10. The photocatalytic activity of $NiO/NaTaO_3$ increases remarkably with doping of La ions [340,341] The reaction scheme for the water splitting on the $NiO/NaTaO_3$:La photocatalyst is shown in Fig. 7.10 [341]. The particle size of the $NaTaO_3$:La crystal (0.1–0.7 nm) is smaller than that of the nondoped $NaTaO_3$ crystal (2–3 nm) and ordered surface nanosteps are created by lanthanum doping. The clue to get a success is to create the small particle size with high crystallinity and to enhance the probability of the reactions of photogenerated electrons and holes with water molecules, rather than recombination. The H_2 evolution sites of the edge (NiO reaction centers) are effectively separated from the O_2 evolution sites of the groove at the surface nanostep structure. This separation is advantageous, especially for water splitting in order to avoid the back reaction. Doping of Ca, Sr, and Ba gives the similar effect as the La doping on the formation of the morphology of $NaTaO_3$ and the enhancement of photocatalytic activity [342]. Time-resolved IR measurements proved that the La doping enhances the lifetime of photogenerated electrons in a conduction band or a shallow trap level. The lifetime enhancement yields in turn the enhancement of photocatalytic efficiency due to reduction of the probability of the recombination [343].

Another interesting example of the influence of the materials engineering upon the efficiency of the photocatalytic concerns layered perovskite structures The band structures of $Sr_2Ta_2O_7$ and $Sr_2Nb_2O_7$ is highly schematically shown in Fig. 7.11 [344], wherein the band structure of NiO used as a co-catalyst is also presented [345]. These crystal structures are composed of NbO_6 and TaO_6 octahedra. O_2 evolution on $Sr_2Ta_2O_7$ runs so easy as that on $Sr_2Nb_2O_7$ because the potentials of their valence bands are highly positive to oxidize water into O_2. Therefore, it is due to the difference in the conduction band level that the photocatalytic activity of $Sr_2Ta_2O_7$ is higher than that of $Sr_2Nb_2O_7$. The high conduction band level causes the driving force for reduction of water to form H_2, which provides $Sr_2Ta_2O_7$ decomposition of pure water without co-catalysts.

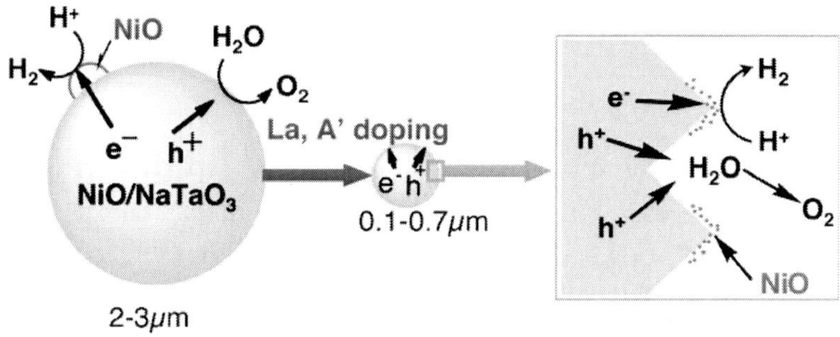

Figure 7.10 Mechanism of highly efficient water splitting over $NiO/NaTaO_3$:La photocatalyst. Reprinted with permission from Ref. [341]. Copyright (2003) American Chemical Society.

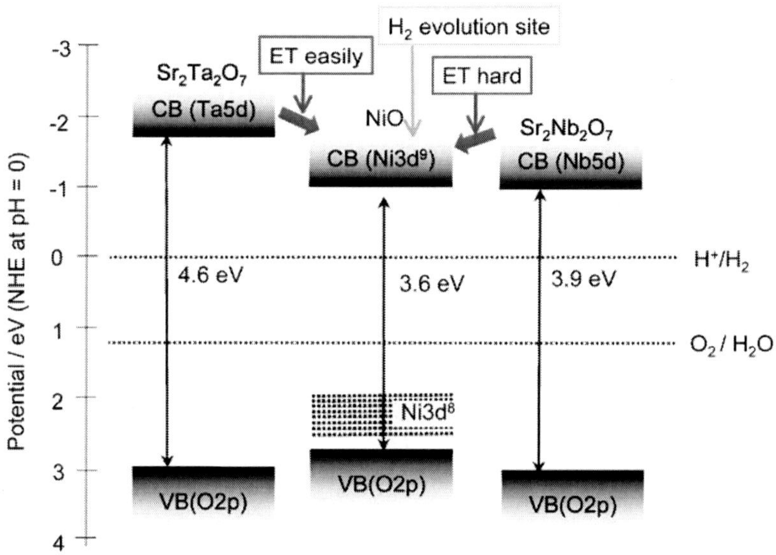

Figure 7.11 Band structures of $Sr_2M_2O_7$ (M = Nb and Ta) photocatalysts and NiO co-catalyst. Reprinted with permission from Ref. [344]. Copyright (2003) American Chemical Society.

In general, pretreatment of H_2 reduction and subsequent O_2 oxidation is indispensable for obtaining high activities for NiO-loaded photocatalysts. The pretreatment is necessary for $NiO/Sr_2Nb_2O_7$ as usual but not for $NiO/Sr_2Ta_2O_7$. In the case of the nontreated $NiO/Sr_2Ta_2O_7$ photocatalyst, it is possible that the photogenerated electrons in a conduction band of $Sr_2Ta_2O_7$ can transfer to a conduction band of NiO because of the suitable potential difference as shown in Fig. 7.11. In contrast, this appears to be difficult for $Sr_2Nb_2O_7$ because the potential difference in the conduction band between $Sr_2Nb_2O_7$ and NiO is not so pronounced. In such a case, pretreatment would be necessary as well as in the case of $NiO/SrTiO_3$. Therefore, the effect of the conduction band level dominates the photocatalytic activities of the $Sr_2M_2O_7$ system.

The numerous examples above demonstrated the series of ideas that one can exploit to develop various catalytic systems for water decomposition. Despite that the light energy conversion via water splitting is in the stage of research it is sought to be advanced and promising. It is developing very fast and new photocatalysts are being discovered one after another.

7.6 Conclusions

In this chapter we became familiar with the first artificial method of hydrogen production, namely, water splitting under the action of light or in other words artificial photosynthesis. To date it is apparently the most promising method due to its ecological purity. But to create a commercial system for artificial photosynthesis is an enormously difficult task. It is possible however to expect that rapidly developing semiconductor nanotechnologies yield a breakthrough in this important area.

Chapter 8

Conventional Methods of Hydrogen Production

8.1 Introduction

In previous two chapters we have considered relatively new and potential methods of commercial production of hydrogen. These methods, namely bioproduction of hydrogen and artificial photosynthesis (water splitting under the action of light), demand employment and development of high technologies, especially nanotechnologies.

It should be said that methods of conventional hydrogen production do not belong to high technologies, but, the fact cannot be ignored that they are well-developed for commercial technologies and are being used to date.

8.2 Steam Reforming

This method of hydrogen production requires either natural gas or hydrocarbons as raw materials. Natural gas is an abundant gas mixture consisting primarily of methane, typically with 0–20% higher hydrocarbons. Bulk hydrogen is usually produced by the steam reforming of methane or natural gas [346]. This process runs in two stages. First at high temperatures (700–1100 °C), steam (H_2O)

Hydrogen Energy for Beginners
Alexander Gavrilyuk

ISBN 978-981-4364-77-5 (Hardcover), 978-981-4364-78-2 (eBook)
www.panstanford.com

reacts with methane (CH_4) to produce syngas which is another gas mixture that contains varying amounts of carbon monoxide and hydrogen:

$$CH_4 + H_2O \rightarrow CO + 3H_2 + 191.7 \text{ kJ/mol [347]} \quad (8.1)$$

In a second stage, further hydrogen is generated via the lower-temperature water gas shift reaction, performed at about 130°C:

$$CO + H_2O \rightarrow CO_2 + H_2 - 40.4 \text{ kJ/mol [347]} \quad (8.2)$$

The oxygen atoms are detached from the water (steam) to oxidize CO to CO_2. This oxidation also provides energy to maintain the reaction. Additional heat required to drive the process is generally supplied by burning some portion of the methane.

Steam reforming generates carbon dioxide (CO_2). Since the production is concentrated in one facility, it is possible to separate the CO_2 and dispose it off properly, for example, by injecting it in an oil or gas reservoir, although this is not currently done in most cases.

This is a well-developed commercial technology that has been used for a long time for hydrogen production.

Its disadvantage is that the produced hydrogen is contaminated by CO_2 and requires careful rectification. But even after that, hydrogen still contains residual quantities of CO_2. But the main disadvantage is that CO_2 released into atmosphere gives an additional income to its pollution.

8.3 Partial Oxidation

The partial oxidation reaction occurs when a substoichiometric fuel–air mixture is partially combusted in a reformer, creating a hydrogen-rich syngas. A distinction is made between thermal partial oxidation and catalytic partial oxidation.

The general reaction equation is as follows [348]:

$$C_nH_m + n/2 \cdot O_2 \rightarrow n \bullet CO + m/2 \cdot H_2 \quad (8.3)$$

8.4 Electrolysis and Thermolysis

Hydrogen is produced on an industrial scale by the electrolysis of water. This can be facilely done with a few volts in a simple apparatus like a Hofmann voltameter. A Hofmann voltameter is an apparatus for electrolyzing water, invented by August Wilhelm von Hofmann in 1866. It consists of three joined upright cylinders, usually glass (Fig. 8.1). The inner cylinder is open at the top to allow addition of water and an ionic compound to improve conductivity, such as a small amount of sulfuric acid (H_2SO_4). Water is split into hydrogen and oxygen by the action of electric current. As it was already discussed, water is decomposed into hydrogen and oxygen under the applied voltage >1.23 V. Pure water doesn't conduct electricity but on adding sulfuric acid it does. A platinum electrode is placed inside the bottom of each of the two side cylinders, connected to the positive and negative terminals of a source of electricity. When current is run through Hofmann's voltameter, gaseous oxygen forms at the anode and gaseous hydrogen at the cathode. Each gas displaces water and collects at the top of the two outer tubes. Larger scale production usually relies on high-pressure and high-temperature systems to improve the energy efficiency of electrolysis [348].

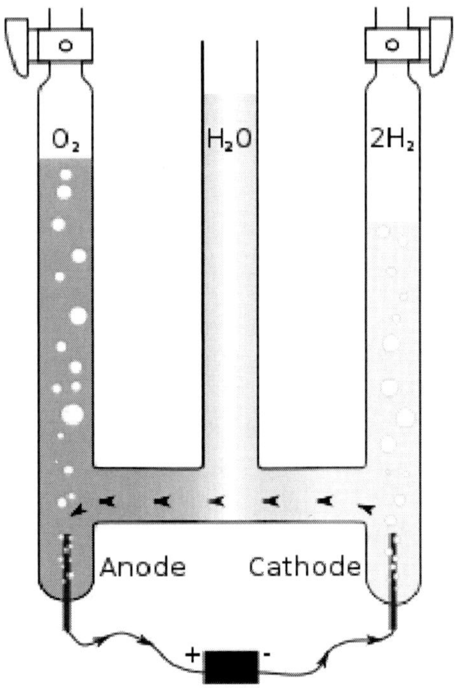

Figure 8.1 Hoffman voltammeter (Image: IIVQ) [348].

Experimental processes include electrolysis at very high temperatures (800°C), therefore, much of the energy required to release hydrogen is supplied as heat instead of electricity. Various catalytic agents are being studied to improve the efficiency of high-temperature electrolysis.

Water spontaneously dissociates at around 2500°C, but this thermolysis occurs at temperatures too high for usual process piping and equipment. Catalysts are required to reduce the dissociation temperature [348].

If electrolysis occurs in pure water, H^+ cations accumulate at the anode and OH^- anions accumulate at the cathode. This can be proved by plunging a pH indicator to the water: the water near the anode is acidic whereas the water near the cathode is basic. The negative hydroxyl ions that approach the anode mostly combine with the positive hydronium ions (H_3O^+), forming water. The positive hydronium ions that approach the negative cathode mostly combine with negative hydroxyl ions to form water. Relatively few hydronium (or hydroxyl) ions reach the cathode (or anode). This can cause a concentration overpotential at both electrodes.

Overpotential is an electrochemical term which refers to the potential difference (voltage) between a half-reaction's thermodynamically determined reduction potential and the potential at which the redox event is experimentally observed.

Pure water is a fairly good insulator since it has a low dissociation constant $K_w = 1.0 \times 10^{-14}$ at room temperature and thus pure water conducts current poorly, the value of its specific conductivity being $\sim 0.055\ \mu S \cdot cm^{-1}$. Unless a very large potential is applied to provide an increase in the dissociation of water, the electrolysis of pure water proceeds very slowly limited by the overall conductivity [348].

If a water-soluble electrolyte is added, the conductivity of the water is enhanced considerably. The electrolyte dissociates into cations and anions; the anions move towards the anode and neutralize the buildup of positively charged H^+ there; similarly, the cations move toward the cathode and neutralize the buildup of negatively charged OH^- there. This provides the continuous flow of electricity.

An electrolyte must be chosen carefully, since an anion from the electrolyte is in competition with the hydroxide ions to give up an electron. An electrolyte anion with less standard electrode potential than hydroxide will be oxidized instead of the hydroxide, and no oxygen gas will be produced. A cation with a greater standard electrode potential than a hydrogen ion will be reduced instead, and no hydrogen gas will be produced.

The following cations have lower electrode potential than H^+ and are therefore suitable for use as electrolyte cations: Li^+, Rb^+, K^+, Cs^+, Ba^{2+}, Sr^{2+}, Ca^{2+}, Na^+, and Mg^{2+}. Sodium and lithium are frequently used, as they form inexpensive, soluble salts.

If an acid is used as the electrolyte, the cation is H^+, and there is no competitor for the H^+ created by disassociating water. The most commonly used anion is sulfate (SO_4^{2-}), as it is very difficult to oxidize this ion to the peroxodisulfate ion, the standard potential for oxidation being −2.05 V [348].

Strong acids, such as sulfuric acid, and strong bases, such as potassium hydroxide (KOH) and sodium hydroxide (NaOH), are frequently used as electrolytes due to their strong conducting abilities.

A solid polymer electrolyte, such as Nafion, can also be used. When a special catalyst is applied on each side of the solid electrolyte, it can efficiently split the water molecule with as little potential as 1.8 V [348].

Water electrolysis does not convert 100% of the electrical energy into the chemical energy of hydrogen. The process requires more extreme potentials than what would be expected based on the cell's total reversible reduction potentials. This excess potential accounts for various forms of overpotential by which the extra energy is eventually lost as heat. For a well-designed cell the largest overpotential is the reaction overpotential for the four electron oxidation of water to oxygen at the anode. Reaction overpotential is an activation overpotential that specifically relates to chemical reactions that must formally precede electron transfer. An effective electrocatalyst to facilitate this reaction has not been developed. Platinum alloys are the default state of the art for this oxidation.

Developing a cheap and effective electrocatalyst for this reaction would be a great advance.

The simpler two-electron reaction to produce hydrogen at the cathode can be electrocatalyzed with almost no reaction overpotential by platinum or, in theory, a hydrogenase enzyme. If other less-effective materials are used for the cathode then another large overpotential must be paid for.

The energy efficiency of water electrolysis varies widely—some report 50–80%. These values refer only to the efficiency of converting electrical energy into hydrogen's chemical energy. The energy lost in generating the electricity is not included. For instance, when considering a power plant that converts the heat of nuclear reactions into hydrogen via electrolysis, the total efficiency may be closer to 30–45%, although the inefficiencies of power plants in turning heat into electrical energy is not usually included in efficiency, so the former measure of 50–80% is probably a more realistic efficiency.

However the research and development in the field of electrolysis has recently been greatly affected by the work performed by the team from Massachusetts Institute of Technology. In 2008, a group, lead by Daniel Nocera, announced the development of an electrocatalyst composed of the abundant metal cobalt and phosphate [349]. The team demonstrated a new method of splitting water into hydrogen and oxygen, predicting that it will unleash a "solar revolution". However, the research had little to do with solar power, but it had far-reaching implications as a storage medium, making the use of renewable but intermittent energy sources like the sun and the wind more practical for everyday use. The work follows several decades of investigations in a field sometimes known as "artificial photosynthesis," because it attempts to mimic one of nature's best tricks—storing energy in chemical bonds. In this case, energy from the sun can be stored in hydrogen fuel when it is split from water; a catalyst is being developed to do so in an efficient and cost-effective manner.

Nocera's catalyst inspires expectations that everyone with a roof could someday have a simple way of making hydrogen from water.

8.5 Which Is the New Catalyst for Water Electrolysis?

This new scheme promises to do what existing commercial electrolyzers do, but in a much simpler and efficient way. The setup in this scheme is easy to build, operate, and maintain, and the catalyst identified in Ref. [349] uses cheap and abundant materials—cobalt and phosphorous—instead of the rare and expensive metals used as catalysts in the existing systems. This is important because the system should be cheap enough to be installed in every house. The new material, however, is inexpensive as it can act as an electrode during water electrolysis, requires neutral pH solution, and can generate hydrogen and oxygen by making use of small amounts of electricity. Electric power generated by solar panels during the day can be used efficiently to generate hydrogen and oxygen by electrolysis. The products of this electrolysis, hydrogen and oxygen, can be then used at night to produce electric currents inside fuel cells (Fig. 8.2). The catalyst consists of a combination of cobalt and phosphate that is deposited on an indium–tin oxide electrode. This unique mix appears to catalyze the water electrolysis process, although the exact mechanism responsible for this effect is largely unknown. It was announced that with the new electrode the water electrolysis process can produce the same amount of oxygen by using only 1 V, rather than 1.6 V in the case of more traditional methods.

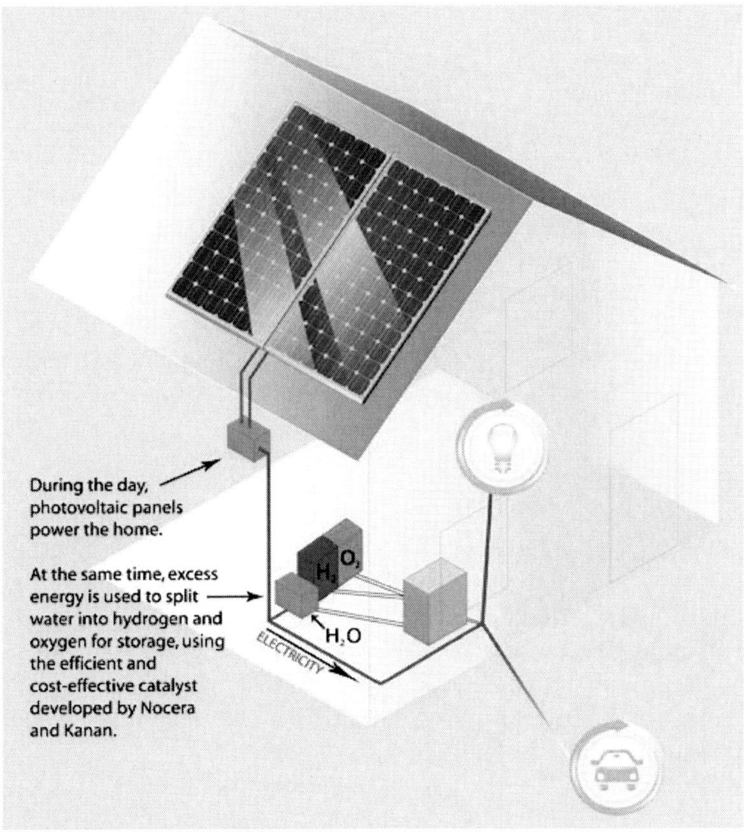

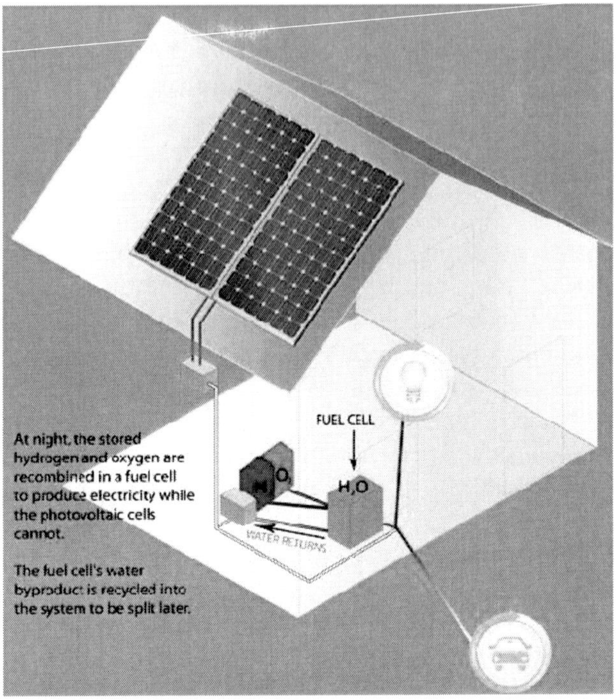

Figure 8.2 Schematic diagram of the energy supply to an individual house during day and night (Image: MIT News Office) [350].

The enhanced efficiency of the water electrolysis makes it possible to produce cheap systems for energy supply in individual houses. This interesting work shows clearly that the potential of electrolysis in energy conversion systems is not exhausted [351].

8.6 Conclusions

In this chapter, we covered the conventional methods of hydrogen production, which are not able to satisfy the requirements of the modern hydrogen energetics. Still there is a potential for further development in traditional electrolysis, which is connected with the search of new and effective catalysts. On the other hand, the idea to combine the cells for electrolysis with a hydrogen fuel cell looks interesting and prospective.

Chapter 9

Hydrogen Fuel Cells

9.1 Introduction

One more interesting example concerning the importance of hydrogen in energy processes is the work of hydrogen fuel cell.

A fuel cell is an electrochemical gadget that converts chemical energy from a fuel into electric energy [352]. Electricity is generated via the reaction between a fuel and an oxidizing agent. The fuel cell requires supply of reactants, whereas the reaction products must be removed out of it. This system operates with the help of an electrolyte. If the necessary reactants are supplied the fuel cells can operate continuously.

Since the reactants are being consumed from an external source, fuel cells represent a thermodynamically open system where the reactants are being replenished. This is different from batteries that store electric energy chemically and hence represent a thermodynamically closed system. A fuel cell works as long as it is being supplied by reactants and reaction products are being removed out of it [352].

In practice, there are many possible combinations of fuels and oxidants. A hydrogen fuel cell uses hydrogen as its fuel and oxygen (usually from air) as its oxidant, but other reactants are also possible. For example, other fuels include hydrocarbons and alcohols, whereas other oxidants include chlorine and chlorine dioxide [352].

Hydrogen Energy for Beginners
Alexander Gavrilyuk

ISBN 978-981-4364-77-5 (Hardcover), 978-981-4364-78-2 (eBook)
www.panstanford.com

In a hydrogen fuel cell, hydrogen and oxygen atoms combine to yield water molecules, that is, the reversed reaction of water splitting is used. This is a downhill reaction accompanied with negative changes of Gibbs free energy, that is, $\Delta G < 0$ and for this reason it runs facilely. The thermodynamic functions of this reaction are known from the water splitting discussed in Chapter 7.

A stream of hydrogen is delivered to the anode side of the membrane electrode assembly (MEA). At the anode side it is catalytically split into protons and electrons. This oxidation half-cell reaction is represented by:

$$H_2 \rightarrow 2H^+ + 2e^-, E^\circ = 0\ V_{SHE} \tag{9.1}$$

The reduction half-cell reaction at the cathode is represented by

$$0.5\ O_2 + 2H^+ + 2e^- \rightarrow H_2O, E^\circ = 1.23\ V_{SHE} \tag{9.2}$$

The overall reaction is as follows:

$$H_2 + 0.5\ O_2 \rightarrow H_2O, E^\circ = 1.23\ V_{SHE} \tag{9.3}$$

9.2 Design of a Fuel Cell

There are plenty of various fuel cells; however, they all have general common features. They are made up of three main functional units that are sandwiched together: the anode, the electrolyte, and the cathode (Fig. 9.1). Two chemical reactions occur at the interfaces of the three different domains resulting in the consumption of fuel, creation of water or carbon dioxide, and generation of electric current, which can be used to power electrical devices, normally referred to as the load [352].

At the anode a specially selected catalyst oxidizes the fuel, usually hydrogen, turning it into a positively charged ion and a negatively charged electron. The electrolyte is a substance specifically designed so that it is permeable to the ions, but not to the electrons. The released electrons travel through a wire creating the electric current. The ions flow through the electrolyte to the cathode. On reaching the cathode, the ions are recombined with the electrons and the two react with a third chemical, usually oxygen, to create water or carbon dioxide [352].

So, again we see the membrane architecture for the energy conversion process where hydrogen plays the main role. The ability of hydrogen to break out easily into the charged carrier is employed.

The electrolyte performs the function of charge separation similar to that for the biological membranes in metabolic processes considered above. Again Mitchell's "proticity" is created via splitting of hydrogen atoms into electrons and hydrogen proton, which drives the vectorial flow of the protons to the cathode.

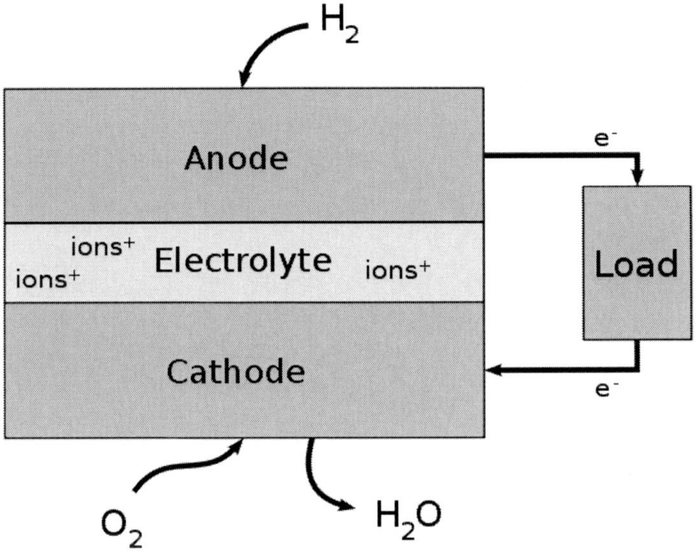

Figure 9.1 Block diagram of a fuel cell (Image: Paulsmith99) [353].

The most important design features of a fuel cell are as follows:

(i) The electrolyte substance that usually defines the type of fuel cell.
(ii) The fuel that is used. The most common fuel is hydrogen.
(iii) The anode catalyst that splits the fuel into electrons and ions. The anode catalyst is usually made up of very fine platinum powder.
(iv) The cathode catalyst that turns the ions into the chemical product like water or carbon dioxide. The cathode catalyst is often made up of nickel.

A typical fuel cell produces a voltage from 0.6 V to 0.7 V at full rated load. Voltage decreases as current increases, due to several factors, such as activation loss, ohmic loss (voltage drop due to resistance of the cell components and interconnects), and mass transport loss (depletion of reactants at catalyst sites under high loads, causing rapid loss of voltage) [352].

To produce and transport the desired amount of energy, fuel cells can be combined in series and parallel circuits, where series yields higher voltage, and parallel allows a higher current to be supplied. Such a design is called a fuel cell stack. The cell surface area can be enhanced to allow stronger current from each cell.

In addition to electricity, fuel cells produce water, heat and, depending on the fuel source, very small amounts of nitrogen dioxide and other emissions. The energy efficiency of a fuel cell is generally between 40–60%, or up to 85% efficient if waste heat is captured for use [352].

9.3 Proton-Exchange Membrane Fuel Cells

In the archetypical hydrogen–oxygen proton-exchange membrane fuel cell (PEMFC) design, a proton-conducting polymer membrane (as in metabolic processes!), the electrolyte, separates from the sides of the anode and the cathode. Figure 9.2 shows the design of PEMFC [354].

On the anode side, hydrogen diffuses to the anode catalyst where it later dissociates into protons and electrons. The protons are conducted through the membrane to the cathode, but the electrons are forced to travel in an external circuit (supplying power) because the membrane has negligible electron conductivity. On the cathode catalyst, oxygen molecules react with the electrons (which have traveled through the external circuit) and protons to form water—in this example, there is only one waste product, which can be either liquid or vapor.

In addition to this pure hydrogen type, there are hydrocarbon fuels for fuel cells, including diesel, methanol, and chemical hydrides. The waste products with these types of fuel are carbon dioxide and water.

Very often Nafion is used as a membrane for the hydrogen fuel cell.

Nafion is a sulfonated tetrafluoroethylene based fluoropolymer-copolymer discovered in the late 1960s by Walther Grot of DuPont. It is the first of a class of synthetic polymers with ionic properties which are called ionomers. Nafion's unique ionic properties are a result of incorporating perfluorovinyl ether groups terminated with sulfonate groups onto a tetrafluoroethylene (teflon) backbone. [355,356]. Nafion has received considerable amount of attention as a proton conductor for proton-exchange membrane (PEM) fuel cells because of its excellent thermal and mechanical stability.

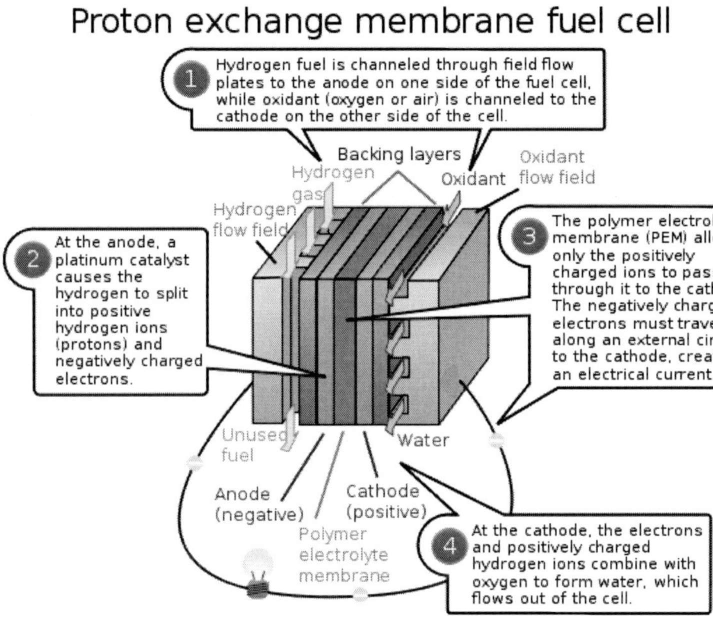

Figure 9.2 Construction of a high-temperature PEMFC: Bipolar plate as electrode with in-milled gas channel structure, fabricated from conductive composites (enhanced with graphite, carbon black, carbon fiber, and/or carbon nanotubes for more conductivity); porous carbon papers; reactive layer, usually on the polymer membrane applied; polymer membrane (Image: Jafet) [357].

The chemical basis of Nafion's superior ionic conductive properties is a matter of continuous research. Protons on the SO_3H (sulfonic acid) groups "hop" from one acid site to another. Pores allow movement of cations but the membranes do not conduct anions or electrons. Nafion can be manufactured with various cationic conductivities [358].

Nafion derivatives are first synthesized by the copolymerization of tetrafluoroethylene (TFE) (the monomer in Teflon) and a derivative of a perfluoro (alkyl vinyl ether) with sulfonyl acid fluoride. The latter reagent can be prepared by the pyrolysis of its respective oxide or carboxylic acid to give the olefinated structure.

9.4 Preparation and Properties of the Nafion Membrane

The resulting product is an $-SO_2F$-containing thermoplastic that is extruded into films. Hot aqueous NaOH converts these sulfonyl fluoride ($-SO_2F$) groups into sulfonate groups ($-SO_3^-Na^+$). This form of Nafion, referred to as the neutral or salt form, is finally converted to the acid form containing the sulfonic acid ($-SO_3H$) groups. Nafion can be cast into thin films by heating in aqueous alcohol at 250°C in an autoclave. By this process, Nafion can be used to generate composite films, coat electrodes, or repair damaged membranes. The structural formula of Nafion is shown in Fig. 9.3.

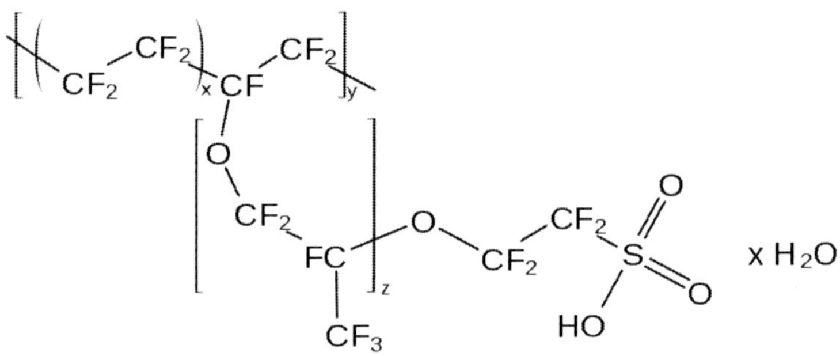

Figure 9.3 Structural formula of Nafion (Image: Roland) [359].

The morphology of Nafion membranes is a matter of continuing study to gain greater control on its properties. Nafion structure may also possess other important properties such as water management, hydration stability at high temperatures, electro-osmotic (EO) drag, as well as the mechanical, thermal, and oxidative stability.

The first model for Nafion, called the cluster-channel or cluster-network model, proposed an equal distribution of sulfonate ion clusters [356] with a 4 nm diameter held within a continuous fluorocarbon lattice (Fig. 9.4). Narrow channels about 1 nm in diameter interconnect the clusters, which explains the transport properties [356,358].

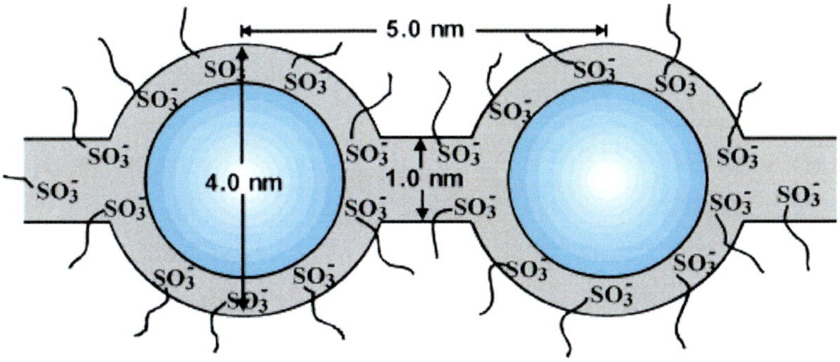

Figure 9.4 Cluster-network model for the morphology of hydrated Nafion. Reprinted from Ref. 361, copyright (1983), with permission from Elsevier.

Advanced morphological investigations have yielded other models, namely, a core–shell model where the ion-rich core is surrounded by an ion-poor shell, a rod model where the sulfonic groups arrange into crystal-like rods, and a sandwich model where the polymer forms two layers whose sulfonic groups attract across an aqueous layer where transport occurs [360]. Consistency between the models includes a network of ionic clusters; the models differ in the cluster geometry and distribution. Although no model has yet been determined as fully correct, some scientists have demonstrated that as the membrane hydrates, Nafion's morphology transforms from the cluster-channel model to a rod-like model [361].

A more recent water-channel model [362] was proposed based the simulations on small-angle X-ray scattering data and solid state nuclear magnetic resonance study. In this model, the sulfonic acid functional groups self-organize into hydrophilic water channels of ~2.5 nm diameter through which small ions can be easily transported, while the hydrophobic polymer backbone with crystallites provides good mechanical stability.

Figure 9.5 shows highly schematically the main features of the water-channel model of Nafion, at a hydration level of 20 vol% (11 wt%) water [358]. The structure contains ~2.4 nm-diameter water channels (cylinders) lined with hydrophilic side groups, Fig. 9.5b. These channels can be considered as cylindrical inverted micelles. On the outside, they are stabilized by the relatively straight helical backbone segments observed by nuclear magnetic resonance [362,363]. With a persistence length of 3–5 nm for the Nafion backbone [364,365] the densely packed cylinders having shell of relatively stiff backbones can be expected to have a persistence length of tens of nanometers (Fig.9.5a).

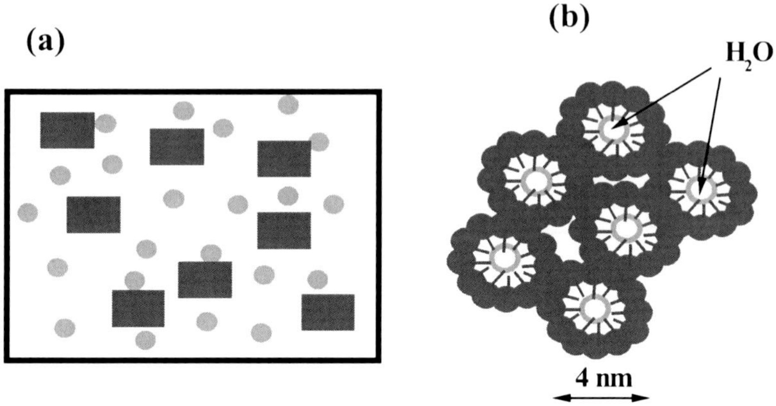

Figure 9.5 Parallel water-channel (inverted-micelle cylinder) model of Nafion. (a) Cross-sections through the cylindrical water channels (light gray) and the Nafion crystallites (dark gray) in the noncrystalline Nafion matrix (white). (b) Schematic diagram of the approximately hexagonal packing of several inverted-micelle cylinders.

It is clear that in this type of fuel cells the membrane must be hydrated, requiring water to be evaporated at precisely the same rate that it is produced. If water is evaporated too quickly, the membrane dries, resistance across it increases, and eventually it will crack, creating a gas "short circuit" where hydrogen and oxygen combine directly, generating heat that will damage the fuel cell. If the water is evaporated too slowly, the electrodes will flood, preventing the reactants from reaching the catalyst and stopping the reaction. Methods to manage water in cells are being developed like EO pumps that focus on flow control. Just as in a combustion engine, a steady ratio between the reactant and oxygen is necessary to keep the fuel cell operating efficiently.

The same concerns temperature: it must be maintained throughout the cell in order to prevent destruction of the cell through thermal loading. This is particularly challenging as the $2H_2 + O_2 \rightarrow 2H_2O$ reaction is highly exothermic, so a large quantity of heat is generated within the fuel cell.

The membrane must also not allow either gas to pass to the other side of the cell, a problem known as gas crossover. Finally, the membrane must be resistant to the reducing environment at the cathode as well as the harsh oxidative environment at the anode. Splitting of the hydrogen molecule is relatively easy in the presence of a platinum catalyst. However, splitting the oxygen molecule is more difficult, and this causes significant electric losses. An appropriate catalyst material for this process has not been discovered, and platinum is the best option.

Nafion was found effective as a membrane for PEM fuel cells by permitting hydrogen ion transport while preventing electron conduction. Solid polymer electrolytes, which are made by connecting or depositing electrodes (usually noble metal) to both sides of the membrane, conduct the electrons through an energy requiring process and rejoin the hydrogen ions to react with oxygen and produce water [355].

The PEMFC is a prime candidate for vehicle and other mobile applications of all sizes down to mobile phones, because of its compactness. However, water management is crucial for performance: too much water will flood the membrane, too little will dry it; in both cases, power output will drop. Water management is a very difficult subject in PEM systems, primarily because water in the membrane is attracted toward the cathode of the cell due to polarization. A wide variety of solutions for managing the water exist, including integration of EO pumps. The pumps are fabricated from silica nanospheres [366] or hydrophilic porous glass; the pumping mechanism is generated by an external electric field applied on an electric double layer (EDL) and generates high pressures (more than 340 atm (34 MPa) at 12 kV applied potentials) and high flow rates (e.g., 40 ml/min at 100 V in a pumping structure less than 1 cm^3 in volume). EO pumps are compact, have no moving parts, and scale favorably with fuel cell design. The EO pump might drop the parasitic load of water management in fuel cells from 20% to 0.5% of the fuel cell power [367].

Furthermore, the platinum catalyst on the membrane is easily poisoned by carbon monoxide (CO) (no more than one part per million is usually acceptable) and the membrane is sensitive to things like metal ions, which can be introduced by corrosion of metallic bipolar plates, metallic components in the fuel cell system, or from contaminants in the fuel/oxidant. The most commonly used membrane is Nafion by DuPont, which relies on liquid water humidification of the membrane to transport protons. This implies that it is not feasible to use temperatures above 80 to 90°C, since the membrane would dry.

PEM systems that use reformed methanol were proposed, as in Daimler Chrysler Necar 5; reforming methanol, that is, making it react to obtain hydrogen, is however a very complicated process that also requires purification from the CO the reaction produces. A platinum–ruthenium catalyst is necessary as some CO will unavoidably reach the membrane. The level should not exceed 10 parts per million. Furthermore, the start-up times of such a reformer reactor are of about half an hour. Alternatively, methanol, and some other biofuels can be fed to a PEM fuel cell directly without being reformed, thus making a direct methanol fuel cell (DMFC). These devices operate with limited success.

9.5 Catalyst Research

Much of the current research on catalysts for PEM fuel cells can be classified as having one of the two main objectives: to obtain higher catalytic activity than the standard carbon-supported platinum particle catalysts used in current PEM fuel cells or to reduce the poisoning of PEM fuel cell catalysts by impurity gases. Examples of these two approaches are given in the following sections.

9.5.1 Increasing Catalytic Activity

As mentioned above, platinum is by far the most-effective element used for PEM fuel cell catalysts, and nearly all current PEM fuel cells use platinum particles on porous carbon supports to catalyze both hydrogen oxidation and oxygen reduction. However, due to their high cost, current Pt/C catalysts are not feasible for commercialization. The U.S. Department of Energy estimates that platinum-based catalysts will need to use roughly four times less platinum than is used in current PEM fuel cell designs in order to represent a realistic alternative to internal combustion engines. Consequently, one main goal of catalyst design for PEM fuel cells is to increase the catalytic activity of platinum by a factor of four so that only one-fourth as much of the precious metal is necessary to achieve similar performance.

One method of increasing the performance of platinum catalysts is to optimize the size and shape of the platinum particles. Decreasing the particles' size alone increases the total surface area of catalyst available to participate in reactions per volume of platinum used, but recent studies have demonstrated additional ways to make further improvements to catalytic performance. For example, one study reports that high-index facets of platinum nanoparticles [i.e., Miller indexes with large integers, such as Pt (730)] provide a greater density of reactive sites for oxygen reduction than typical platinum nanoparticles [366].

A second method of increasing the catalytic activity of platinum is to alloy it with other metals. For example, it was recently shown that the Pt_3Ni (111) surface has a higher oxygen reduction activity than pure Pt(111) by a factor of ten [367]. This dramatic performance enhancement is attributed by the authors to modifications in the electronic structure of the surface, which reduce its tendency to bond to oxygen-containing ionic species present in PEM fuel cells and enhance in turn the number of available sites for oxygen adsorption and reduction.

9.5.2 Reducing Poisoning

The other popular approach to improving catalyst performance is to reduce its sensitivity to impurities in the fuel source, especially CO. Presently, it is not economical to mass-produce pure hydrogen gas by electrolysis or any other means. Instead, hydrogen gas is produced by steam, reforming light hydrocarbons, a process which produces a mixture of gases that also contains CO (1–3%), CO_2 (19–25%), and N_2 (25%) [368]. Even tens of parts per million of CO can poison a pure platinum catalyst, therefore, increasing platinum's resistance to CO is an active area of research.

For example, one study reported that cube-shaped platinum nanoparticles with (100) faces displayed a four-fold enhancement in oxygen reduction activity compared with randomly faceted platinum nanoparticles of similar size [369]. The authors concluded that the (111) facets of the randomly shaped nanoparticles bonded more strongly to sulfate ions than the (100) facets, reducing the number of catalytic sites open to oxygen molecules. The nanocubes they synthesized, in contrast, had almost exclusively (100) facets, which are known to interact with sulfate more weakly. As a result, a greater fraction of the surface area of those particles was available for the reduction of oxygen, boosting the catalyst's oxygen reduction activity.

In addition, researchers have been investigating ways of reducing the CO content of hydrogen fuel before it enters a fuel cell as a possible way to avoid poisoning the catalysts. One recent study revealed that ruthenium–platinum core–shell nanoparticles are particularly effective at oxidizing CO to form CO_2, a much less harmful fuel contaminant [370]. The mechanism that produces this effect is conceptually similar to that described for Pt_3Ni above: the ruthenium core of the particle alters the electronic structure of the platinum surface, rendering it better able to catalyze the oxidation of CO.

9.6 Conclusions

In this chapter, we covered the principles of functioning of the hydrogen fuel cell where chemical energy is converted directly into electrical energy. It should be emphasized that the hydrogen fuel cells are devices whose operation depends highly upon various catalytic processes. On the other hand, solid state catalysis is inevitably connected with nanotechnologies. So, the development of nanosized catalysts, as well as ionic membranes, is the most promising trend in the development of hydrogen fuel cells.

Chapter 10

Hydrogen Storage

10.1 Introduction

In this chapter the fundamentals of the methods for storing hydrogen for subsequent use and the materials employed for this purpose will be considered.

The ideas and methods concerning hydrogen storage are related to many different issues. It is worth emphasizing that in its essence the trend "hydrogen storage" is in fact related rather to materials that serve for hydrogen accommodation than to hydrogen itself, although specific properties of hydrogen also play here a definite role. In fact hydrogen storage is the problem for material science and hence a dream material for hydrogen storage should be elaborated, which is able to meet all requirements to perform this important function.

We shall consider mainly the properties of materials suitable for hydrogen storage that accommodate either atomic or molecular hydrogen (dihydrogen) performing the function of boxes or bags wherein hydrogen should be accommodated and packed with maximal possible density, with maximum convenience to release when it is necessary, and, certainly, with minimal expenses.

It is necessary to stress that hydrogen storage is the crucial point in the hydrogen energy infrastructure. And for this reason we are necessitated to consider here also some economical aspects of this problem that dictates scientific and technological approaches to this extremely important problem.

Hydrogen Energy for Beginners
Alexander Gavrilyuk

ISBN 978-981-4364-77-5 (Hardcover), 978-981-4364-78-2 (eBook)
www.panstanford.com

Most research into hydrogen storage is focused on storing hydrogen as a lightweight, compact energy carrier for mobile applications.

Firstly it makes sense to consider concisely conventional, long-time used and well-established methods including high pressures, cryogenics, and then to switch to modern ones related to material science and high technology such as physical and chemical storage in chemical compounds that reversibly release H_2 upon heating.

10.2 Conventional Methods of Hydrogen Storage

10.2.1 Liquid Hydrogen

At present, liquid hydrogen (LH) is the basic fuel of choice for various hydrogen-fueled cars. It can be used in the same manner as it is being used in space shuttles [371].

The advantage of LH is that it stores approximately 2.6 times the energy per unit mass as gasoline. The disadvantage is that it needs about four times the volume for a given amount of energy, due to its very low density. LH has less energy density by volume than hydrocarbon fuels such as gasoline by approximately a factor of four [371]. This highlights the density problem for pure hydrogen: there is actually about 64% more hydrogen in a liter of gasoline (116 g hydrogen) than there is in a liter of pure LH (71 g hydrogen). The carbon in the gasoline also contributes to the energy of combustion [371].

However LH requires cryogenic storage and boils around 20.3 K (–253°C). Moreover, the production of liquid or highly compressed hydrogen requires high amount of energy (due to the cost of liquidification and of solid cryogenic tanks, since they must be impermeable to tiny H_2 molecules) and may become dangerous in some circumstances [371].

10.2.2 Compressed Hydrogen

Compressed hydrogen is the gaseous hydrogen which is kept under pressure. Compressed hydrogen, in comparison to liquid one, is quite different to store. It is stored in hydrogen tanks at 350 bar and 700 bar and is used in hydrogen vehicles [371]. Compressed hydrogen gas has good energy density by weight, but poor energy density by volume versus hydrocarbons, hence it requires a larger tank to store. Large hydrogen tanks are heavier than the small hydrocarbon tanks used to store the same amount of energy, all other factors remaining equal. Increasing gas pressure would improve the energy density by volume, making smaller, but not lighter container tanks [371]. Compressed hydrogen will require 2.1% of the energy content to power the compressor. Higher compression without energy recovery will mean more energy lost at the compression step. Compressed hydrogen storage can exhibit very low permeation [371].

10.2.3 Underground Hydrogen Storage

Underground hydrogen storage is the practice of hydrogen storage in underground caverns, salt domes and depleted oil/gas fields. Large quantities of gaseous hydrogen have been stored in underground caverns by Imperial Chemical Industry for many years without difficulties. The storage of large quantities of hydrogen underground in solution-mined salt domes, aquifers, or excavated rock caverns or mines can function as grid energy storage which is essential for hydrogen economy. By using a turboexpander the electricity requirements for compressed storage on 200 bar amount to 2.1% of the energy content [371].

10.3 Solid-State Hydrogen Storage

10.3.1 Requirements for Solid-State Hydrogen-Storage Materials

Thus, solid-state storage is now considered as the safest and most effective way for routine handling hydrogen. Many scientific programs have targeted a “dream hydrogen-storage material (HSM)”, highlighting the exclusive importance of this fascinating scientific, technological, and socio-economic issue for our civilization.

The targets were set by the U. S. Department of Energy (the targets assume a 6.5 kg H_2 storage system) [371]. The goals for 2015 are even more demanding: 9.0 wt.% and 81 kg H_2 per m^3, which approach the expectations of the automotive industry [371]. This task is difficult to achieve since that the mass density of elemental hydrogen is only 70.8 kgm^{-3} in its liquid state at 20 K (1 atm), and that 5 kg of hydrogen gas fills a volume of 56 m^3 under ambient conditions. Furthermore, one should bear in mind the additional contribution of the mass of the necessary container.

Briefly, an ideal chemical HSM is to have a low molar weight (to decrease the reservoir/storage mass), be inexpensive, have fast kinetics for absorbing and desorbing H_2 in the 25–120°C temperature range, and store large quantities of hydrogen reversibly. It is important to emphasize that at present no single material fulfills all of these requirements.

An ideal solid HSM, therefore, for practical applications should, for both economic and environmental reasons, obey the five main commandments of hydrogen storage [8,372]:

(i) High storage capacity: minimum 6.5 wt.% abundance of hydrogen and at least 65 gL^{-1} of hydrogen available from the material
(ii) Thermal decomposition temperature for a material charged with hydrogen $T_{dec} \sim$ 60–120°C
(iii) Reversibility of the thermal absorption/desorption cycle: low temperature of hydrogen desorption and low pressure of hydrogen absorption (a plateau pressure of the order of a few bars at room temperature), or ease of nonthermal transformation between substrates and products of decomposition
(iv) Low cost
(v) Low-toxicity of a nonexplosive and possibly inert (to water and oxygen) storage medium

And all these targets are for the hydrogen-storage system, not the HSM. System densities are often around half of those of the working material, thus while a material may store 6 wt.% H_2, a working system using that material may only achieve 3 wt.% when the weight of tanks, temperature and pressure control equipment, etc., is considered [8,372]. Storing hydrogen effectively in a solid-state material would instantly command commercial interest if the amount of reversibly adsorbed/desorbed hydrogen exceeds 6.5 wt.% [8,372] and more than 65 gL^{-1} [8,372].

On the other hand, there is always a possibility for these target values to be reduced if the cost of fossil fuels grows abruptly due to economic or political reasons. Nevertheless, one should seriously perceive the above as a prime target at least for the next twenty years.

Furthermore, one should not forget the irreversibility problem in hydrogen economy. For example, chemical generation of hydrogen from certain materials is highly effective and thermodynamically favorable [8]. However, the recycling of the resulting products back to their initial state might be complicated, energy intensive, and, consequently, energy and cost expensive. Nevertheless, it may happen that under sufficient market pressure, a particular kind of irreversibility (e.g., desorption is easy, but reabsorption is difficult) would then become acceptable. In such a case, a container of (obligatorily) cheap HSMs could be discarded without regeneration, similar to many existing single-use batteries [8].

The most promising modern methods of hydrogen storage based upon high technology and the knowledge of material science are characterized usually as physical or chemical. These methods provide the hydrogen storage via physical or chemical interaction of hydrogen with a number of materials. The fundamentals of these methods is strong interaction of either molecular or atomic hydrogen with materials of the storage medium [8].

Simple atomic mass–based calculations reveal that only the light (i.e., low atomic number) chemical elements can satisfy the criterion (i). Thus, the series of efficient HSMs can only be built from targeted chemical elements from an unfortunately short list: Li, Be, B, C, N, O, F, Na, Mg, Al, Si, and P. Furthermore, due to the toxicity or unfavorable chemical properties of hydrogen's connections with Be, F, Si, and P, the list of chemical elements suitable for hydrogen storages must be reduced to only eight elements. Heavier ones may enter the multiple component system only as a low-abundant additive, presumably for fine-tuning of properties or as a catalyst [8].

In 2010, only two storage technologies were recognized as being susceptible to meet the targets of the Department of Energy of the United States: metal–organic framework, MOF-177 (see below), exceeds 2010 target for volumetric capacity, while cryo-compressed H_2 exceeds more restrictive 2015 targets for both gravimetric and volumetric capacity [371].

10.3.2 Physical Hydrogen Storage in Microporous Materials

This method employs the storing of molecular hydrogen (dihydrogen) accommodated in pores of highly porous materials suitable for the storage. Molecular hydrogen doesn't undergo any chemical transformation and adsorbs and desorbs without dissociation.

The use of hydrogen physisorption on porous materials is one of the main methods being considered for vehicle applications. The objective is storage of large amounts of hydrogen at near-ambient temperatures and "safe" pressures. These materials would also need to satisfy the requirements of reasonable volume, weight and realistic kinetics for charging and discharging the hydrogen gas. The U.S. Department of Energy (DOE) has set targets to develop and verify on-board hydrogen-storage systems achieving 6 and 9 wt.% for 2010 and 2015, respectively [372]. The weak interaction of hydrogen physisorbed within pores needs to be understood in order to optimize porous materials for hydrogen adsorption. The important measurements for this application are

(i) amount adsorbed as a function of pressure
(ii) temperature dependence of adsorption
(iii) the enthalpies of adsorption
(iv) the adsorption/desorption characteristics

Amounts adsorbed can be compared using adsorption isotherms while the temperature dependence can be evaluated using isotherms over a temperature range and also, isobars. Measurements of a series of hydrogen adsorption isotherms over a temperature range provide information for calculation of the enthalpies of adsorption. These characteristics are important in relation to the hydrogen charging and discharging characteristics. Adsorption/desorption kinetic measurements are required because of the need for fast charging and discharging of hydrogen.

Pores are divided into three categories on the basis of their size: micropores, of dimensions below 2 nm, mesopores, between 2 nm and 50 nm, and macropores, which are greater than 50 nm [373]. The porous materials being considered for physisorbed molecular hydrogen storage are predominantly microporous. They are also often in the subcategory of ultramicroporous, in which the pore dimensions are less than 0.7 nm; close to the size of a single hydrogen molecule [374]. Microporous materials comprise a very large domain within the research area related to HSMs. The narrow width of their micropores leads to an increase in the density of hydrogen in the pore network compared to the gas phase. This occurs primarily at relatively low, sub-ambient, temperatures [374].

Van der Waals (attractive/repulsive) forces take place between the solid surface and the gas molecules (Fig. 10.1). At these length scales the adsorption potentials of the opposing pore walls overlap, which significantly increases the density of the adsorbed hydrogen, compared to the gas phase, at any given temperature and pressure (Fig. 10.1) [375]. The ideal pore size for maximal attraction of an adsorbate is the same as its diameter to provide optimal interaction with all the surrounding adsorbent walls, regardless of geometry, thus maximizing the total van der Waals force on the adsorbate.

Figure 10.1 illustrates the mechanism of the enhancement of the adsorbtion potential in micropores and gives a clue in the strategy for creation of an ideal material for the hydrogen storage: it should be a material with the maximum possible specific surface possessing the microporous structures with pore sizes maximally suitable to a hydrogen molecule sizes.

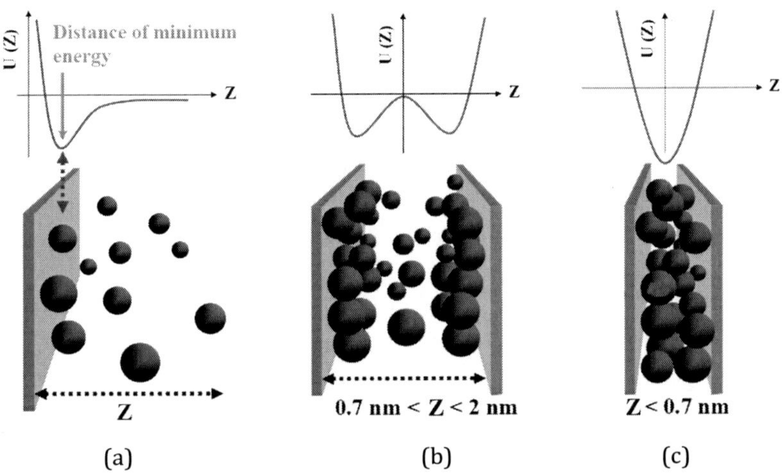

Figure 10.1 The difference in the adsorption potential value for adsorbed molecules on (a) a free surface and (b and c) in slit-shaped pores. In (b), the adsorption potential and the density is higher than in flat surface, which increase futher in (c) as the pore size decreases. As per IUPAC classification, a macropore has a width more than 50 nm; a mesopore has a width between 2 nm and 50 nm; and a micropore has a width less than 2 nm. Adapted from Ref. [375].

For many years there has been significant and intensive research on the use of many various microporous materials for gas storage, but in the last few years the interest in this area has increased dramatically. The discovery and exciting development the technology of carbon nanotubes (CNTs) and their perspective to be used as a possible storage medium has attracted significant attention of many workers [374,375]. Progress proceeds also with the synthesis of microporous organic polymers, as well as new hybrid material classes, and new candidate materials are likely to emerge as the search, as well as the interest in these materials for a range of other applications, continues [374].

In this section, we cover the main groups of adsorbents currently being considered as storage media.

The adsorption of hydrogen by microporous materials is, as a rule, fast, which is practically advantageous for hydrogen storage, but low temperatures are required to achieve significant capacities at useful storage pressures. In practice, low temperature storage requires cooling, which unavoidably adds to the weight of a storage unit and is therefore a significant disadvantage. The hydrogen adsorption capacity of microporous materials at ambient temperature is currently low for practical use [376,377]. However, one may hope to enhance the interaction of hydrogen with the materials to realize this possibility in the future. This is a quickly developing domain for research and development; the situation being changed day by day. It may occur that when this book is published, the requirements of the DOE of the United States (US DOE) will be achieved in some HSMs.

However, a number of factors should be also considered in order to assess the suitability of a material for widespread use as a hydrogen store. The important physical and chemical properties include those related to safety, such as the pyrophoricity of the material and its toxicity, as well as those that affect the operation and hence the design of hydrogen-storage systems, such as the thermal conductivity of the material in an activated state. Other practical considerations include the cost of the material and its raw constituents, and their natural abundance [374].

In makes sense to start our consideration of the microporous materials from metal–organic framework (MOF) materials, since namely these materials are very deductive both due to the huge variety of problems arising on the way of HSMs engineering as well as due to the strategy and the methods used to salve these problems. And, perhaps, at least at present time, these materials are valued as the most promising among the materials for physical hydrogen storage.

10.3.3 Metal–Organic Frameworks

MOFs represent an intensively investigated class of synthetic porous materials for hydrogen storage.

MOFs are highly crystalline inorganic–organic hybrid structures that contain metal clusters or ions (secondary building units, or SBUs) as nodes and organic ligands as linkers (molecular bridges) [378–382]. As such, they demonstrate a series of performances inherent to the building blocks, such as geometric rigidity, chemical functionality, or chirality [382]. Their preparation technology is relatively simple and quite reproducible: it makes it possible to control the design for manufacturing targeted products from the vast number of MOFs that are potentially accessible [382]. One simple outcome of the use of molecular, rather than monatomic, bridges is the extension of the length between the metal centers, which can lead to the formation of large void regions (see Fig. 10.2).

In order to facilitate the process of structure prediction, the concept of SBUs as structural entities was adopted from zeolite structure analysis. These are simple geometric figures representing the inorganic clusters or coordination spheres that are linked together by the (typically linear) organic components to form the product framework.

Considerations of the geometric requirements for a target framework and implementation of the design and synthesis of such a framework have been termed reticular synthesis [382], where the term "reticular" means having the form of a (usually periodic) net [382]. This process requires both an understanding of the local coordination patterns of the metal and organic units and foreknowledge of what net topologies they will adopt [383]. The accumulation of knowledge that has given rise to such precepts has primarily resulted from their crystalline nature that allows a high degree of structural characterization to be achieved through X-ray diffraction methods [383].

Guest molecules (solvent) occupying the pores can be removed by solvent exchange and heating under vacuum giving rise to porous structure of MOFs without destabilizing the frame; hydrogen molecules being adsorbed onto the surface of the pores by physisorption [382].

Compared to other traditional porous systems as zeolites and porous carbon materials, MOFs have very high number of pores and surface area which allow higher hydrogen uptake in a given volume. For this reason, research interests on hydrogen storage in MOFs have been growing continuously since 2003 when the first MOF-based hydrogen storage was achieved [382]. MOFs have many favorable features, such as high porosity, reproducible and relatively facile syntheses, amenability to scale-up, and chemical modification for targeting desired properties. Enormous number of geometric and chemical variations of MOFs along with different combinations of secondary basic units and linkers create a terrific number of possibilities for researchers in their search by varying materials of metal ions and linkers to discover a “magic” combination that will provide the maximum hydrogen uptake.

An outstanding property of MOFs that has triggered their study as hydrogen-storage candidates is their enormous surface areas. Many MOFs are now reported in the literature to have surface areas greater than 1000 m^2g^{-1}, which is higher than that demonstrated for zeolites. In particular, the dinitrogen isotherm measured for MOF-177 at 77 K exhibits the highest uptake of N_2 for any material to date, and gives rise to a monolayer-equivalent surface area of 4500 $m^2\ g^{-1}$ [384].

Single-crystal X-ray diffraction studies reveal the framework has cavities of 10.9–11.8 Å diameter connected in all directions by channels. The initial study of the hydrogen uptake by this material at 77 K revealed that while it did not reach saturation at 1 atm, it adsorbed comparatively less than other related materials displaying smaller surface areas [385]. Supercritical hydrogen, as expected, adsorbs more weakly than dinitrogen at 77 K, and pore-filling is not possible at reasonable pressures. Thus, both pore geometry and surface structure must be optimized to increase the uptake in these materials.

As MOFs are composed of a great assortment of chemical moieties that are characterized to a higher degree (by crystallography), they hold more promise for establishing which structural factors are most effective in adsorbing hydrogen.

The synthesis of MOFs is commonly carried out by solvothermal methods: heating a mixture of organic linker and metal salt in a solvent system that usually contains formamide functionality. The materials are generally prepared in "one-pot" solvothermal syntheses under mild conditions. Powder X-ray diffraction is a simple and effective technique to establish the purity of the crystalline fraction, which often approaches 100% of the sample. Synthetic reproducibility with high purity is essential for the establishment of structure–property relationships [382]. The details of the synthesis are presented elsewhere (see, e.g., Ref. [378] and references therein).

10.3.3.1 Structures

The archetypal MOF is $Zn_4O(BDC)$, where BDC = 1,4-benzenedicarboxylate, which is commonly known as either MOF-5 or IRMOF-1.6 [383] (Fig. 10.2). This material consists of zinc oxide clusters joined by benzene linkers. The metallic clusters are known generically as SBUs and in MOF-5 the structure forms a highly porous cubic network. Thousands of different MOFs have been studied for their hydrogen adsorption properties; there number increasing dramatically every year.

For adsorbents with thick or dense walls, hydrogen uptake values will be low, as it was demonstrated for slit and cylindrical pores in nanostructured carbon materials [386].

So, the requirements to the wall of an adsorbent sound: the walls of an adsorbent should be composed of light elements, be as thin as possible (the best is, one atomic layer thick), and be highly segmented. A simple calculation demonstrating the advantage of segmenting graphene sheets has recently been reported [386]. The incision process that isolates groups of aromatic rings from each other yields an increased surface area as more edges become exposed. The main idea is to separate these isolated units, which can be afforded in MOFs by reticular synthesis.

However, many MOFs reported have spacings between their walls (or links) that are too large for the effective adsorption of hydrogen. For example, the distance between phenylene faces in MOF-5 is much larger than the 2.89 Å kinetic diameter of H_2 [382].

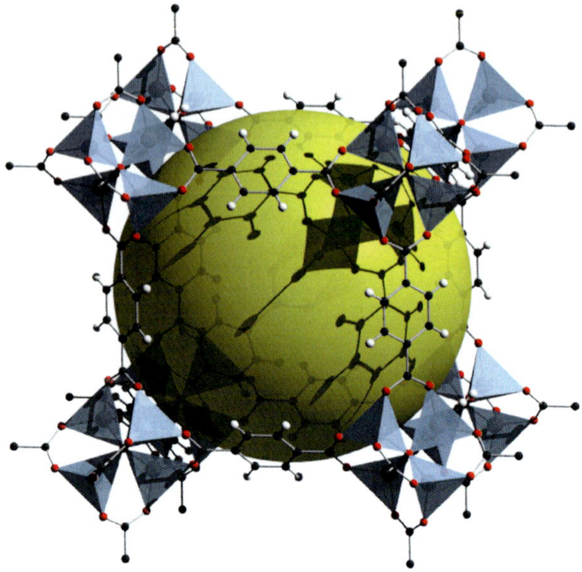

Figure 10.2 (a) The MOF-5 structure as ZnO_4 tetrahedra (blue polyhedra) are connected by BDC linkers (O, red and C, black) to give an extended 3-D cubic framework with interconnected pores of 8 Å aperture width and 12 Å pore (yellow sphere) diameter. Yellow sphere represents the largest sphere that can occupy the pores without coming within the van der Waals size of the framework (Image: Tony Boehle) [387].

An unused void volume at the center of these pores would result in the lower volumetric packing density, even at monolayer coverage, whereas to approach the targeted guidelines for H_2 storage it is ultimately required that the density of the adsorbed hydrogen would be that of its liquid state [382].

To improve this situation in MOFs, smaller pores are necessary. This arrangement can be afforded by the use of shorter links and the investigation of new topologies. A rigid linear dicarboxylate such as acetylenedicarboxylate is required to stabilize a smaller pore analogue of MOF-5 [382]. This is an attractive target for future syntheses, but it is not necessary to only consider simple cubic topologies of this type. By simply distorting the primitive cubic net—for example, by shearing along the face or body diagonals—the pores become elliptical and their shortest dimension is reduced [382].

There are a vast number of smaller pore structures with various topologies that have been published, and their H_2 adsorption is the subject of investigation. While MOFs with constricted pores may exhibit smaller gravimetric capacities, gains may be made in volumetric capacity [388,389]. It should be taken into account, that MOFs with very short linkers that may at first not appear to be porous from dinitrogen adsorption studies, may be composed of apertures that are too small to admit N_2, but are large enough to allow the passage of H_2 [390]. From a design perspective, however, it must always be borne in mind that as the framework density increases, the gravimetric H_2 capacity decreases and therefore a compromise must be found.

10.3.3.2 Impregnation

Another idea for optimization of pore size is to insert another adsorbate surface within large-pore MOFs. This can be achieved by either impregnation with a nonvolatile guest or catenation by another identical framework.

Some large molecules such as C_{60} and Reichardt's dye can be included into MOF-177 from the solution phase [384]. This yields firstly reduction of the free diameter of the pores, and, secondly, may create additional adsorptive sites (Fig. 10.3). If impregnation with reactive species is used the more adsorptive sites are created that could enhance hydrogen uptake at room temperature. However, the doping level might be achieved, where the additional mass added by the guest is not compensated by additional hydrogen uptake. The guest should also not block the existing adsorptive sites on the framework. Another requirement is that the guest must be well-bound or have very low vapor pressure to avoid desorption along with the hydrogen. These considerations imply that lightweight and reactive, but large and spindly, molecules or complexes are required. It is easy to understand that all these requirements are contradictory.

10.3.3.3 Catenation

Framework catenation is another interesting idea in MOF research [382]. The intergrowth of two or more frameworks is possible for many topologies and is often observed when long linkers are used [382]. Catenation can be realized either in the form of interpenetration, where the frameworks are maximally displaced from each other, or interweaving, where they are minimally displaced and exhibit many close contacts [382]. The main purpose of catenation is a reduction of the free diameter of the pores, therefore, that could improve hydrogen uptake. It was reported that in a series of Zn_4O carboxylates the interwoven IRMOF-11 material showed the greatest hydrogen uptake at 77 K [385]. Fourfold catenated IRMOFs have been reported to adsorb approximately 1 wt.% hydrogen at room temperature and 48 bar [391].

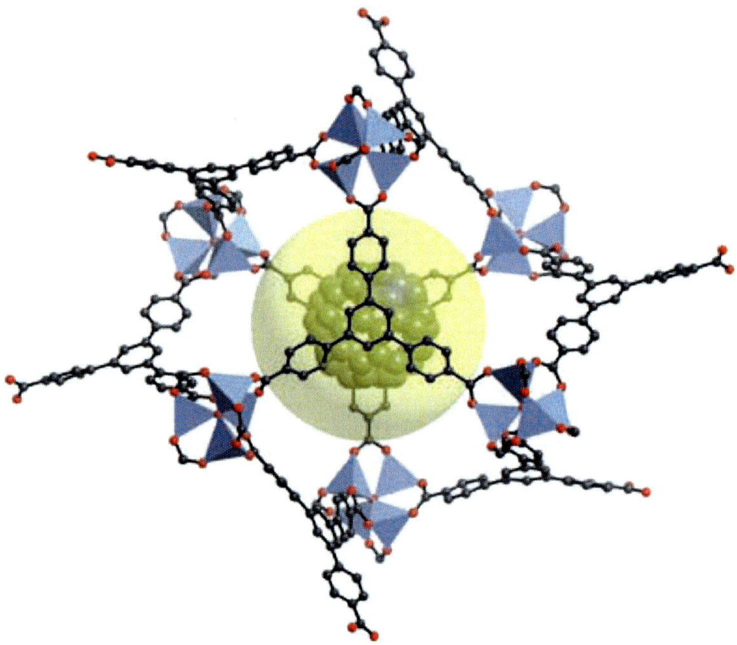

Figure 10.3 The large pore (yellow sphere) in MOF-177 is 11.8 Å in diameter, large enough to include C_{60} molecules that may provide additional surface sites for the physisorption of H_2. Reprinted with permission from Ref. 392. Copyright (2004) American Chemical Society.

Interpenetration rather than interweaving may help to maximize the exposed surfaces of catenated frameworks. Although interweaving can lead to reinforcement of the individual frameworks, by improving the rigidity and favoring stability in the absence of guests, which occurs due to an effective thickening of the walls, the number of framework atoms and their associated adsorptive sites that are exposed to the pore volume is reduced.

Conversely, interpenetration decreases the pore size without blocking any adsorptive sites and in most cases also leads to a more convoluted void region. At the same time there is no an appreciable impact on the diffusion rate of supercritical hydrogen through the solid unless the apertures constrict to the size of the H_2 kinetic diameter.

Although at the first glance it seems that interpenetrated frameworks should be preferential over interwoven ones, this type of catenation is unlikely to occur in the evacuated state. Most catenated frameworks to date are formed via interweaving, since those that are really interpenetrated have free solvent molecules that separate the individual frameworks where void regions do exist [382]. It is unavoidable that during the evacuation process attractive interactions between catenated frameworks yield their interweaving. To minimize the proportion of framework atoms that are involved in close contacts, incorporation of specified chemical features could be carried out. For example, the inclusion of a hydrogen bond donor–acceptor group at the midpoint of a linker could lead to localized interframework contacts only at the linker midpoints, thus leaving essential distance between the SBUs. This arrangement would be optimal in the situation where the primary adsorptive sites are on the SBUs and blocking of them results from full interweaving [382].

10.3.3.4 Open metal sites

The next interesting idea to enhance the hydrogen uptake is to use coordinative unsaturation that occurs in metal clusters that which additional terminal ligands bound to them [382]. Commonly these are solvent molecules coordinated to the metal by a Lewis acid or a Lewis base interaction. In the most favorable cases, if the binding of the ligand is sufficiently weak to allow its release, it may be released with heating as neutral guest species without deterioration of the framework; an open metal site being exposed to the void region. MOFs composed of metal clusters with lesser rigidity or constructed of metal ions that don't fulfill the electronic requirements to sustain an open metal site often collapse after loss of the terminal ligands, but may rearrange to a new form displaying permanent porosity [393].

Another strategy for constructing MOFs with open metal sites is to embed these within the linker. This has been achieved using the equatorially chelating ligand N,N'-phenylenebis (salicylideneimine) dicarboxylate, which can be used to construct a framework isoreticular with MOF-69 [394] after binding a metal ion at its center [395]. The solvent molecules axially coordinated to the linker-bound metal should be removable after framework construction, since the metal is not a key component of the framework. The prospects of exposing open metal sites in a porous material are exciting indeed. To date, these have been divalent cations that are too electron deficient to be expected to form metal hydrides or hydrogen's complexes. Their coordinative unsaturation, however, may lead to stronger physisorption and is concurrent with a decrease in the density of the framework as a result of the loss of the terminal ligands. Evidence for stronger binding of H_2 to open Ni^{2+} sites has been presented recently for an open-framework phosphate [396]. The possibility of incorporating other metals with coordinative unsaturation that will reversibly chemisorb hydrogen is plausible and this is an admirable target for future syntheses.

10.3.3.5 MOFs of light metals

An important strategic requirement in exploring MOFs composed of other metals is the necessity to reduce the framework density to enhance gravimetric hydrogen capacity. Thus, the MOFs formed by light main group metal cations such as Li+, Na+, Mg^{2+}, and Al^{3+} are of great interest. One could expect that while the monovalent cations will most likely have frameworks that are susceptible to hydrolysis, the large polarizing power of Mg^{2+} and Al^{3+} should lead to relatively strong coordination bonds. Oxophilicity and predominantly octahedral coordination are well-known characteristics of the two cations, and novel solvent systems will probably need to be employed to avoid formation of dense oxide/hydroxide phases. The only aluminum MOF to date was prepared hydrothermally by using fluoride ions as a mineralizer. The aluminum hydroxide terephthalate MIL-53 (Fig. 10.4) displays an uptake of 3.8 wt.% at 77 K and 16 bar, but an unusual amount of hysteresis upon desorption [382]. Magnesium MOFs are also very rare, in fact few systematic studies have been reported [382].

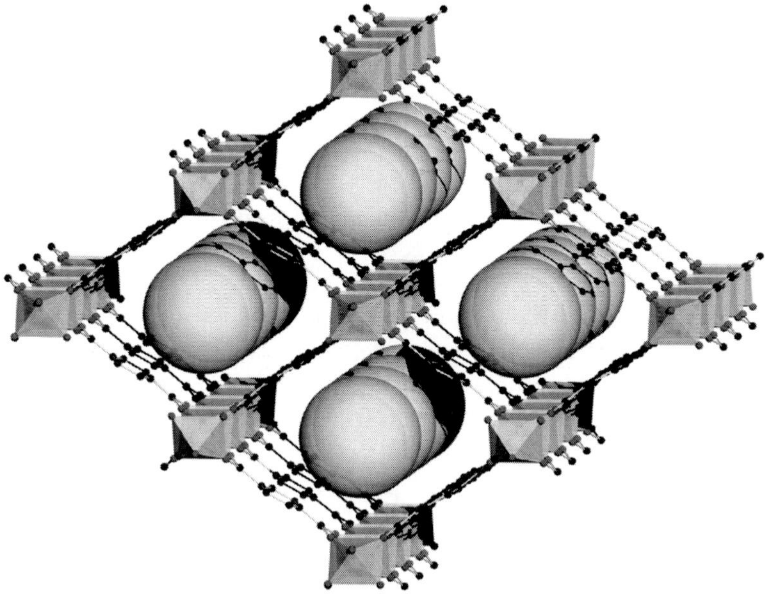

Figure 10.4 MOF MIL-53, M(OH)(BDC) (M = Al^{3+} or Cr^{3+}) (Image: Tony Boehle) [397].

10.3.3.6 Functionalized linkers, SBU adsorptive sites, and the energetics of physisorption

The stimulus for creation MOFs with constricted pores is driven by the need to enhance the heat of adsorption of hydrogen in the adsorbents. In an ideal case, the adsorbent for this purpose should have both large gravimetric and volumetric capacities at ambient temperature, and along with this also quick uptake and release of dihydrogen [382].

Since the van der Waals forces are weak the hydrogen capacity of even materials with a very high surface area occurs small at room temperature. Energy of interaction of ~20 $kJmol^{-1}$ has been amounted for porous materials to maintain high capacity at ambient temperatures.

Many workers undertook serious attempts, both theoretical and experimental, to estimate the interaction energy of hydrogen with nanostructured carbon materials. This value is agreed to be ~ 5 $kJmol^{-1}$, being estimated both from high-level calculations of the gas-phase interaction of H_2 with substituted and expanded aromatic compounds [398] as well as variable-temperature adsorption studies on various carbon materials [399,400]. Most experimental evidence regarding desorption of hydrogen from activated carbon materials at elevated temperatures have referred this to chemisorbed hydrogen at dangling bonds [401].

Thus, the influence of chemical substitution or nanostructuring of carbon materials on the physisorptive interaction energy is rather small. Organic linkers with aromatic backbones such as phenylene, naphthalene, bipyridine, and biphenylene have been employed in MOF structures to enhance the rigidity of the frameworks. These molecular units mimic the nanostructure of activated carbon materials and nanotubes which are primarily composed of sp^2-hybridized carbon atoms, and hence it is believed that adsorption energetics should be similar for both classes of materials. This situation would be especially true for MOFs where only a small proportion of the framework is composed of metal, such as those with small nuclearity SBUs. Ab initio calculations indicate that the energy of interaction of hydrogen with substituted benzene is marginally enhanced by the addition of electron-donating groups [400]. The effects of these modifications remain untested in MOFs. Chemical modification with the retention of topology has been demonstrated for a large series of Zn_4O-based MOFs [402], and initial studies reveal that the linker length and width appear to have an influence on the hydrogen uptake at 77 K [385] and room temperature [403]. As the linker size have a profound impact on the framework pore size, it is not clear at present whether nets in uptake are mainly achieved by the surface energetics of the linker or the resulting geometric changes in the pore.

Further alteration of the linker energetics, such as inclusion of heteroatoms in the aromatic backbone also requires investigation. And MOFs have the additional advantage of a heterogeneous surface structure with local dipoles that may enhance the attraction of hydrogen [384]. These are mainly associated with the SBUs where ionic and dative bonding occur. MOFs composed of $Zn_4O(CO_2)_6$ SBUs such as MOF-5 and MOF-177 can be considered as an assembly of six-membered rings. Evidence for stronger binding of H_2 in MOFs was obtained by inelastic neutron scattering (INS) experiments [403].

Thus, the composition of such metal-containing ring systems which are incorporated in the SBUs influences hydrogen binding. To understand the details of the hydrogen interaction at the SBUs is particularly important for larger clusters as they begin to resemble an array of metal oxide nanoparticles in a porous organic matrix. Furthermore, the use of alternative metals generates new SBUs and possibly new framework topologies, the search for new MOFs with enhanced uptake is limitless [382].

10.3.3.7 Characterization of hydrogen uptake in MOFs

The uptake in MOFs is typically characterized by adsorption isotherms measured at 77 K up to 1 atm (the operating range of a typical commercial adsorption apparatus), but data collected at room temperature and high pressure have also been reported. The high-pressure conditions are more relevant for practical applications, especially as the pressure limits in commercial hydrogen tanks increase. As the low-temperature hydrogen isotherms can be measured using essentially the same apparatus as that used to determine the MOF surface area, these data are also very useful at this early stage of exploration.

As always, care must be taken to ensure the accuracy of the results, especially when gravimetric methods are used. As H_2 gas has the lowest molecular weight and physisorbs weakly, ultrahigh-purity gas must be used and the adsorption apparatus adequately tested for leaks. Ideally, the results should be confirmed by a complementary method, such as volumetric adsorption [385].

More laborious measurements are required to quantify the heat of adsorption in the most promising compounds. Spectroscopic investigation of the hydrogen–framework interaction has been performed using INS studies. This technique provides a sensitive probe of the adsorptive sites for molecular hydrogen as it makes it possible to observe the hindered rotational transitions of bound molecules. Neutrons are the ideal scattering probes for such experiments as X-rays are insensitive to the low electron density of hydrogen atoms.

Very interesting and important observations were made for MOF-5 loaded with varying amounts of hydrogen [403]. These spectra revealed at least two distinct adsorptive sites to exist on the framework. They were assigned to the inorganic and organic components. MOFs have again the advantage over nanostructured carbon materials and amorphous adsorbents in that they have a high degree of long-range order. If such adsorptive sites are indeed localized, they should be periodically arranged. It therefore may be possible to detect the adsorbed hydrogen molecules at very low temperatures by a diffraction experiment. A report has recently appeared describing the modeling of electron density, measured by synchrotron powder X-ray diffraction of adsorbed H_2 [404].

Computational modeling studies of the hydrogen interaction with MOFs and their chemical components is also important for our understanding of physisorption in these materials. The simplicity and high symmetry of some of the structures should make them ideal object for study, to the level that has been achieved for zeolites [405,406]. Grand Canonical Monte Carlo simulations of argon and methane in MOFs have been reported, and these display good agreement with measured isotherms [407–409]. The physisorption of hydrogen in MOF-5 has also been examined, in addition to more detailed calculations for H_2 with models of its inorganic and organic moieties [398,410–412]. The Monte Carlo study indicates that binding at the $Zn_4O(CO_2)_6$ clusters is ~ 1.5 kJmol^{-1} higher in energy than at the phenylene groups, but a broad range of interaction energies was determined for their system at 300 K corresponding to occupation of a broad distribution of sites [412]. Further computations of the hydrogen dynamics in MOFs will be indispensable in providing future directions for materials optimization.

An important issue for the hydrogen storage in MOFs is the H_2 sorption properties in order to control the characteristics of H_2 binding. Again here INS spectroscopy of the rotational transitions of the adsorbed hydrogen molecules is usually informative. INS is an experimental technique commonly used in condensed matter research to study atomic and molecular motion as well as magnetic and crystal field excitations [413]. It distinguishes itself from other neutron scattering techniques by resolving the change in kinetic energy that occurs when the collision between neutrons and the sample is an inelastic one. Results are generally communicated as the dynamic structure factor (also called inelastic scattering law) $S(q,\omega)$, sometimes also as the dynamic susceptibility $\chi(q,\omega)$ where the scattering vector q is the difference between incoming and outgoing wave vector, and hω is the energy change experienced by the sample (negative that of the scattered neutron). When results are plotted as function of ω, they can often be interpreted in the same way as spectra obtained by conventional spectroscopic techniques; insofar as INS can be seen as a special spectroscopy.

Neutrons are scattered inelastically far more strongly by hydrogen than by any other element, which facilitates the observation of rotational tunnel splitting of the librational ground state of the H_2 molecule [414]. This splitting is akin to the ortho–para transition for free H_2, which was covered in Chapter 1, and this mode is forbidden in optical spectroscopy [415]. This splitting is an extremely sensitive measure of the rotational potential energy surface, a feature that has made it possible to determine fine details of hydrogen adsorption by INS.

Hydrogen gas uptake by MOF-5 at 78 K was measured by introducing small amounts of H_2 into a chamber containing the guest-free form of the material and monitoring the weight change as a function of increasing doses of H_2 (Fig. 10.5) [375]. The measured sorption isotherm shows a type I behavior, for which is typical that the saturation is reached at low pressures followed by a pseudoplateau at higher pressure of H_2 with a maximum uptake of 45 mg of H_2 per gram of MOF-5 [4.5 wt.% = 17.2 H_2 per $Zn_4O(BDC)_3$ formula unit]. The observed sharp uptake of H_2 at lower pressure indicated favorable sorption interactions between the MOF-5 framework and H_2 molecules. Adsorbed H_2 molecules could also be reversibly desorbed from the pores by reducing pressure.

The H_2 sorption was evaluated in conditions that mimic a typical application environment, namely, room temperature and pressures considered safe for mobile fueling. A different sorption apparatus was constructed, in which a 10 g sample of guest-free MOF-5 was charged with H_2 up to 20 bar and the weight change monitored as a function of H_2 uptake and release [403]. MOF-5 showed substantial H_2 uptake that increased linearly with pressure, giving 1.0 wt.% at 20 bar. These findings demonstrate the potential of MOFs for H_2 storage applications. The isotherm at ambient temperature is expected to be approximately linear as observed because the material is noticeably undersaturated with H_2 in the pressure range explored and, in principle, at higher pressures can take up more H_2, up to at least the amount observed at 78 K.

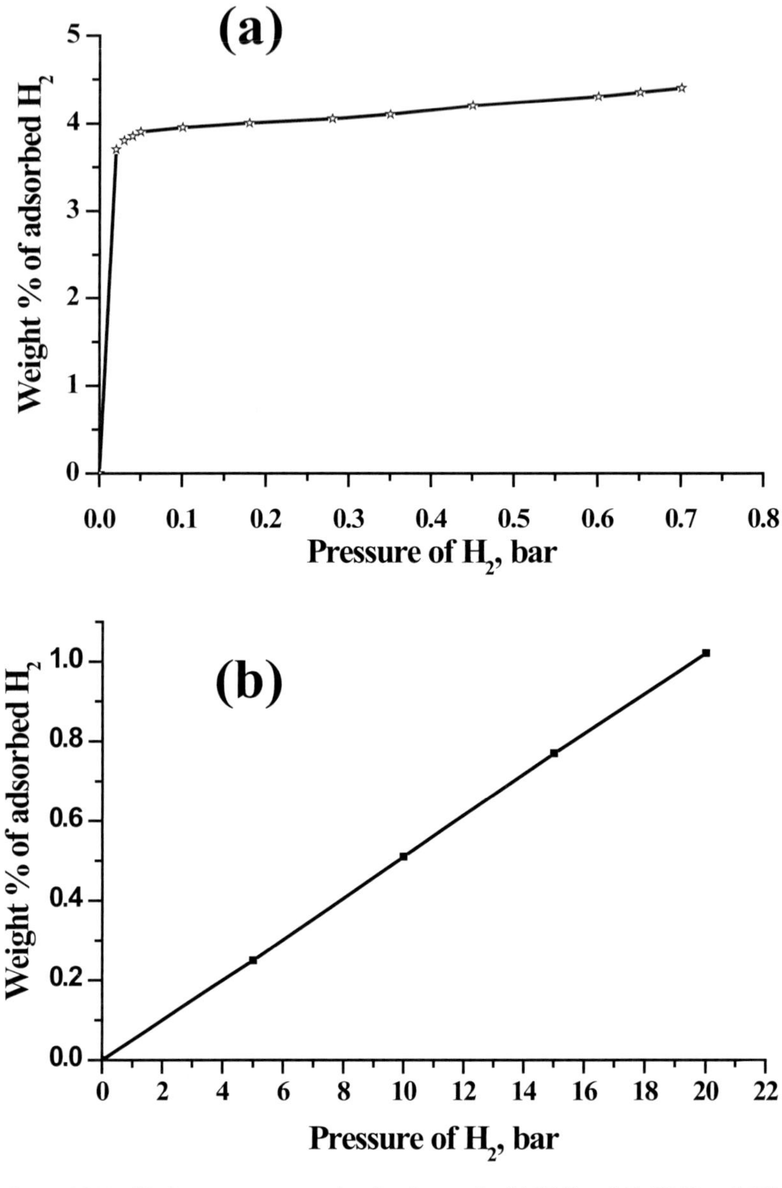

Figure 10.5 Hydrogen gas sorption isotherm for MOF-5 at (A) 78 K and (B) 298 K.

As noted above, 78 K sorption data suggested that 17.2 H_2 per formula unit is close to saturation. The barrier to rotation for the binding site near Zn is somewhat greater than those on the BDC, but also slightly lower than that for the extra-framework Zn^{2+} transition was observed at ~8 meV [403]. Various factors could contribute to this difference, including different degrees of accessibility of the Zn and the strong electrostatic field in the zeolite supercage. The barriers found for MOF-5 are noticeably higher (0.40 and 0.24 kcal mol^{-1}) than those found in carbons, including single-walled nanotubes (SWNTs) [403], in which it is ~0.025 kcal mol^{-1}. Moreover, the rotational band in the case of carbons has a width of nearly 2.5 meV compared with 0.5 meV in MOF-5 case. This value corroborates the much lower mobility (and hence stronger interaction) for H_2 in MOF-5 than in carbons. The INS results for hydrogen in MOF-5 point to the importance of the organic linkers in determining H_2 uptake levels.

Accordingly, using the same experimental setup and details outlined above for the room-temperature measurements on MOF-5, H_2 sorption was examined in IRMOF-6 and IRMOF-8 where it was found that the specific H_2 uptake is approximately doubled and quadrupled, respectively, for IRMOF-6 and -8 relative to that found for MOF-5 at the same (room) temperature and pressure (10 bar). Specifically, for IRMOF-8, the H_2 uptake under those conditions is 20 mg of H_2 per gram (2.0 wt.%)—a capacity well above those found for "active" carbon (0.1 wt.%) and "graphitic" carbon (0.3 wt.%) [403]. The percent uptake found for MOF-5, IRMOF-6, and IRMOF-8 at room temperature and 10 bar is equivalent to ~1.9, 4.2, and 9.1 H_2 per formula unit, respectively. These capacities are well below the saturation level as determined at 78 K for MOF-5 (17.2 H_2 per formula unit), which points to the potential for higher uptake at ambient temperature and higher pressures. The capacity of these structures for hydrogen at room temperature is comparable to the highest capacity achieved for CNTs at cryogenic temperatures, although the capacity of those materials is very sensitive to preparation conditions and appears to saturate at lower pressures [403]. It could be anticipated that further increases in performance can be expected, with new IRMOFs having similar but larger organic linkers.

To either increase the operating temperature of a microporous hydrogen store, or increase the maximum storage capacity at near ambient temperature, an increase in the strength of the hydrogen-surface interaction is required.

For this reason, one should think to use the structures with the complexes of transition metals. Molecular hydrogen is known to form so-called Kubas complexes with nearly all transition metals that are dihydrogen complexes coordination complexes containing intact H_2 as a ligand (Fig. 10.6) [416]. The prototypical complex is $W(CO)_3(PCy_3)_2(H_2)$. This class of compounds represents intermediates in metal-catalyzed reactions involving hydrogen. Hundreds of dihydrogen complexes have been reported. Most examples are cationic transition metals complexes with octahedral geometry [416].

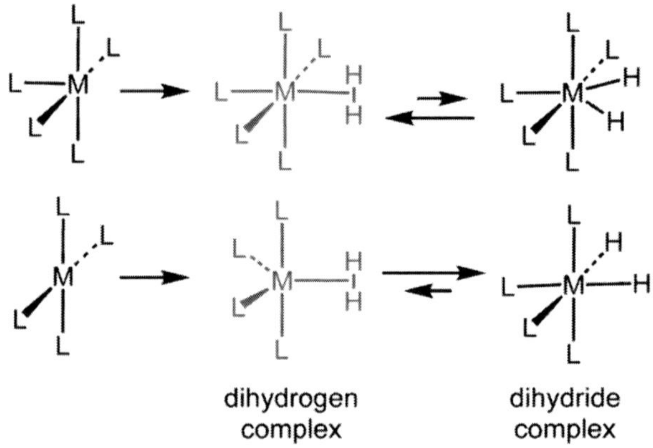

Figure 10.6 An example of formation and equilibrium structures (Kubas complexes) of metal dihydrogen and dihydride complexes (L = ligand) (Image: Smokefoot) [416].

The formation of Kubas structures implies the possibility of enhancing the hydrogen-surface interaction via a similar mechanism in MOFs and offers the promise of developing the latter with much improved storage properties.

10.3.3.8 Conclusions concerning the metal–organic frameworks

The series of multifaceted considerations elucidated above showed that significant research will be required to satisfy many of them simultaneously. A material with high gravimetric and volumetric hydrogen capacity at practical conditions should have a high surface area with pores of appropriate dimension for hydrogen and a large heat of adsorption. The strategies of impregnation, catenation, and the inclusion of open metal sites are just a few possibilities to be tested. To date, nearly 5000 2D and 3D MOF structures have been reported in the literature, but only a fraction of these have been examined for their porosity and far fewer of them have been tested for their hydrogen-storage capacity. This fact, combined with the advantages of MOFs such as ease of synthesis and functionalization of their organic linkers and inorganic SBUs are favorable circumstances for achieving the DOE target for on-board hydrogen storage [382].

Varying several factors such as surface area, pore size, catenation, ligand structure, spillover, and sample purity can result different amount of hydrogen uptake in MOFs.

In 2006, chemists at the University of California, Los Angeles, and the University of Michigan have achieved hydrogen-storage concentrations of up to 7.5 wt.% in MOF-74 at a low temperature of 77 K [417,418]. In 2009, researchers at University of Nottingham reached 10 wt.% at 77 bar (1,117 psi) and 77 K with MOF NOTT-112 [419]. Most articles about hydrogen storage in MOFs report hydrogen uptake capacity at a temperature of 77 K and a pressure of 1 bar because such condition is commonly available and the binding energy between hydrogen and MOF is large compare to the thermal vibration energy which will allow high hydrogen uptake capacity.

10.3.4 Zeolites

Zeolites are microporous aluminosilicates formed from AlO_4 and SiO_4 tetrahedra. They have a range of practical applications that exploit their ion exchange, molecular sieving and catalytic properties. The term zeolite is often used to describe compounds with similar structures that are formed by elements other than Al and Si, including P, Ga, Ge, B, and Be, but these materials are also known as zeotypes. These materials form a variety of different crystallographic structures.

Their ordered crystalline nature gives them uniform cavities and channels size in the microporous range (see Fig. 10.7), which is an argument to nominate them as candidate-materials for hydrogen storage. The framework structures are relatively rigid and they have high specific surface areas and large pore volumes.

Thus, the strategies in the research and development of these structures are similar to those in MOFs. One should create highly porous material with size of approximately dihydrogen diameter and additionally provide sufficient energy of interaction between hydrogen molecule and the material surface.

These strategies can be achieved by altering the Si/Al ratio of the framework, or the equivalent stoichiometry in the case of zeotypes. The anionic nature of zeolite frameworks allows the presence of cations within their structure. Exchange of these cations can also significantly alter their properties.

However, despite zeolites are highly porous, with high surface areas, the hydrogen-storage capacities reported in the literature to date being fairly low. In their study of low pressure (<0.1 MPa) hydrogen uptake by a range of porous solids, it was concluded that zeolites are less likely to be effective as HSMs, in comparison to MOFs or porous carbons, due to their limited pore volume [374]. The values of hydrogen uptakes are ranging within the limits 1.81–2.55 wt.% at 77 K and various pressures that is far less than that in MOFs.

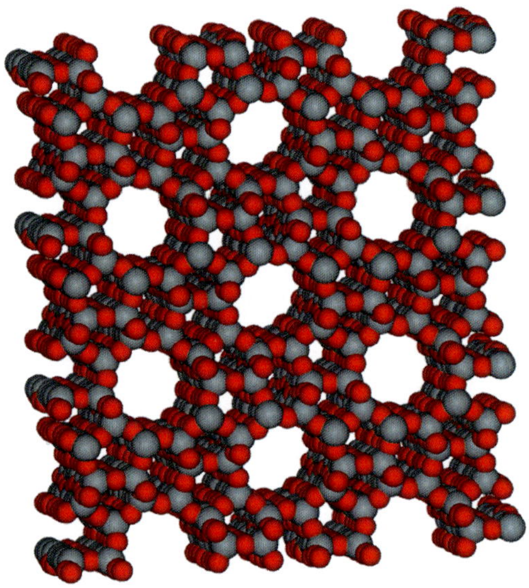

Figure 10.7 The microporous molecular structure of a zeolite (Image: Benjamin Mills) [420].

The results of the molecular mechanics simulations predict that the maximum storage capacities for a range of zeolites are in the region of 2.65 and 2.86 wt.%, whereas in the other research it was found that, theoretically, the hydrogen content could reach (4.8 ± 0.5) wt.% for sodalite (SOD) structures [421,422]. In the study [421] a lower figure of 1.92 wt.% was determined for SOD. In later work [422] lower values for the saturation uptake were found using Grand Canonical Monte Carlo (GCMC) simulations, although the higher figures are supported in the work [423], where it was found an uptake of 4.45 wt.% for Mg–X. However, even these higher capacities are low in comparison to the challenging US DOE storage system targets.

Hydrogen storage at higher temperatures has been proposed through the mechanism of encapsulation [424–426], a process by which, in the case of zeolites A, X, and Y, hydrogen molecules enter the sodalite cages. However, this does not appear to be a mechanism capable of delivering practical storage capacities.

Enthalpies of adsorption, a measure of the strength of the hydrogen-adsorbent interaction for zeolites can have values in the range 3.5–18 kJ mol^{-1}, with the largest of these values being found for (Mg, Na)-Y zeolite [427,428].

However, zeolites have a number of significant practical advantages over other microporous adsorbents. They possess high thermal stability in comparison to MOFs and organic polymers, for example. Thorough degassing of the framework can therefore be performed at 350°C (623 K) with no decomposition [374].

Their crystalline nature allows easy characterization of the host material and they have a well-defined and understood pore size in comparison with activated carbon. Current industrial production of synthetic zeolites also demonstrates the feasibility of their synthesis in large quantities, although this is also the case for carbons. Unfortunately, it seems unlikely that zeolites will be useful as practical hydrogen-storage media for onboard applications because they do not show the required capacities for practical storage applications [429]. Zeolites are, however, certainly valuable as model systems for the further investigation of the interaction of hydrogen with microporous materials [430], and so their continued study is both likely and worthwhile. Hydrogen storage in zeolites was covered in detail in [431] and this review is recommended to interested readers.

10.3.5 Carbons

Several types of microporous carbons can be nominated as potential materials for hydrogen storage, including activated carbons, CNTs and nanofibers and, more recently, microporous templated carbons [432]. Carbon seems to be very attractive as a host material for this purpose because of its low molar mass. It has also chemically stability and can be synthesized in a number of different forms. From a practical point of view, porous carbons are already commercially produced in large quantities for a broad range of applications and are relatively inexpensive [372].

10.3.5.1 Activated carbon

Activated carbon is a porous form of carbon that can be synthesized via both chemical and physical activation methods. Depending on the raw material and the activation method and conditions, the resultant carbon will have a specific, although not necessarily clearly defined, pore structure. Activated carbons can be predominantly macro-, meso-, or microporous, but the latter are the main focus for hydrogen storage [372]. Activated carbons tend to have slit-shaped pores [433,434], but also exhibit a relatively wide pore size distribution in contrast to the crystalline adsorbents, such as the zeolites or MOFs, which, as it was shown above, have a well defined pore size and pore geometry. There have been many studies of hydrogen adsorption by activated carbons and, according to [432], these show that gravimetric storage capacities can reach 5.5 wt.% at 77 K. However, the pore structure of carbons can take many different forms, depending on the chosen activation method and the raw material used for the synthesis, and so a range of capacities have been reported in the literature. Carbons are also challenging to characterize accurately, as demonstrated by the variability in the results of a recent interlaboratory study [435].

In addition to the experimental characterization difficulties, modeling work on these materials is significantly limited by the lack of knowledge regarding their microstructure, which is fundamentally different to crystalline microporous adsorbents. The latter can be characterized crystallographically using powder diffraction but activated carbons are X-ray and neutron amorphous. The structure itself therefore has to be modeled, or an idealized slit-pore model used. However, estimated maximum theoretical capacities appear to agree reasonably well with those determined experimentally. The use of activated carbon monoliths for hydrogen storage is sometimes advocated [436] because microporous activated carbon in this form has a greater bulk density than the same carbon in powder form.

Equivalent monoliths formed from MOFs have not yet been reported in the literature, at the time of writing, and it is possible that activated carbon monoliths have significant advantages in terms of volumetric storage capacity. An experimental comparison would be worthwhile, however, because it may show that the higher gravimetric capacities reported for the best performing MOFs are counterbalanced by lower practical volumetric capacities in real storage devices due to lower bulk material density.

10.3.5.2 Carbon nanostructures

Hydrogen carriers based on nanostructured carbon (such as carbon buckyballs and nanotubes) have been proposed. Despite initial claims of greater than 50 wt.% hydrogen storage, to date it is generally accepted that less than 1 wt.% is practical [437].

After Kroto and Smalley discovered Fullerene, one of carbon allotropes (a cluster of 60 carbon atoms: C60). A fullerene is any molecule composed entirely of carbon, in the form of a hollow sphere, ellipsoid, or tube [438,439]. Figure 10.8 shows the structure of so called buckminsterfullerene, that is a spherical fullerene molecule with the formula C60 [435]. Fullerenes, due to their elegant design, excellent mechanical properties, their hollow cage and high energy barrier for breaking a C–C bond have been considered as possible nanocage carriers for small molecules. Fullerene hydrogen storage is an attractive alternative for use as hydrogen storage.

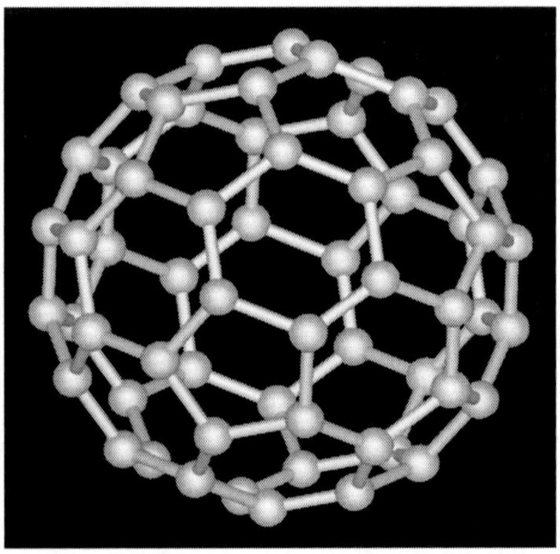

Figure 10.8 Buckminsterfullerene C_{60} (Image: Saperaud) [440].

In 1991, Iijima [441] discovered thin and long straw-shaped CNTs (Fig. 10.9). The nanotubes range in length from a few tens of nanometers to several micrometers, and outer diameter from about 2.5 nm to 30 nm. A carbon atom in nanotubes forms a hexagonal honeycomb lattice of sp^2 bond with three other carbon atoms. As the inner diameters of the tubes are extremely thin down to about several nanometers, the tubes are called nanotubes.

Figure 10.9 Carbon nanotubes (Image: Arnero) [442].

CNTs are formed by rolled graphite sheet, with a least inner diameter of 10.15 nm up to several nm and a length of 10–100 μm [443]. Tubes formed by only one single graphite layer are called SWNTs (Fig. 10.9). For tubes contains multiple concentric graphite layers are called multiwalled nanotubes (MWNTs, Fig. 10.10).

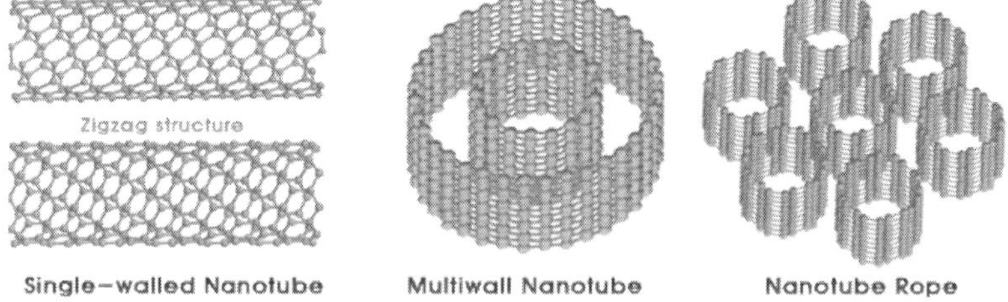

Figure 10.10 Structure of carbon nanotubes.

10.3.5.3 Synthesis

Techniques have been developed to produce nanotubes in sizeable quantities, including arc discharge, laser ablation, high-pressure carbon monoxide (HiPco), and chemical vapor deposition (CVD). Most of these processes take place in vacuum or with process gases. CVD growth of CNTs can occur in vacuum or at atmospheric pressure. Large quantities of nanotubes can be synthesized by these methods; advances in catalysis and continuous growth processes are making CNTs more commercially viable [372].

Their ultrahigh surface-to-volume ratio makes CNTs highly suited for investigating fundamental aspects of gas adsorption. For instance, the interior region of nanotubes offers a deep well for physical adsorption, which provides an ideal example of a microporous material. For this reason, the hydrogen sorption properties of carbon nanostructures, such as nanotubes and nanofibers, have been investigated extensively in recent years.

Although the issue has not yet been resolved completely, it seems fairly safe to conclude that both nanotubes and nanofibers are unlikely to be the hydrogen-storage panacea that some authors have suggested. In addition to nanotubes and nanofibers, fullerenes and carbon nanohorns have also been investigated, either experimentally or theoretically, for their hydrogen-storage properties [372].

10.3.5.4 Three-dimensional carbon nanotubes for hydrogen storage

Researchers in Germany have designed a 3-D carbon-based nanotube matrix that can store and release hydrogen extremely efficiently. According to the authors the material could form the basis of a sustainable hydrogen economy [444]. A computer-based approach to design a 3-D carbon nanotube structure was used that can store more hydrogen at room temperature than any other carbon-based material [444]. The resulting material would be lightweight, cheap to make and non-toxic, which means that it would be suitable for large scale production.

The US Department of Energy's target for HSMs by 2015 is 6 wt.%, that is, 6% of the combined weight of the material after gas adsorption should be stored hydrogen. This new nanotube material has a total hydrogen uptake of 5.5 wt.% at room temperature which suggests it could be a promising candidate.

Inspired by natural sponges, a computer model was designed that placed CNTs in the hole positions of a theoretical sponge network (Fig. 10.11). Usually cylinders are put next to each other that is a way of packing them. But there is another way to put them so they cross over each other, like the holes in a sponge. Using modeling the arrangements was found the nanotubes could be arranged in to fit this criteria.

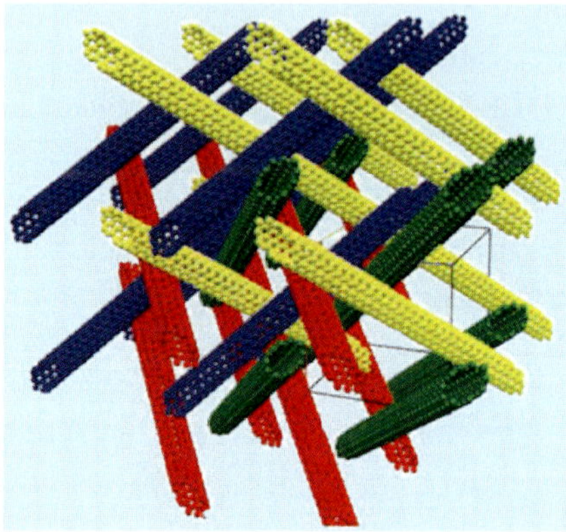

Figure 10.11 Hydrogen initially attaches at the points where the nanotubes touch. This does not occur when the nanotubes are packed in parallel with the entire edge of each tube touching others. Reprinted from Ref. 443 with permission from Cambridge University Press, copyright 2002.

The CNTs were arranged in parallel, but with none of them intersecting. Each nanotube is in contact with others, but not all the way along the edge of the nanotube. After hydrogen adsorption simulations, the team found that these structures adsorbed a significant amount of hydrogen. The way the hydrogen nanotubes were put together provides some points of contact and it was proposed that at these points hydrogen preferentially adsorbed. Putting cylinders next to each other in a parallel fashion doesn't have the same effect [444].

The next step is to make the structures experimentally.

10.3.5.5 Graphene–organic frameworks

A thin flake plain carbon, just one atom thick, became world famous last year. The discovery of the super material graphene gave Andre Geim and Konstantin Novoselov the Nobel Prize in Physics 2010

[445]. Graphene has a wide range of unusual and highly interesting properties. As a conductor of electricity it performs as well as copper. As a conductor of heat it outperforms all other known materials.

Graphene—carbon formed into sheets a single atom thick—now appears to be a promising base material for capturing hydrogen, according to recent research [446,447]. The findings suggest stacks of graphene layers could potentially store hydrogen safely for use in fuel cells and other applications. Graphene has become something of a celebrity material in recent years due to its conductive, thermal and optical properties, which could make it useful in a range of sensors and semiconductor devices. The material does not store hydrogen well in its original form, according to a team of scientists studying it at the National Institute of Standards and Technology (NIST) Center for Neutron Research, USA. But if oxidized graphene sheets are stacked a top one another like the decks of a multilevel parking lot, connected by molecules that both link the layers to one another and maintain space between them, the resulting graphene-oxide framework (GOF) can accumulate hydrogen in greater quantities [Fig. 10.12].

Inspired to create GOFs by the MOFs that are also under scrutiny for hydrogen storage, the team is just beginning to uncover the new structures' properties. It was found that GOFs can hold at least a hundred times more hydrogen molecules than ordinary graphene oxide does. The easy synthesis, low cost and non-toxicity of graphene make this material a promising candidate for gas storage applications.

The GOFs can retain 1% of their weight in hydrogen at a temperature of 77 K and ordinary atmospheric pressure—roughly comparable to the 1.2% that some well-studied MOFs can hold [447].

Another of the team's potentially useful discoveries is the unusual relationship that GOFs exhibit between temperature and hydrogen absorption. In most storage materials, the lower the temperature, the more hydrogen uptake normally occurs. However, it was discovered that GOFs behave quite differently. Although a GOF can absorb hydrogen, it does not take in significant amounts at below 50 Kelvin (–223°C). Moreover, it does not release any hydrogen below this "blocking temperature"—suggesting that, with further research, GOFs might be used both to store hydrogen and to release it when it is needed, a fundamental requirement in fuel cell applications [447].

Some of the GOFs' capabilities are due to the linking molecules themselves. The molecules the team used are all benzene-boronic acids that interact strongly with hydrogen in their own right. But by keeping several angstroms of space between the graphene layers—akin to the way pillars hold up a ceiling—they also increase the available surface area of each layer, giving it more spots for the hydrogen to latch on [447].

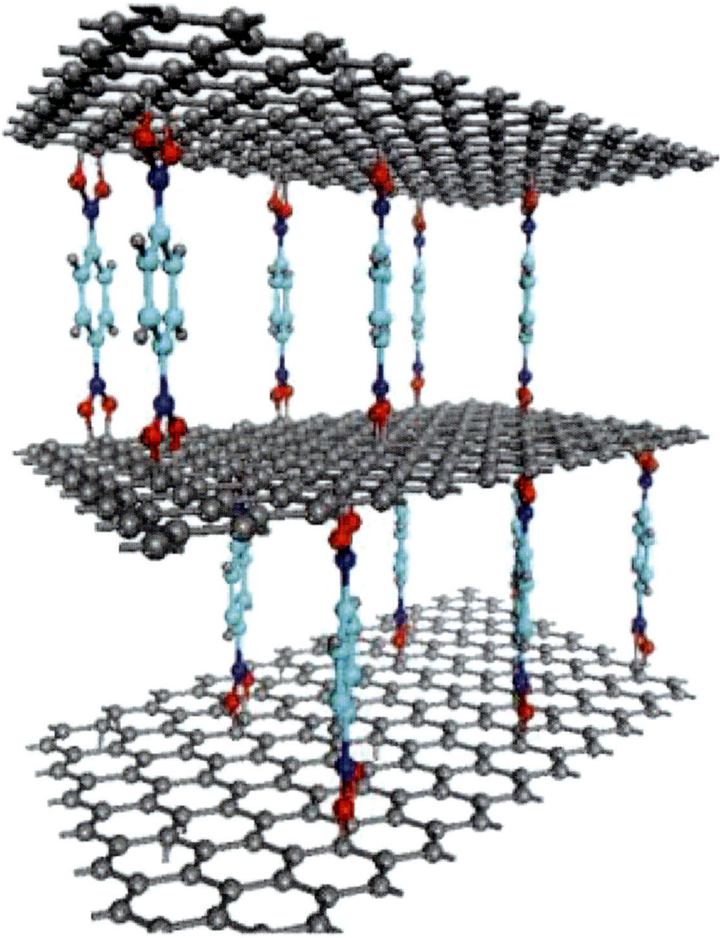

Figure 10.12 A GOF is formed of layers of graphene connected by boron-carboxylic "pillars." GOFs such as this one are just beginning to be explored as a potential storage medium for hydrogen (Credit: NIST) [446].

According to the team, GOFs will likely perform even better once their parameters will be explored the in more detail. The strategy will concern attempts to optimize the performance of the GOFs and explore other linking molecules as well [447].

10.3.5.6 Templated carbons

Templated carbons are a form of microporous or mesoporous carbon typically synthesized by introducing a carbon precursor, such as sucrose or acetonitrile, into the pores of an inorganic template. Carbonization and the subsequent removal of the template results in a pore structure that is relatively well defined compared to their activated carbon counterparts [448,449]. Such materials can exhibit impressive hydrogen adsorption properties, with the largest capacity reported to date being 6.9 wt.% for a zeolite-templated carbon at 77 K and 20 bar [422]. The uptake was Type I, with an estimated saturation capacity of 8.33 wt.%. Unlike activated carbons, these materials show evidence of microstructural ordering through the appearance of Bragg peaks in their X-ray powder diffraction patterns.

Another type of templating uses carbide precursors and can produce microporous carbons with very well defined pores sizes [450,451]. According to Ref. [451] the hydrogen uptake of these so called carbide-derived carbons (CDCs) can reach 4.7 wt.% at 60 bar and 77 K. Although this is not particularly high, the CDCs offer the opportunity to further tune and optimize the pore size and continued investigation of the interaction of hydrogen with these materials would undoubtedly be valuable.

10.4 Interstitial Hydrides

10.4.1 Fundamentals of the Formation of Hydrides

Usually many metals exhibit predisposition towards reactions with dihydrogen forming metallic hydrides. For various metals these reactions may occur at various temperatures and pressures.

Figure 10.13 shows schematically consecutive steps of possible hydrogen–metal interaction arising when the adsorption of molecular H_2 occurs on the metal surface. As seen in the figure, the final stage of the interaction is formation of hydrides. Binary, or higher, hydrides are formed from metallic elements or compounds that react with gaseous hydrogen. Molecular hydrogen dissociates into atomic hydrogen on the surface of the host material and enters the bulk via diffusion between interstitial sites in the host lattice.

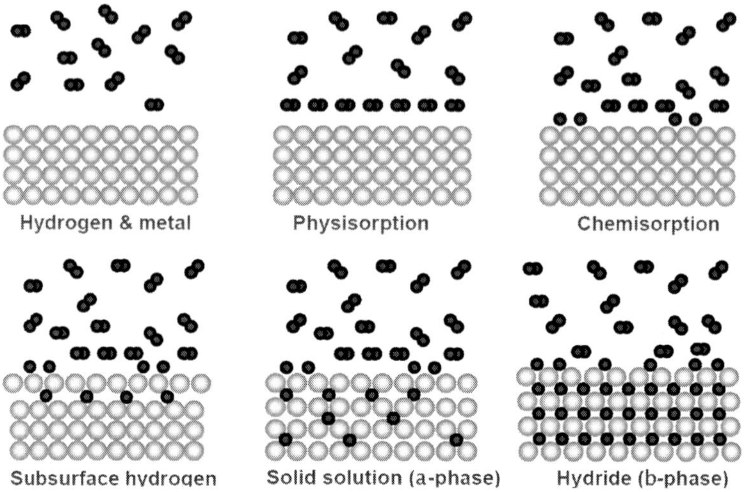

Figure 10.13 Hydrogen absorption in metals. The six distinct states of hydrogen absorption in metals. On the upper left-hand side the metal and hydrogen gas are in two separate phases. Hydrogen molecules can be found in the physisorbed state at low temperature. Hydrogen molecules interact with the electrons at the metal surface, dissociate and bind to the metal atoms at the surface and form the chemisorbed state. Hydrogen can the jump in a subsurface layer and diffuse into the metal host lattice as a solid solution (a-phase). Finally, the hydrogen–hydrogen interaction (volume expansion) becomes important and the hydride phase (b-phase) is formed. Reprinted with permission from Ref. [452]. Copyright (2008) American Chemical Society.

For a researcher it is important to be aware of what is known about the state of hydrogen in metals and of the dynamics of its entry into and release from a metal.

There are some fundamentals that help to describe the formation of hydrides as the result of the hydrogen adsorption on the metal surface and subsequent absorption of the detached atomic hydrogen within the metal bulk. We consider these fundamental clues that facilitate general understanding of the hydrogen processes in metals.

Dihydrogen dissociates first on the metal surface and then dissolved in metals; at low concentrations, exhibits a linear relation between the concentration and the half power of the pressure, p, of the molecular gas [452]. This is evidence that hydrogen dissociates when it enters the metal. Much experimental information and theoretical considerations demonstrate that in transition metals the electron accompanying the proton of the dissociated hydrogen enters the s and d-bands of the host metal changing the density of states at the Fermi surface and causing shifts of the energy bands. The Fermi electrons concentrate about the proton, closely screening the positive charge and producing what is effectively a neutral atom [453]. These electronic reactions produce both short-range and long-range interactions between the dissolved hydrogen species.

Hydrogen occupies interstitial sites in the metal lattice, as shown by neutron diffraction in many cases. In so doing the hydrogen "atom" forces the nearest-neighbor metal atoms farther apart from each other, causing an increase of volume of the nearest-neighbor shell of atoms [454]. The free surface of the metal specimen sweeps out to produce a larger volume increase than the volume increase of the inner shell of nearest-neighbor atoms. This is the physical reason for the increase of the lattice parameter of the metal upon absorption of hydrogen. At equilibrium a uniform distribution of dissolved hydrogen in a metal causes zero hydrostatic stress in the lattice, although each hydrogen atom is individually under a hydrostatic stress. There is no force or pressure tending to push the hydrogen atoms together into one interstitial site [453,454].

In palladium (Pd) the interstitial sites having octahedral symmetry, of which there is one site per Pd atom, are the lower-energy sites so that these are preferentially occupied by hydrogen. Unlike the case with substitutional solid solutions, adding an interstitial solute atom to a metal does not increase the number of lattice sites for that solute species. The consequence is that interstitial solutes obey Fermi–Dirac statistics, and it follows that an infinitely large thermodynamic activity would result from a complete filling of all octahedral sites in a lattice from an electronic point of view. Therefore, as the molecular gas fugacity is increased in an attempt to fill all octahedral sites in Pd with hydrogen a "spill-over" necessarily occurs into the higher-energy class of interstitial sites, namely, those of tetrahedral symmetry. The fugacity (*f*) of a real gas is an effective pressure which replaces the true mechanical pressure in accurate chemical equilibrium calculations.

Because hydrogen changes the local and the global electronic structure of the host metal, one may expect that hydrogen changes the cohesive forces between metal atoms. Several theoretical investigations show that the cohesive interactions are in fact decreased in many transition metals including Pd [453]. The consequence is that the effort required to cause brittle cracking of a metal is decreased by hydrogen [453]. In general, the mechanical response to applied stresses is variously affected by hydrogen in complex ways.

Dissolved hydrogen interacts attractively or repulsively with chemical and with structural singularities in the lattice. The interactions with the former, that is, impurity atoms, are best treated.

The same is true for structural imperfections when the hydrogen is exactly at the imperfection. However, at distances such that linear elasticity theory can be applied and the stress field produced by the imperfection can be calculated, the thermodynamics of stressed bodies can be used [454]. For example, a dislocation whether of edge, screw, or mixed type has a stress field, as does a particle of a second phase within a matrix. The thermodynamics shows that a stress field that has a positive (i.e., tensile) hydrostatic component of stress, σ_h, decreases the chemical potential of an interstitial solute, whereas a negative hydrostatic component of stress increases the chemical potential of the solute. Therefore, at equilibrium, that is, uniform chemical potential everywhere, dissolved hydrogen exhibits a higher concentration on the tensile side of an edge dislocation and a lower concentration on the compressive side. In cases where σ_h is zero, squared terms in the stress become significant and cause an increase in the chemical potential. Shear stresses also affect the chemical potential so that the shear stress field about a pure screw dislocation lowers the chemical potential of dissolved hydrogen. The relations between solute chemical potential and mechanical stress are valid whether the stress is internally generated, as by dislocations, second phase particles, grain boundaries, stacking faults, etc., or by externally imposed mechanical forces upon the solid body [453].

The appearance of one solid phase within another solid phase usually involves a change of density, and therefore stress fields are generated in the two phases which change the equilibrium compositions of the co-existing phases from those characteristic of stress-free phases. Thus, as hydrogen increases in concentration in Pd and the high-concentration phase, termed the β-phase (see 10.13), is generated the larger specific volume of the growing β-phase causes compressive stresses upon it and a complex state of stress in the environing α-phase (Fig. 10.14). Because at room temperature the Pd atoms do not diffuse and only the hydrogen atoms can move, a coherence between the α and β lattices is maintained across the interface when the β-particle is small. The coherent equilibrium involves equilibrium concentrations different from those obtaining when the coherency is destroyed by the nucleation of interfacial dislocations. Thereby the difference arises between coherent and incoherent metal-hydrogen phase diagrams, and also the kinetic phenomena of hysteresis in hydrogen absorption an evolution into and from Pd.

Hydrogen has a very large mobility in transition metals especially in Pd, α-Fe, V and Nb. Its diffusivity is proportional to the product of the mobility and a thermodynamic parameter, $d(\ln a)/d(\ln x)$, where a is the thermodynamic activity and x the concentration of the hydrogen [453]. This term can become significant at high concentrations, and in the neighborhood of the critical concentration and temperature of a miscibility gap system (of which Pd-H is an example) the thermodynamic term becomes very small.

Another reason for a decrease of the diffusivity of hydrogen in metals is the attractive interaction with chemical and structural defects. This is the so-called trapping phenomenon [455,456] which is very important at low hydrogen concentrations. Thus, dislocations in a metal will slow down the diffusion of hydrogen when the dislocations are static. However, if the dislocations are caused to move by applied, or by internally generated shear stresses, hydrogen can be transported with the moving dislocations by virtue of the attractive interaction discussed above.

Diffusive hydrogen fluxes are proportional to the gradient of the chemical potential of the dissolved hydrogen. Since mechanical stress affects the chemical potential, it is clear that solely gradients of stresses can produce diffusive fluxes even in the absence of concentration gradients, and stress gradients can be as important, or more so, than concentration gradients for inducing diffusion of hydrogen. Temperature gradients and gradients of electrical potential can also produce fluxes of hydrogen. Deductions about the state of ionicity of the dissolved hydrogen cannot be drawn from measurements of the motion of dissolved hydrogen in an electric field because of the existence of momentum transfer between electrons and electron holes and the dissolved hydrogen [453].

The absorption of hydrogen into the lattice of a metal from the molecular gas must be preceded by adsorption of the molecule upon the surface, dissociation into adsorbed atoms, surface diffusion of the adsorbed atoms, and jumping of these atoms into the sub-surface region (Fig. 10.13). Each of these steps is strongly affected by the atomic-scale topography of the surface and by the quantity and distribution of impurity species and any one of these steps may be kinetically controlling. If hydrogen is presented to the metal surface by electrochemical deposition or by means of a partially dissociated and/or ionized hydrogen gas, the dissociation step is avoided. In all cases the adsorbed species may encounter each other, recombine and return to the environing phase as hydrogen molecules. Hence the transition of adsorbed atoms into the lattice is limited by the kinetics of the recombination reaction. To mitigate this limitation, recombination inhibitors ("poisons") may be added to the environing phase; the exact mechanisms by which these agents act are complex and not fully understood.

Further penetration from the sub-surface plane of the lattice is controlled by diffusion whose boundary condition is the sub-surface concentration, c_o, of hydrogen. This kinetically controlled concentration may be conceptually related to the input fugacity, *fi*, of the source of the hydrogen. It is equal to the pressure of an ideal gas which has the same chemical potential as the real gas. f_i equals the fugacity f_e of molecular hydrogen gas that at equilibrium would produce c_o in a metal at the given temperature.

Sieverts law:

$$c_o = c = s\,fe^{1/2}, \tag{10.1}$$

where s is a constant relates f and the hydrogen concentration, c, in the metal at equilibrium [453]. For a gas of molecular hydrogen, there is a correlation between f_e and the gas pressure, p. However, f_i is a kinetically determined quantity without any relation to environing pressure. It is incorrect to believe that a hydrogen-charging system characterized by large f_i (because large c_o values are produced) subjects that metal lattice to a hydrostatic pressure given by the relation between p and f_e.

The input of hydrogen into the sub-surface layer expands that layer, as discussed in a previous section. Since that layer is coherent with the underlying metal which at the moment does not contain hydrogen and is therefore not expanded, a very large compressive stress is produced below the input surface; its magnitude decreases with increasing distance into the metal, and the stress becomes tensile further in. The distribution of both of the normal and the shear stresses is a function of the time of diffusion, but the largest values occur just below the input surface when the value c_0 is generated there [457].

Because the shear stresses produced may occur very large, dislocations are generated. Some of these dislocations form entangled arrays near the input surface which increase the hardness of the metal and decrease the in-diffusion of hydrogen by trapping the atoms. Other dislocations are caused to glide, by the shear stresses, deeper into the specimen, carrying trapped hydrogen with them and forming pile-ups at grain boundaries or being annihilated at grain and interphase boundaries.

The merging of dislocations into grain boundaries causes the grain boundary to change its atomic configuration and rotation of the grains to occur with respect to each other [453]. Rotation of grains at the specimen surface can produce localized surface elevations and depressions, and emergence of edge dislocations can produce slip lines and serrations on the surface. Because of the dependence of f_i upon the atomic topography of the input surface, different metal grains achieve different c_0 values and because dislocation dynamics are complicated functions of grain boundary configuration and of the orientation of the grain lattice with respect to the stress field, adjoining grains achieve different hydrogen contents at any one instant of time [453]. This phenomenon produces additional mechanical stresses between grains. This, as well as dislocation pile-ups at grain boundaries can produce microcavities the growth of which can produce linkage of the cavities and ultimately separation between grains [453]. Such separations, or cracks, fill with molecular gas at a pressure consistent with Sieverts law for the local hydrogen concentration. If the cracks communicate with the external surface of the specimen they provide channels for the loss of hydrogen from that region of the specimen [453].

Therefore, as the molecular gas fugacity is increased in an attempt to fill all octahedral sites in Pd with hydrogen a "spill-over" necessarily occurs into the higher-energy class of interstitial sites, namely, those of tetrahedral symmetry.

10.4.2 Palladium as an Example for Metal Hydrides

After consideration of some fundamentals of the metal hydration it makes sense to start this section, where concrete metal hydrides will be covered, with the description of one most famous hydrophilic metal, namely, Pd, since exactly this metal can serve as a model material which fully exhibits various features of hydrogen-metal interaction.

Pd is a unique material with a strong affinity to hydrogen owing to both its catalytic and hydrogen absorbing properties. Pd has the potential to play an important role in many aspects of the envisioned hydrogen economy, including hydrogen purification, storage, detection, and fuel cells. Pd exhibits a number of exceptional qualities which enable its application in a vast of hydrogen technologies since it has the ability to absorb large volumetric quantities of hydrogen at room temperature and atmospheric pressure, and subsequently forms palladium hydride (PdH_x) [458]. As shown in Fig. 10.14, the absorbed hydrogen atoms occupy interstitial octahedral sites of the face-centered cubic lattice [459].

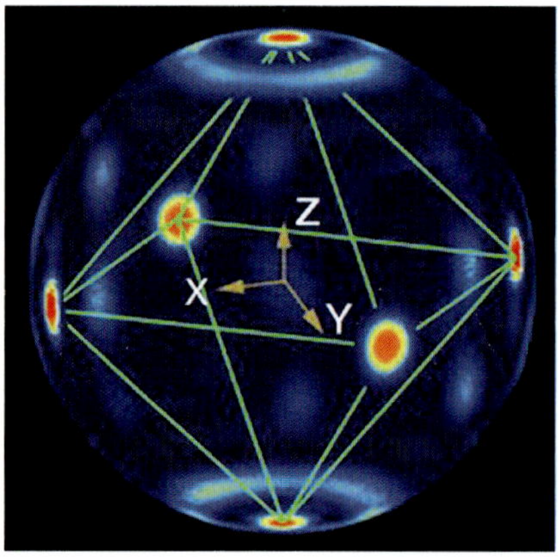

Figure 10.14 A reconstructed neutron holographic image showing the positions of six Pd atoms surrounding a hydrogen nucleus in an octahedral interstitial site of a face-centered cubic lattice. Reprinted from Ref. [459], copyright (2005), with permission from Elsevier.

The temperatures and equilibrium pressures required for the formation of various metal hydrides are compared in Table 10.1 [460].

Table 10.1 Metal hydrides and their hydrogen-storage properties

Metal	Hydride	Wt.% H	P_{eq}, T
Pd	$PdH_{0.6}$	0.56	0.02 bar, 298 K
Mg	MgH_2	7.60	1 bar, 573 K
$LaNi_5$	$LaNi_5H_6$	1.37	2 bar, 298 K
ZrV_2	$ZrV_2H_{5.5}$	3.01	10^{-8} bar, 323 K
FeTi	$FeTiH_2$	1.89	5 bar, 303 K
Mg_2Ni	Mg_2NiH_4	3.59	1 bar, 555 K
TiV_2	TiV_2H_4	2.60	10 bar, 313 K

This table clearly shows both advantages and drawbacks of Pd as a HSM. The main advantage is that Pd can facilely form a hydride under ambient conditions, whereas the main drawback is the low uptake of hydrogen as compared with the other metals in the table. Thus, Pd is a unique metal which can facilely absorb H_2 under low pressures and temperatures but it is a heavy element and the hydrogen uptakes are relatively low. Thus, it is possible to exclude this element from perspective ones since it is not capable of meeting the criteria of the US DOE which means that it would be never used as a pure material for hydrogen storage as such.

Nevertheless, indeed the situation with Pd is not so pessimistic. It will be shown below that Pd is an indispensable element in the systems of hydrogen storage and has its own personal area of application performing unique function in combination with other HSMs.

The very attractive hydrogen absorbing property of Pd has undergone extensive study in both the gas phase [461–463] and under electrochemical conditions [464–467].

Other than Pd, most metals require energy input to overcome an activation barrier for dissociation, which requires the application of high hydrogen pressures or elevated temperatures (see Table 10.1).

On Pd surfaces, the dissociative adsorption of H_2 molecules occurs with little or even no activation energy barrier [468,469]. These unique dissociative properties of Pd enable it to serve as a catalyst to facilitate hydrogen absorption and desorption in other metal hydrides [470].

Further, its inherent selectivity for hydrogen, fast sorption kinetics, and reversibility of hydride formation allow Pd-based alloy membranes to attain a high hydrogen gas quality at 99.99999% purity.

Hydrogen sensors are also being investigated that exploit the changes in the properties of Pd upon exposure to hydrogen [471].

Carbon-supported Pt is commonly used as an electrocatalyst in low temperature fuel cells that utilize hydrogen or small organic molecule feedstock [472]. The high cost of Pt and the limited world supply are barriers to widespread use [473,474]. Pd is of great interest as a substitute material for Pt, not only due to its chemical similarity to Pt, but also because of its significantly reduced cost (approximately 1/5 that of Pt) and 50 fold greater abundance.

The extensive use of Pd for hydrogen storage remains impractical, however, due to its relatively high price. Nevertheless, with the application of appropriate cost-saving methods such as alloying with less expensive metals and utilizing nanoscale materials, Pd has the potential to play a significant role in many aspects of the hydrogen economy, from purifying hydrogen to harnessing its energy via fuel cells.

10.4.2.1 Hydrogen purification with the help of palladium

Hydrogen may be produced from either hydrocarbons or water by employing a wide series of techniques. The most environmentally friendly production technique involves the electrolysis of water, which uses off-peak power or power from renewable sources such as wind turbines or solar cells. Currently, the majority of H_2 is produced in the form of *syngas* that is generated by the gasification of hydrocarbons, which generates large amounts of CO and CO_2.

A high purity hydrogen gas stream is crucial to the performance of fuel cells since certain species poison the catalysts even at very low concentrations [475,476]. The application of a highly selective Pd-based membrane is currently the most advanced method to purify the gas [476]. A major drawback for membranes that are comprised of pure Pd is hydrogen embrittlement [477].

The absorption of hydrogen into Pd occurs in two distinct phases. The α-phase appears at low concentrations of hydrogen (solid solution), and the β-phase forms at high concentrations of hydrogen (metal hydride). Associated with the α/β-phase transition is an expansion of the face-centered cubic crystal lattice. This large expansion of the lattice, due to hydrogen embrittlement, can cause the cracking of the membrane. In general, β_{min} decreases and α_{max} increases in conjunction with elevated temperature. As a result, the width of the co-existence region or miscibility gap ($\beta_{min} - \alpha_{max}$) decreases with rising temperatures until a critical threshold T_c = 570 K (bulk Pd) is reached. Above T_c, the miscibility gap is no longer observed. For this reason, membranes that are comprised of pure Pd are held above 570 K in order to avoid hydrogen embrittlement.

It has been shown that the size of the Pd particles has an effect on this phase transition [478]. The miscibility gap and T_c are decreased with the use of nanoparticles [479,480]. The hysteresis in the PC isotherm of Pd nanoparticles (2.6 ± 0.4 nm in diameter) disappeared at 393 K. The absence of hysteresis, in addition to the absence of a plateau region in the equilibrium pressure, implies that the critical state of the α- and β-phases has been attained. Thus, a remarkable decrease in T_c was recognized in Pd nanoparticles. Hydrogen embrittlement may also be controlled at ambient temperatures by alloying Pd with metals which have larger atoms than Pd itself (see [458] and references therein]. In PdAg, for example, which is the most common hydrogen extraction metal alloy used by industry, the lattice is pre-expanded by the silver atoms and thus is less influenced by hydrogen than a pure Pd lattice. An optimal hydrogen permeation value is achieved at a silver content of 23 wt.%. The isotherms for PdAg alloys do not show a well-defined plateau as in the case of Pd, and the lattice expands continuously in this region. The α/β-phase transition results in a volumetric expansion of only 0.38 % for this alloy. The same effect has been seen with Pd–Cd nanostructures. A composition of Pd–Cd (15 at.%) was found to eliminate the α/β-phase transition with optimal hydrogen solubility and, consequently, increased the kinetics of the absorption process [458].

In heterogeneous catalysis, it is observed that poisons (e.g., sulfur, arsenic, and carbon monoxide) deactivate catalyzed reactions through irreversible adsorption at active sites. Recently, Pd-based sorbents were shown to effectively remove the contaminants mercury, arsenic, selenium, and phosphorus from high temperature syngas [458 and references therein]. Much of the work done with respect to the poisoning species of Pd-based hydrogen permeable membranes has suggested that conducting hydrogen purification at elevated temperatures is likely to decrease the problematic effects described above.

10.4.2.2 Palladium as an important assistant in hydrogen storage

The critical properties that a HSM should possess include that it be (i) lightweight, (ii) inexpensive and readily available, (iii) accommodating for high volumetric and gravimetric densities of hydrogen, and exhibit (iv) rapid sorption kinetics, (v) easy activation, (vi) low temperature dissociation or decomposition, (vii) appropriate thermodynamic properties, (viii) long-term cycling stability, and (ix) high reversibility. Pd should not be considered as a sole HSM since it is somewhat expensive and has a low gravimetric hydrogen density [458]. However, it does satisfy all of the other requirements that define a HSM, which enable it to "assist" in the hydrogenation and dehydrogenation of other materials. Pd acts not only as a catalyst to facilitate the uptake and dissociation of hydrogen in other metal hydrides, it can also protect the surface from corrosion, allowing for the study of air-sensitive samples in ambient conditions [458]. At temperatures of below 250 K, hydrogen absorption and desorption is effectively hindered and it becomes frozen within the sample. This allows for the absorption of a desired hydrogen concentration at room temperature, followed by rapid cooling to cryogenic temperatures to facilitate analysis of the sample.

Inexpensive, lightweight, and high capacity metal hydrides such as MgH_2 (7.6 wt.%), are considered as viable candidates for practical hydrogen-storage applications [481–483]. There are, however, two major disadvantages inherent to MgH_2: (i) the sorption kinetics involved are quite sluggish; and (ii) MgH_2 has a high thermodynamic stability requiring temperatures that exceed 300°C for the desorption of hydrogen [458,481]. Although hydrogenation is most commonly carried out in a high-pressure hydrogen atmosphere at elevated temperatures, metals such as Pd and its alloys may be electrochemically charged with hydrogen at ambient temperatures in contrast to magnesium and its alloys that may not be charged via this method as they dissolve in acidic solutions or form magnesium hydroxide in alkaline media [458].

However, these issues can be resolved through the use of a Pd capping layer. It was shown that a film of Mg_xTi_{1-x} that is capped with a thin (3–4 nm) layer of Pd can be electrochemically charged with hydrogen. The cumulative hydrogen-storage capacity of this material has been shown to approach 1750 mAh/g, which corresponds to 6.4 wt.% [458].

Even though many metals absorb hydrogen exothermally, it is necessary to overcome a certain activation energy barrier, which is contingent on surface elements, in order to dissociate hydrogen molecules. Since many metals have slow hydrogenation kinetics, Pd can be employed as a catalyst to dissociate hydrogen molecules, and thus decrease the hydrogenation time. The activation energies of H_2 sorption for bulk MgH_2, mechanically milled MgH_2, and nanocatalyst doped MgH_2, are shown in Fig. 10.15 [484]. The activation barrier is drastically lowered via nanocatalyst doping. In addition, the collision frequency between H_2 molecules and transition metal nanoparticles may increase with the reduction in the size of the catalyst.

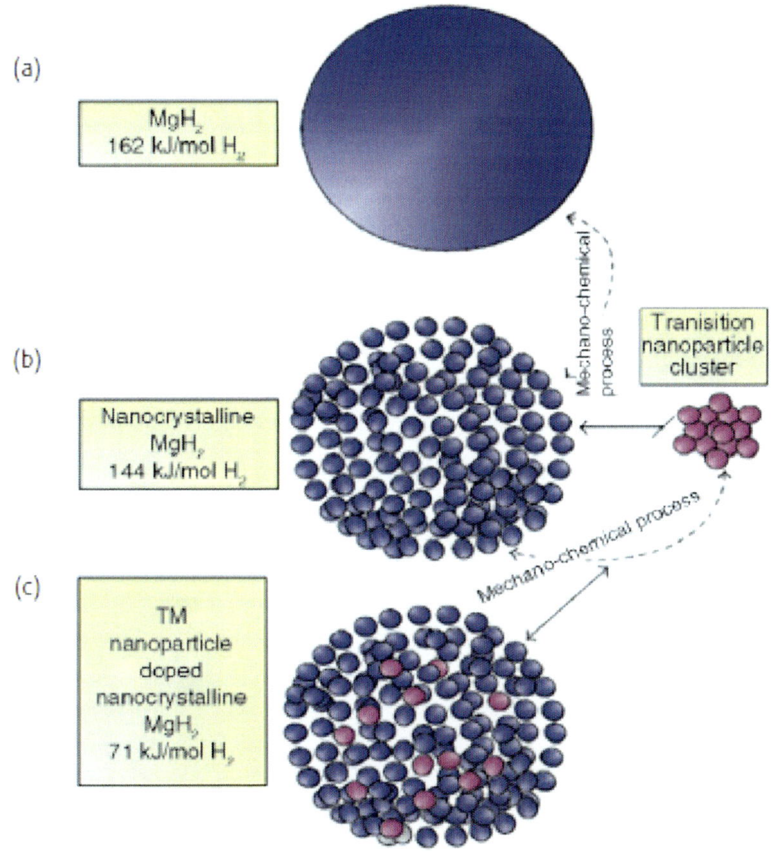

Figure 10.15 Conceptual model of MgH_2 cluster: (a) plain, (b) nanocrystalline, and (c) nanocatalyst-doped materials. Reproduced from Ref. [484]. Copyright © 2008 Michael U. Niemann et al.

Figure 10.16 displays the kinetics of hydrogen absorption to Mg with different particle sizes (a) and to Pd-doped Mg nanoparticles (b). Pure Mg takes three times as long to attain the equivalent hydrogen capacity as Pd-doped Mg [485]. A computational study has revealed that an optimal hydrogenation catalyst would not only have superior dissociation capabilities, but also exhibit low potency dissociated hydrogen atom binding with the metal. Pd offers an attractive compromise for these properties, which may be further enhanced when it is alloyed with other metals [458].

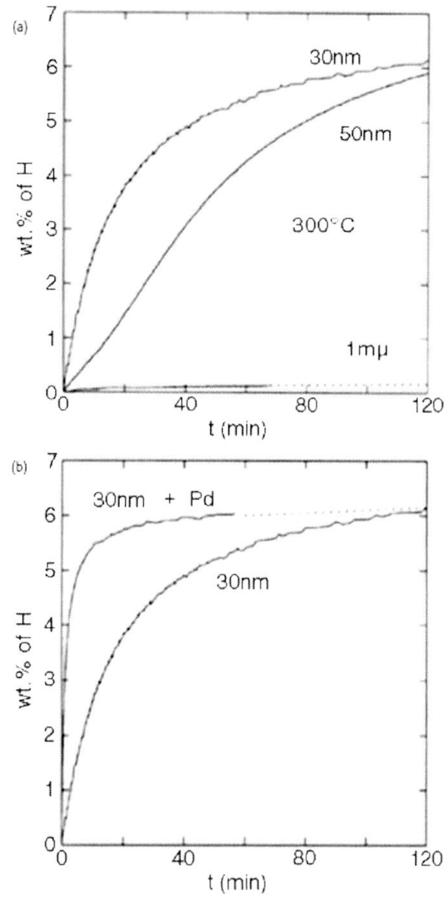

Figure 10.16 The effect of grain size on hydrogen absorption into ball-milled Mg and (b) a comparison of nanosized Mg with and without a Pd catalyst. Reprinted from Ref. [485], copyright (1999), with permission from Elsevier.

As we have seen before, another promising class of HSM is carbon. Carbon offers the advantages of being lightweight and inexpensive, while exhibiting onboard reversibility and rapid sorption kinetics. In contrast to the chemisorption in metal hydrides, the phenomenon of physical adsorption is essentially the accumulation of undissociated hydrogen molecules upon a surface of microporous carbon. It has been found that hydrogen uptake in carbon is proportional to its inherent surface area and/or pore volume. High hydrogen adsorption capacities (3–8 wt.%) may be obtained only at very low cryogenic temperatures or high pressures. Operating within the cryogenic domain poses certain challenges such as maintaining the storage vessel at low temperatures.

The storage of hydrogen using carbon materials at practical temperatures relies on a chemical reaction between carbon and hydrogen atoms, which involves the cleavage of the H_2 molecule. This may be accomplished through the use of transition metal catalysts [458].

The metallic doping of carbon materials results in the following phenomena, which are collectively termed as the hydrogen spillover mechanism: (i) H_2 molecules dissociate on the metal catalyst particles; (ii) H atoms migrate from the metal catalyst particles to the carbon support material; and (iii) H atoms diffuse on the substrate [458]. The hydrogen spillover mechanism is illustrated in Fig. 10.17 [486]. When spillover occurs, the hydrogen capacity of the combined system (carbon + metal) will typically exceed that of the individual components. An advantage that is provided by the use of Pd as the catalyst is that a portion of the hydrogen is stored within the catalyst itself (as PdH_x). Hence, the capacity of the overall material is likely to be higher than that of a nonabsorbing metal [458].

However, a catalyst that exhibits an excellent compromise between hydrogen dissociation and migration from the metal should produce much better results. Because the strength of the C–H interaction tends to be much lower than that of the Pd–H, alloying Pd with metals which have a lower affinity for hydrogen should enhance the effects of spillover. This was recently validated when PdCd nanoparticles, which were dispersed on activated carbon, generated a spillover enhancement that was far greater than the Pd nanoparticles alone, due to the weak strength of the Cd–H bonds of surface resident Cd atoms [458].

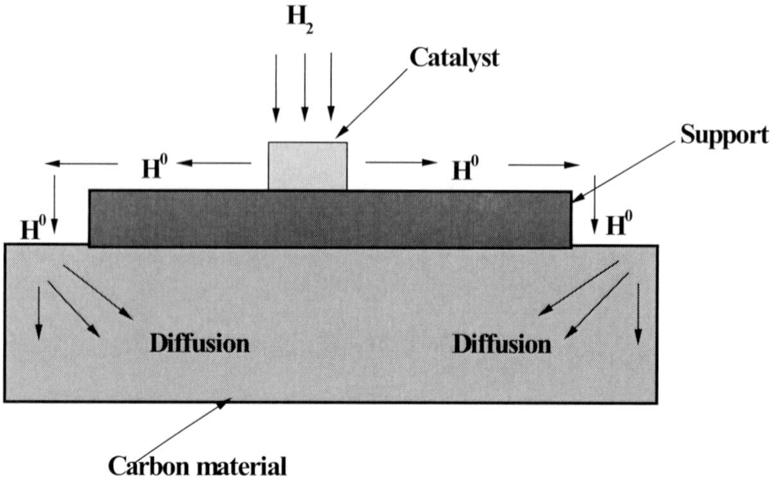

Figure 10.17 Spillover mechanism: (i) dissociative adsorption of H_2 molecules on a metal catalyst, (ii) migration of H atoms to the support, and (iii) diffusion on the bulk of the carbon material.

10.4.2.3 Hydrogen detection with the help of palladium

Various types of sensor have been studied and implemented for the sensing of oxidizing and reducing gases and namely H_2 sensors have gained great interest due to the undesired and/or unanticipated presence of hydrogen within certain commercial/industrial environments. Although hydrogen is attractive as a carbon-free fuel, it is colorless, odorless, an asphyxiant (through displacement of oxygen), and it is highly flammable in air. A potential hydrogen gas sensor must satisfy four requirements for wide-scale deployment, namely: sensitivity, selectivity, specificity, and cost. Most approaches utilize Pd as a catalyst, since Pd has a great absorptive affinity for H_2. The mechanical, electrical, magnetic, and optical properties of the metal undergo dramatic changes on exposure to hydrogen [458].

There are two primary classes of hydrogen selective sensors that are based on Pd. The most popular of these exploits the increased electrical resistivity of Pd, which is built up due to the intensified electron scattering that is initiated by the incorporation of hydrogen [458].

The alternate class is an optical sensor that consists of a layer of Pd, which is coated onto an optically active material that translates the concentration of hydrogen into an optical signal [458]. The key principle that underlies hydrogen sensing is the conversion of the α-phase (conductive) of Pd to the β-phase (less conductive).

The nanoeffect on the α/β-phase transition has been utilized successfully in many cases. It was discovered that Pd nanowires have the capacity for detecting H_2 much more rapidly than Pd films. A non-mechanism was elucidated for enabling hydrogen switches and sensors, which was based on Pd mesowire arrays. Rather than the observation of an increase in resistivity upon exposure to hydrogen, a decrease was registered. This mechanism involves the closing of nanoscopic gaps within the nanowires, which are caused by the dilation of the Pd grains as they undergo hydrogen absorption [487].

10.4.2.4 Fuel cell catalysis

Pt is the most commonly employed electrocatalyst in low temperature fuel cells. The disadvantages of Pt include its high cost, limited world supply, and vulnerability to CO which is present in the H_2 feed stream or is produced as an intermediate during the oxidation of small organic molecules. Although Pt exhibits the highest activity among pure metals in cathodic oxygen reduction reactions, it is still not a satisfactory candidate for conducting this sluggish reaction on its own. Numerous studies have been undertaken toward the identification of alternative catalyst materials that might replace Pt, either partially or completely.

Although Pd is of great interest as a substitute material for Pt, it does not appear to be active enough for utilization as a catalyst in most oxidation or reduction reactions. Hence, Pd alloys and composites thereof have been widely investigated for these purposes. Pd–Pt bimetallic catalysts have been found to outperform Pt in many reactions, and Pd is shown to be a far superior catalyst than Pt in formic acid oxidation. For the oxygen reduction reaction, Pd-alloys have also demonstrated improved performance when compared to Pt, especially in the presence of methanol. The transition from Pt-based to Pd-based catalysts has commenced and will likely result in the development of more cost effective materials with enhanced activity toward the commercialization of fuel cells.

10.4.2.5 Conclusions concerning palladium

This section has described the significant roles of Pd-based nanomaterials in hydrogen purification, storage, detection, and fuel cells. It is anticipated that Pd-based nanomaterials will be vital to both enable and develop a hydrogen economy. The extent of the application of these nanomaterials will be largely contingent on cost and availability. Critical targets in Pd-based nanomaterials research should therefore be three-fold: (i) to advance our understanding to a point where it will be possible to successfully tailor high-performance Pd-based nanomaterials by design; (ii) to develop novel synthesis methods that allow precise control over product morphology and composition and which will serve to optimize the catalytic and hydrogen absorbing efficiencies of Pd-based nanomaterials; and (iii) to design and fabricate robust nanoscale Pd entities that can survive under the difficult conditions of high temperatures and various oxidizing or reducing environments.

In regard to hydrogen absorbing applications (e.g., purification, storage, and detection), there are two common, albeit detrimental issues that should be encompassed within the research focus; specifically, the development of countermeasures to negate both the onset of hydrogen embrittlement and the effects of poisoning species.

Hydrogen solubility plays an essential kinetic role in certain applications (e.g., storage, fuel cell catalysts) that exploit hydrogen dissociation, where surface interactions are of greater concern than those involved with bulk phase absorption. In these cases, the adsorption strength is a crucial factor. While research on HSMs has slowed somewhat, due to discouraging results combined with failures to meet stipulated requirements, it is now more imperative than ever to investigate the use of Pd-based catalysts to assist in the realization of this goal. Combinatory studies, including both experimental and computational methods, to facilitate further elucidation of the dynamics of the alloying effect should help us to reach a deeper understanding of the electronic and geometric nuances that are at play in this process.

Intensive research has been recently conducted on the metal hydrides for improving adsorption/desorption properties for hydrogen storage. The development of more advanced computational techniques for the detailed study of hydride formation and the desorption of hydrogen may be a beneficial avenue of research that will serve to support experimental work and/or lead to a better understanding of bimetallic or trimetallic Pd-based hydride systems. In conclusion, the issues of cost, hydrogen embrittlement, poisoning species, hydrogen solubility, and hydrogen adsorption strength are critical areas for investigative research toward the development of functional Pd-based nanomaterials to enable technologies that will drive the hydrogen economy.

We paid so much attention to Pd since it is a very lucky material that to demonstrate numerous processes and problems that arise in metal hydrides. It would be easier then to describe these processes in other hydrides which are being investigated in the hydrogen-storage systems.

10.4.3 Hydrides for Hydrogen Storage

As it was considered above molecular hydrogen dissociates into atomic hydrogen on the surface of the host material and enters the bulk via diffusion between interstitial sites in the host lattice. Unfortunately, binary hydrides (MHx, where M is a metallic element and x is the hydride stoichiometry) occur generally to be either too unstable or too stable for use as practical storage materials. The former means that the pressures at which the host metal reversibly absorbs and releases hydrogen are too high at practical storage temperatures and the latter means the pressures are too low. In terms of formation or decomposition enthalpies, ΔH is too high in the former case (either small negative numbers or positive) and too low in the latter (a relatively large negative value) [374].

The highlight: however, when two or more metallic elements are combined, particularly one that forms a stable hydride and one that does not, the resultant alloy or intermetallic tends to form a hydride of intermediate stability.

The study of metallic hydrides began nearly 150 years ago with the discovery of the hydrogen-absorbing properties of Pd. Work beginning in the 1960s resulted in the later commercialization of nickel–metal hydride (Ni–MH) batteries in which the negative electrode material forms an intermetallic hydride. Although this has been the most commercially successful application of interstitial hydrides, metal hydride technology is also exploited in a number of other application areas, including gas separation and purification, temperature sensing, thermal compression, and refrigeration or cryocooling [374]. For mobile storage applications, the gravimetric capacity of many interstitial hydrides is relatively low, but some of these compounds show remarkable and practical hydrogen absorption and desorption characteristics, and they are likely to play an important role in a future hydrogen economy.

In this section, we look at a number of the different types of interstitial hydride, and cover intermetallic compounds, solid solution alloys, modified binary hydrides, and amorphous and nanostructured hydrides, including mechanically milled materials, amorphous alloys produced using other methods, and quasicrystals.

10.4.3.1 Intermetallic compounds

The host materials in this group are ordered stoichiometric compounds typically formed from two metallic components, A and B. The A and B components tend to form hydrides AHx and BHy, with enthalpies of formation ΔH_A and ΔH_B, representing a stable and an unstable hydride, respectively. As mentioned above, the resultant intermetallic hydride $A_mB_nH_z$, where m and n are integers and z is a real number, will then tend to have an enthalpy of formation, ΔH_{AB}, where $\Delta H_A < \Delta H_{AB} < \Delta H_B$. Varying the ratio n/m then shifts the value of ΔH_{AB} in either direction. The components A and B can generally be fully or partially substituted by other elements of relatively similar size or chemistry. Hydrogen-absorbing intermetallics form a number of different groups, which can be distinguished by their stoichiometries, including AB_5, A_2B_7, AB_3, AB_2, AB, and A_2B compounds [374].

Around 1970 the reversible hydrogenation of $LaNi_5$ was discovered that [488] yielded an increase in the amount of metal hydride research which in turn ultimately led to the commercialization of Ni–MH battery technology, and the intensive research into metal hydride-based hydrogen storage that proceeds today. A vast number of A and B element combinations are now known to form reversible hydrides, although relatively few do so at useful temperatures and pressures. In this section we will cover some of the combinations and stoichiometries that are of practical interest for hydrogen storage.

Note that different stoichiometries can often be found within the binary phase diagram of any given pair of constituents but, because of the dramatic effect that the ratio of elements has on the hydriding properties, it is likely that only one of these has a formation enthalpy that gives the material hydrogen absorption and desorption capabilities in a practical temperature and pressure regime. As an example, let us consider La and Ni, which combine to produce the well known hydride-forming compound $LaNi_5$. La forms a very stable hydride, LaH_2, with ΔH = −104 kJ mol^{-1} H, while Ni forms a very unstable hydride (ΔH = −3 kJ mol^{-1} H) [489]. The calculated values for the La–Ni–H system are −13.3, −23.0, −40.3, −60.6 and −82.5 kJ mol^{-1} H for $LaNi_5$, $LaNi_2$, LaNi, La_2Ni, and La_5Ni, were reported respectively [489]. The dependence on the ratio of the two constituent metal elements can be seen clearly and matches the trend observed experimentally: the lower the Ni content of the host intermetallic, the more stable the associated hydride compound. Some intermetallics, principally Mg-based compounds such as Mg_2Ni and Mg_2Cu, form stoichiometric hydride complexes upon hydrogenation and can therefore be categorized as complex transition metal hydrides rather than interstitial intermetallic hydrides. Furthermore, there are interesting cases, such as the intermetallic compound $LaMg_2Ni$ that forms a mixed hydride ($LaMg_2NiH_7$) consisting of NiH_4 tetrahedra and 'interstitial' H^- ions, which blur the boundary between the interstitial intermetallic hydrides and their complex counterparts [374]. Many of the intermetallics can be under or over-stoichiometric. In each of the following sections we will briefly summarize the main groups of interstitial hydride-forming intermetallics that are of interest for hydrogen storage, and we therefore cover AB_5, AB_2 and AB compounds.

10.4.3.2 AB_5 compounds

The archetypal AB_5 intermetallic is $LaNi_5$ that facilely forms a hydride under moderate hydrogen pressures and ambient temperatures. It has an enthalpy of formation of −15.7 kJ mol^{-1} H and an enthalpy of decomposition of −15.1 kJ mol^{-1} H [490]. Hydrogen capacity exceeds $LaNi_5H_6$, which yields a reversible gravimetric capacity of ~1.25 wt.% [491]. An essential drawback of this material $LaNi_5$ and some other binary AB_5 compounds, such as $CaNi_5$, is that they can disproportionate and lose their reversible capacity during hydrogen cycling. However, there is a way to resist this disproportionation by modification: these compounds can be modified via partial substitution. The most effective substituent for this purpose is considered to be Sn that dopes $LaNi_5$, with a composition of $LaNi_{5-x}Sn_x$, where x = 0.2 [492]. To reduce the cost of La-containing materials so called mischmetal (Mm), a naturally occurring mixture of rare earths, can substitute the La as the A component. Along with the $LaNi_5$, $LaNi_{5-x}Sn_x$, $MmNi_5$ and $CaNi_5$ intermetallics, there is a large variety of hydride-forming AB_5 compounds that can be composed of many other combinations of the elements, where the A element is either one or more of the lanthanides or Ca, as in the compounds mentioned above, or Y and Zr, and the B elements any of a range of elements including Co, Al, Mn, Fe, Cu, Sn, Si, and Ti, as either full or partial constituents [491,493].

It can be seen clearly that the gravimetric storage capacity of these materials is noticeably lower than the current US DOE target for mobile hydrogen-storage applications. On the other hand, the AB_5-based intermetallics demonstrate fairly good cycling properties along with robust resistance to gaseous impurity contamination, long term cycling stability and a high volumetric storage density, and, because of this, they are an important and encouraging example of a practically effective reversible HSM.

10.4.3.3 AB_2 compounds

AB_2 compounds can be formed from the combination of many different elements. The A elements are usually from group 4 (Ti, Zr, Hf) or the lanthanides (La, Ce, Pr, and so on), whereas the B element can be a transition or non-transition metal, with a preference for V, Cr, Mn and Fe [490]. In the work [494] the most promising elements for electrochemical applications of AB_2 compounds were claimed as A = Mg, Zr and Ti, and B = V, Cr, Mn and Ni, despite off many more can be used as partial substituents.

These intermetallics crystallize, like the AB_5 compounds, and exhibit properties depending on the elemental composition which allows charging reversibly a given AB_2 material with hydrogen. The parameters can be tuned via partial elemental substitution.

It should be stressed that AB_2 intermetallics have already been used practically. In the 1980s, a fleet of Daimler vans and automobiles were operated using AB_2-based hydrogen stores containing the non-stoichiometric compound $Ti_{0.98}Zr_{0.02}CrO_{.05}V_{0.43}Fe_{0.09}Mn_{1.5}$ [495], which exhibits very fast kinetics and good long term cycling stability [496]. This example of the compound containing various substituents shows how many possibilities have the researcher that creating various hydrogen-storage systems.

Although, high material costs and a gravimetric storage capacity of 1.8 wt.% do not, in view of the current US DOE targets, make it particularly perspective for widespread use, despite this it is another example, along with the AB_5-based compounds, of a conventional interstitial metallic hydride that works well in real applications and satisfies many of the required performance criteria.

10.4.3.4 AB compounds

In comparison to the variety of different AB_2 and AB_5 compositions elucidated in the literature, the number of AB compounds suitable for hydrogen storage is limited. It is only TiFe that is of practical interest for storage use, which achieves a total gravimetric capacity of 1.86 wt.% at H/M = 0.975, with a reversible capacity of 1.5 wt.% [491]. The hydrogen absorption properties of TiFe were widely investigated and it has reversible hydrogen-storage characteristics in a useable range. It has an ordered body-centered cubic structure, and exhibits two distinct plateaus in its hydrogen absorption isotherm. Partial substitution can again be used to improve the hydrogen absorption behavior, for example, substitution of Fe with Mn and Ni, which lower the pressure of the first ambient temperature isotherm plateau, and therefore stabilize the hydride with respect to pure TiFe. TiFe and $TiFe_{0.85}Mn_{0.15}$ demonstrate good long term cycling stability, they are low cost, and tend not to be pyrophoric [491], unlike the AB_5 compounds. A pyrophoric substance is a substance that ignites spontaneously in air. Unfortunately their activation is relatively difficult and they are significantly more sensitive to gaseous impurities than AB_5 intermetallics. As with most of the intermetallic hydrogen-storage compounds, their gravimetric capacities are low in comparison to the current US DOE targets, but they are nonetheless favorable to many other interstitial hydrides.

10.4.3.5 Solid solution alloys

Solid solution alloys formed by dissolving one or more hydrogen-absorbing metallic elements in another also show interesting hydrogen-storage properties. Unlike the intermetallics, these materials do not necessarily have stoichiometric or near-stoichiometric compositions. They can be formed from a number of host solvents, including Pd, Ti, Zr, and V. From the previous consideration, it is clear that Pd-based alloys suffer from low gravimetric capacities and the high cost of Pd, and Ti and Zr-based solid solutions tend to be too stable. At the same time, vanadium-based alloys demonstrate favorable absorption properties, and distinct advantages in gravimetric capacities over some intermetallics. Although pure vanadium is very expensive, the use of lower cost ferrovanadium looks more promising and Fe containing V-based solid solution alloys appears feasible HSMs [490]. Typical gravimetric hydrogen capacities can approach 4 wt.% for Ti–V–Fe ($Ti_{43.5}V_{49.0}Fe_{7.5}$) and Ti–V–Cr–Mn alloys, although the latter requires temperatures in the range 520–745 K [497]. The pressure-composition isotherms for these materials exhibit two plateaus, of which only the higher pressure plateau region can be used for hydrogen-storage applications. Therefore, in a practical pressure range, the reversible capacity is significantly lower than the maximum capacity claimed above at around 2.5 wt.% [374]. In the study of Ti–V–Fe alloys, the "available hydrogen quantity" (AHQ) was defined as being the difference between the hydrogen content at 100 kPa on the absorption and desorption isotherms at 253 and 573 K, respectively [498]. It was found the AHQ to be 2.4 wt.% for $Ti_{43.5}V_{49.0}Fe_{7.5}$ [499], a reversible capacity being of 2.3 wt.% for $Ti_{0.32}Cr_{0.43}V_{0.25}$ along with that the alloy exhibits good cyclic stability, maintaining a reversible capacity of approximately 2 wt.% over the course of 1000 hydriding and dehydriding cycles.

10.4.3.6 Modified binary hydrides

MgH_2 is the binary hydride that occurred to be by far the most attention as a potential storage material. Nevertheless, there are two other binary hydrides that should be also mentioned, namely AlH_3 and PdH_x. The former has high gravimetric storage capacity, but it is effectively non-reversible within a realistic hydrogen pressure range for a practical storage unit and therefore requires off-board regeneration. This process is economically and energetically costly, which makes the chances of this material very low. Pd, meanwhile, as it was already mentioned, is impractical as a storage material because of its relatively high operating temperature, low gravimetric capacity, and high cost.

However, research into nanoscale Pd is being actively pursued due to the interesting differences in its hydrogen absorption behavior compared to the bulk material. As it was shown above, the hydrogen solubility of the hydride phase and equilibrium hydrogen pressures are reduced for small particles [458]. The plateau region also slopes and narrows, and the critical temperature, T_{crit}, is significantly reduced in Pd clusters compared to the bulk. Along with this, there is also a reduction in the enthalpies of hydride phase formation and decomposition [458]. From a practical point of view, the cost of Pd would seem to be prohibitive for widespread use. Despite this, it is possible that nanoscale Pd could find use as a catalyst in practical storage materials. But what is more important, Pd hydride can play the role of a model system for the investigation of size effects [458] and therefore intensive investigations of this system could provide further valuable knowledge concerning the effects of particle size reduction in other interstitial metal hydride systems.

Returning to magnesium, MgH_2 is a perspective potential storage material; it has a gravimetric capacity of 7.66 wt.%. However, it has a serious shortcoming: in addition to its high thermodynamic stability ($\Delta H \sim -75$ kJ mol^{-1} H_2), the kinetics of hydride formation and decomposition of the bulk material are too slow for practical use. The attempts are carrying out to improve the kinetics significantly using mechanical sometimes along also without catalytic additives. The positive effects of ball-milling pure MgH_2 have been reported (improvement in the kinetics at 573 K), but the temperatures required for desorption are still too high for practical applications [500,501]. However, alternative methods of producing nanoscale magnesium have also been developed [502], and include an electrochemical synthesis [503], which makes it possible to obtain an average particle size of approximately 5 nm. Concerning to the catalytic enhancement of the hydrogen absorption and desorption properties of MgH_2, the most successful catalyst found to date appears to be Nb_2O_5. The hydrogen desorption rates obtained using various oxide additives are shown in Fig. 10.18 [504]. The reasons for the enhanced absorption and desorption rates are not yet understood. The results of a number of different additives, including AB_5 and instead induce further MgH_2 particle or grain size reduction during the milling. In addition to the kinetic enhancement, there is some theoretical evidence that a reduced particle size can alter the thermodynamics of this system. Ab initio Hartree–Fock and density functional theory calculations, reported by Wagemans et al. [505], show that the stability of Mg hydride is reduced for very small cluster sizes below approximately 1.3 nm.

Although this dimension is significantly smaller than the grain size achieved by mechanical milling, whether with or without an additive, these calculations demonstrate that there may be some potential for nanoscale magnesium hydride to provide hydrogen-storage capabilities in a practical temperature range, although further work in this area is required.

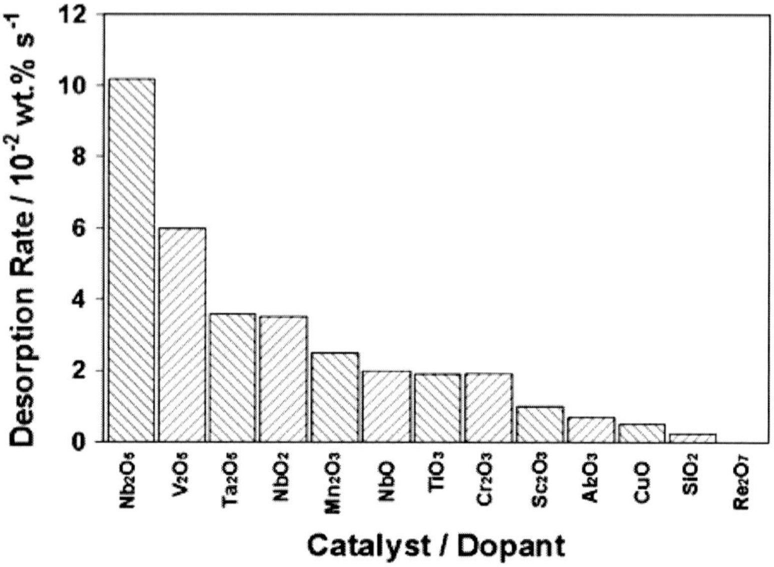

Figure 10.18 The catalytic effect of various transition metal oxides on the hydrogen desorption rate for MgH_2. Reaction rates were determined between 20% and 80% of the maximum hydrogen-storage capacity in each case. Reprinted with permission from Ref. [504]. Copyright (2006) American Chemical Society.

10.4.3.7 Amorphous and nanostructured alloys

In this section primarily three types of amorphous and nanostructured alloy will be covered. Firstly, we concern compounds that are synthesized or processed through the use of mechanical milling. This can have different effects on a material, depending on a number of factors. One possible effect is amorphization, which is the loss of crystallographic order by an originally crystalline material. There are, however, a number of other ways to synthesize or process similarly amorphous alloys and so, secondly, we will discuss amorphous materials that can be produced using alternative routes. Thirdly, we will look at an interesting class of materials known as quasicrystals.

10.4.3.7.1 *Mechanical milling and alloying*

A significant area of research into new HSMs involves the use of mechanical milling, as introduced in the section above for the modification of MgH_2, for which it is well suited. The use of the technique originates from a powder processing method developed to produce oxide dispersion strengthened alloys [500,506]. In materials processing there are several different types of milling [506]. For our purposes, the approaches applied to HSM synthesis or modification can be broadly separated into three areas. Firstly, the synthesis of nanostructured or amorphous hydrogen absorbing alloys using either elemental metals or the combination of elemental metals with crystalline intermetallics or alloys. Secondly, the nanostructuring or amorphization of existing crystalline alloys or intermetallics through milling and, thirdly, the doping of an existing crystalline alloy or intermetallic with a catalytic additive. The latter is essentially the same as the approach used widely for MgH_2.

In each case mentioned, the materials can benefit from the effects of the mechanical milling process, which includes the introduction of nanoscale structural features. Such nanostructuring can lead to size effects that are known to dramatically alter the behavior of materials compared to bulk samples [507,508] and the approach has a broad range of applications in materials science. Materials that have undergone milling can possess a smaller grain size, amorphous and disordered structural features, metastable and high pressure phases, an increased proportion of grain boundaries, a higher density of defects and an increased surface area [506]. All of these features can affect the hydrogen absorption properties of the material in comparison with unmilled, single phase, homogeneous crystalline samples [509–511]. If hydrogen diffusion occurs more rapidly through grain boundary regions, a smaller grain size will enhance the rate at which hydrogen is absorbed or desorbed. Also, amorphous phases of some compounds can show favorable absorption characteristics [512]. In addition, increased defect densities can enhance hydrogen diffusion rates. The activation process, which in many metallic absorbers involves the introduction of high dislocation densities into the host structure, can also be accelerated by milling the material. All of these factors can therefore potentially contribute to the improvement of the hydrogen-storage properties of a material through milling. Although the process does not always have the desired effect, there is significant experimental evidence for the improvement of the hydrogen-storage properties of materials through mechanical milling and alloying [499,500,513]. The different approaches mentioned above have been applied to a range of interstitial hydrides and hydride-forming HSMs. Catalytic additives, such as Pd and Ni, have been milled with TiFe to improve the difficult activation process for this compound [514,515]. Many amorphous and crystalline hydrogen-storage compounds that have been synthesized from the milling of elemental metals include $Ni_{1-x}Zr_x$ amorphous alloys, nanocrystalline $LaNi_5$, amorphous and crystalline TiFe, Ti–V–Mn alloys , and nanocrystalline Mg_2AlNi_2 are reported [374]. The modification of crystalline alloys and intermetallics by milling, meanwhile, has included work on $LaNi_5$, other AB_5-based compounds, Ti–Cr and Ti–Cr–V alloys, and the Mg–Ni alloys that can form complex transition metal hydrides [374].

A further approach to the use of mechanical alloying is reactive milling in which the materials are processed in a hydrogen atmosphere [516,517]. As intimated above, the milling process can also be damaging to the hydrogen sorption properties, as well as beneficial, with hydrogen absorption sometimes prevented entirely as a result of milling [518], or favorable isotherm behavior being lost along with a significant amount of storage capacity [519,520]. However, increases in capacity as a result of milling have also been reported [521].

10.4.3.7.2 *Amorphous alloys*

In addition to the use of milling methods, amorphous alloys can also be produced using alternative synthesis routes. The methods include the rapid quenching of a melt, thermal evaporation, sputtering, electrodeposition and ion implantation [522]. Many amorphous materials produced using these techniques can absorb significant amounts of hydrogen. Those of interest for hydrogen storage were summarized in [522] as being of the form Al_yB_y where A is an early transition metal or rare earth and B is a late transition metal, together with $Pd_{1-y}Si_y$.

Examples of the former include $Ti_{1-y}Ni_y$, where $0.40 \leq y \leq 0.67$, $Zr_{1-y}Ni_y$, where $0.30 \leq y \leq 0.90$, $Zr_{1-y}Fe_y$, where y is 0.24 and 0.25, and $Y_{1-y}Fe_y$, where $0.25 \leq y \leq 0.55$. These compounds have maximum observed hydrogen-to-metal atom ratios (H/M) of 1.25, 1.4, 2.46, and 1.8, respectively. Unlike the crystalline intermetallic hydrides, these amorphous alloys tend to exhibit no plateau in their hydrogen absorption and desorption isotherms. This would seem to be a disadvantage in storage applications, in which the relatively large uptakes over small pressure changes in the plateau region are exploited. However, it has been argued that because this indicates the absence of hydride phase formation, the decrepitation that is associated with this process does not occur, which may lead to greater long term cycling stability. However, it is worth noting that the metastable state of these amorphous materials means that they can undergo recrystallization if subjected to too high a temperature. For storage purposes, it would seem that these amorphous alloys would primarily be of interest if they absorb significantly more hydrogen than their crystalline counterparts, but this is only the case for some of these materials, such as $Ti_{1-y}Cu_yH_x$ and $Pd_{1-y}Si_yH_x$, although there appears to be some discrepancies in the literature. See the review of Eliaz and Eliezer [523] for further discussion of these points. Amorphous alloy systems are discussed in more detail by Fukai [524], with a focus on their thermodynamics, structure and the distributions of site energies in these materials, which differ significantly from ordered hydrides. It is worth noting that hydrogen absorption can itself induce amorphization in some originally crystalline materials [524], although this phenomenon does not necessarily have an application in the field of hydrogen storage.

10.4.3.7.3 *Quasicrystals*

Quasicrystals are an interesting class of materials that show long range order but no translational symmetry. Since the first report in 1984, over one hundred different quasicrystals have been synthesized [525,526] and they have recently been reported to occur in nature [526]. Several quasicrystals have been investigated for their hydrogen-storage properties. The chance to achieve enhanced hydrogen-storage capacities in these materials originates from the large number of potential interstitial sites that exist in their structure. It was found [527] that a Ti-based quasicrystal of composition $Ti_{61}Zr_{22}Ni_{17}$ had a hydrogen-storage capacity of 2.8 wt.%, following electrochemical hydrogenation. The capacity of the $Mg_{44}Al_{15}Zn_{41}$ material, meanwhile occurred to be only 0.6 wt.% at 100 bar (10.0 MPa) and temperatures of 573 and 673 K [528]. Furthermore, the material was found to decompose into a $MgZn_2$ phase and MgH_2 following hydrogenation. Although this latter result is not particularly encouraging, further research is likely necessary to determine whether other quasicrystalline compounds could serve as effective HSMs.

10.4.3.8 Complex hydrides

In the case of complex hydrides, atomic hydrogen is bound either ionically or covalently into the bulk of the storage material. It is then released via the decomposition of the host into two or more components. These materials are generally formed from alkali or alkaline earth metals and [AlH_4], [NH_2], and [BH_4] anionic hydrides. As with many of the materials covered in this chapter, complex hydrides were first synthesized many years ago but it is only through recent work that their potential use as practical reversible HSMs has been realized. This has occurred, most notably, in the case of sodium alanate ($NaAlH_4$), due to the discovery of the catalytic enhancement by Ti-doping of the hydrogenation/dehydrogenation process [529], but also by the discovery of the reversible hydrogen sorption properties of the Li–N–H system [530]. The term complex hydride has become an umbrella term encompassing the alanates, nitrides and borohydrides that are currently being considered for hydrogen storage [531,532]. In this section, we cover each of these in turn and also look at the A_2BHx compounds that are classed as complex transition metal hydrides.

10.4.3.8.1 *Alanates*

The prototype HSM in this group is sodium alanate, $NaAlH_4$ [374]. Its structure consists of sodium atoms surrounded by [AlH_4] tetrahedra. During the dehydrogenation process this phase decomposes into an intermediate Na_3AlH_6 phase with an associated release of gaseous hydrogen:

$$3NaAlH_4 \rightarrow Na_3AlH_6 + 2Al + 3H_2 \tag{10.2}$$

A second reaction step then results in further hydrogen evolution:

$$Na_3AlH_6 \rightarrow 3NaH + Al + 3/2\ H_2 \tag{10.3}$$

Reaction (10.2) occurs between approximately 210°C and 220°C (483 K and 493 K), and reaction (10.3) around 250°C (523 K). The dehydrogenation of NaH does not occur until approximately 425°C (698 K) and does not, therefore, play a practical role in the reversible hydrogen-storage process. The two stage reaction results in an isotherm with two plateaus. At 210°C (483 K), plateau pressures of 15.4 and 2.1 MPa were determined [374]. The irreversibility of this reaction, the instability of the hydride, and the slow desorption kinetics showed seemingly that this material has no chance to be considered as a promising storage material. However, unexpectedly in the mid 1990s, the remarkable positive impact of the addition of $TiCl_3$ on the reversibility and the kinetics of the hydrogenation process was discovered [529]. Later, the technology was developed further [374,533]: it was shown that further improvements could be achieved through mechanical mixing of the alanate and the dopant compounds. To date many other effective dopants have been inferred, including $ScCl_3$, $CeCl_3$, and $PrCl_3$ [374]. Ti-doped $NaAlH_4$ readily desorbs hydrogen at temperatures in the region of 120°C (393 K) and can be rehydrogenated at 170°C (443 K) in 15 MPa of hydrogen [374]. Although the role of the Ti catalyst is not yet fully understood, this material is nominated as a candidate for storage applications.

Other alanates that are being considered for hydrogen storage include $LiAlH_4$, $KAlH_4$, $Mg(AlH_4)_2$ and $Ca(AlH_4)_2$, which have gravimetric hydrogen capacities of 10.54, 5.71, 9.27, and 7.84 wt.%, respectively. Continued research in the area will hopefully reveal new alanates, new alanate phases or mixed alanate combinations, as well as new catalysts, particularly if significant progress can be made in understanding the role of the Ti dopant.

10.4.3.8.2 *Nitrides, amides, and imides*

The potential of the Li–N–H system as a potential storage material was first reported in [530]. The system forms three stoichiometric ternary compounds: lithium imide (Li_2NH), lithium amide ($LiNH_2$) and lithium nitride hydride (Li_4NH). The latter, however, is not involved in the proposed reversible hydrogen-storage process. This begins with lithium nitride (Li_3N), which is hydrogenated to form a combination of lithium imide and lithium hydride (LiH).

Further hydrogenation results in the formation of lithium amide and further lithium hydride, and the complete process is represented by the following reaction,

$$Li_3N + 2H_2 \rightarrow Li_2NH + LiH + H_2 \rightarrow LiNH_2 + 2LiH \qquad (10.4)$$

Dehydrogenation of the imide requires high vacuum and temperatures above 600 K [530]. These are unsuitable conditions for reversible hydrogen storage, but the reaction between the imide and the amide is reversible under more moderate conditions of both temperature and pressure. The mixed hydride/imide on the right-hand side of reaction (10.4) gives a high theoretical hydrogen capacity of 10.4 wt.%, but the reversible capacity between the imide and amide is approximately 6.5 wt.% (one H_2 molecule released from $LiNH_2$ + 2LiH).

A number of other similar materials have been studied as hydrogen-storage media, including the ternary compounds $Mg(NH_2)_2$, $RbNH_2$, $CsNH_2$ and Ca–N–H, and the quarternary and higher systems Li–Ca–N–H, Li–Al–N–H, Na–Mg–N–H, Na–Ca–N–H, Mg–Ca–N–H and Li–Mg–Ca–N–H [532]. It can be seen that these materials have the potential to provide high gravimetric storage capacities and are therefore of great interest. However, the Li–N–H system suffers from a number of drawbacks, including the high hydrogenation and dehydrogenation temperatures, and air or moisture sensitivity [534]. Besides contributing to the degradation of samples during long term cycling, the evolution of ammonia is also an issue because it is a very effective poison for PEM fuel cell membranes [535].

10.4.3.8.3 *Borohydrides*

Borohydrides have the highest gravimetric hydrogen-storage capacities of any of the complex hydrides. $LiBH_4$ contains 18.5 wt.% hydrogen, and releases it through one of the following two reactions [536],

$$LiBH_4 \rightarrow Li + B + 2H \tag{10.5}$$

or,

$$LiBH_4 \rightarrow LiH + B + 3/2H_2 \tag{10.6}$$

However, the decomposition temperature is too high for practical purposes.

According to [532], $LiBH_4$ releases three of its four hydrogen atoms upon melting at 280°C (553 K), with an enthalpy of decomposition of –88.7 kJ mol^{-1} H_2. LiH is very stable and its dehydrogenation occurs only above a temperature of 727°C (1000 K). Nevertheless, the dehydrogenation reaction of $LiBH_4$ is a reversible process, although rehydrogenation requires elevated pressures and temperatures of 35.0 MPa at 600°C (873 K) and 20.0 MPa at 690°C (963 K). As for most of the complex hydrides, the hydrogenation mechanism of LiH and B is not yet understood, but it is thought to result either from the intermediate reaction of LiH with B, to form a compound that is subsequently filled with hydrogen, or from the reaction of B and H to form diborane, which then spontaneously reacts with LiH to give the full borohydride [532].

A number of other alkali metal and alkaline earth metal borohydrides have high gravimetric and volumetric hydrogen-storage capacities. For example, the theoretical gravimetric capacities of $NaBH_4$, KBH_4, and $Mg(BH_4)_2$ are 10.6, 7.4, and 14.8 wt.%, respectively, with volumetric capacities of 113.1, 87.1, and 146.5 kg m^{-3}, respectively.

However, as for all of the complex hydrides, their high theoretical capacities mean that further work on these materials would be valuable. It was discussed in Section 10.4.3 the destabilization of complex hydrides via mixing with other compounds, focusing on the use of this strategy for $LiBH_4$ with various additives including hydrides (MgH_2), magnesium salts (MgF_2, MgS_2, and $MgSe_2$), elemental metals (Al), alloys, oxides (TiO_2) and carbon [537]. This offers an interesting route for the modification of borohydrides, as well as other complex hydrides, and for hydrogen-storage applications, and is likely to be the subject of continued research.

10.4.3.9 Complex transition metal hydrides

A number of complex transition metal hydrides have been known for some time to reversibly desorb and absorb hydrogen. Examples of these materials include Mg_2FeH_6, Mg_2NiH_4 and Mg_2CoH_5 [538]. Some of the host metallic constituents form stable intermetallics, such as Mg_2Ni and Mg_2Cu, but most do not. The complex transition metal hydrides include compounds with impressive storage capabilities, including a H/M ratio of 4.5 in the case of $BaReH_9$ and a volumetric capacity of approximately 150 g L^{-1}, twice that of LH, as is the case for Mg_2FeH_6 [538]. Mg_2NiH_4 is the material in this group that has probably received the most attention as a candidate storage material, due in part to its gravimetric hydrogen capacity of 3.6 wt.%. Its structure is now known to consist of tetrahedral NiH_4 complexes [539–541]. Complex transition metal hydrides generally suffer from the same problem as the other complex hydrides currently being considered for hydrogen storage, namely the high absorption and desorption temperatures. Mg_2NiH_4 has an enthalpy of formation of -32.3 kJ mol^{-1}

H and requires temperatures above 520 K for hydrogen desorption, and higher temperatures still for absorption. Progress has, however, been made on the destabilization of this material through mechanical milling [539]. Orimo and Fujii [513,539] attributed the decrease of the hydrogen desorption temperature in mechanically milled Mg_2Ni to the presence of a large proportion of grain boundaries following the milling process. However, the hydrogen-storage capacity was concomitantly decreased to 1.6 wt.% in the milled material.

Although these materials have not recently received as much attention as the alanates, amides and imides, and borohydrides, they are prominent examples of another type of non-interstitial hydride. It seems likely that there are many more, as yet undiscovered, multinary complex hydride compounds that could potentially serve as effective HSMs. Therefore, although they do not necessarily exhibit gravimetric capacities in excess of the US DOE targets, continued work into new complex hydride compounds, such as the complex transition metal hydrides, may yet prove fruitful.

10.4.3.10 Other materials

In the last three sections we have covered a range of potential reversible storage materials grouped into the porous adsorbents, interstitial hydrides and complex hydrides. In this section we look at materials that do not fit readily into these categories, which include the clathrates and ionic liquids. We shall also look at materials that show enhanced hydrogen storage through the mechanism of hydrogen spillover. Although the latter are principally porous adsorbents, the proposed storage mechanism is fundamentally different to molecular physisorption; hence its separate treatment.

10.4.3.10.1 *Clathrates*

Clathrate hydrates are a novel proposed solution to the hydrogen-storage problem. They are inclusion compounds formed from hydrogen-bonded networks of water molecules and have long been known due to their occurrence in natural gas and oil pipelines [542]. The guests can include species such as methane, nitrogen and argon, but the discovery of molecular hydrogen encapsulation [543] has led to the study of clathrates for solid-state hydrogen storage [544]. Although pure hydrogen clathrate hydrates are stable only at high pressures or low temperatures, recent work has shown that they can be stabilized under near-ambient conditions using a promoter such as tetrahydrofuran (THF) [544,545]. There are three clathrate hydrate structures, namely sI, sII, and sH, and the latter two have been studied for their hydrogen-storage properties.

Recently, hydrogen storage in the sH phase has been proposed, with estimated capacities in the region of 1.4 wt.% [546–548]. The stabilization of sH has been achieved using alternative promoters, including methyl tert-butyl ether (MTBE) and 1,1-dimethylcyclohexane (DMCH) [547,548]. The reduced capacities of the stabilized clathrates, and hence the source of some of the controversy, originates from the occupancy of a proportion of the cages in the clathrate structures by the promoter molecules rather than hydrogen. In the case of sII, it was argued that reducing the quantity of THF used to stabilize the clathrate allowed hydrogen to occupy some of the larger cages occupied by THF at higher concentrations, thus allowing significantly higher hydrogen-storage capacities to be achieved [546]. As for many of the materials covered in this chapter, hydrogen storage using both clathrate hydrates and alternative organic clathrates is the subject of ongoing research.

One significant disadvantage of clathrates is the slow rate of clathrate formation.

10.4.3.10.2 *Ionic liquids*

Ionic liquids have been attracting significant amounts of attention due to their use in catalysis, as so-called green solvents, and in a vast array of other applications [547,548]. They are defined as materials composed of cations and anions which melt at or below 100°C (373 K) [547], and their negligible vapor pressures mean that they are environmentally friendly in comparison to the volatile organic solvents that they can potentially replace in many industrial and chemical processes. The potential of imidazolium ionic liquids for hydrogen storage, with a volumetric hydrogen capacity of up to 30 g L^{-1} [549], has been recently reported. However, the dehydrogenation temperature of the Pd/C-catalyzed material was in the region 230 to 300°C (503 to 573 K) and the hydrogenation time was approximately 100 hours. Although this performance does not appear particularly encouraging, it is worth bearing in mind that, according to [547], there are over one million simple ionic liquids, and many more binary and ternary systems. Therefore, further work is certainly required if a meaningful conclusion on the suitability of this interesting class of materials for hydrogen-storage applications is to be drawn.

10.5 Conclusions

In this chapter various approaches to hydrogen storage were covered. As it was shown, the solid-state hydrogen storage is expected to be the most promising issue to solve this very difficult problem. To do this one needs to develop new high technologies in which nanomaterials and nanosystems would have the most valuable contribution. To date, various ideas, materials, and technologies have been tested for appropriateness to create a commercial hydrogen-storage system. As it was shown, the problem concerns the bewildering number of various materials in which the very different ideas are used to achieve the targeted objectives.

Understanding of the approaches and methods used in hydrogen-storage research is an important condition for success in this field. For this reason one should be able to combine different knowledge obtained from various methods and materials for hydrogen storage. The author hopes that this chapter could be helpful to achieve these goals.

Chapter 11

Production of Atomic Photochemical Hydrogen and Photoinjection of Hydrogen in Solids

11.1 Introduction

Hydrogen atoms, due to their small size, can be easily, and often in great quantity, inserted into various solids involving their radical transformations. Inner cavities, pores, structural channels, and interlayers facilitate the accommodation of hydrogen atoms in many solids. Special interest is paid to the process of injection of hydrogen produced by some external influence at room or even at low temperatures, such as an electric field, irradiation by light, electron and ion beams, X-rays, and g-rays.

Among the solids that are adjustable for insertion of hydrogen atoms maximally valent transition metal oxides (TMO) possess a special position. Due to the peculiarities of the oxide structure, hydrogen atoms can be easily inserted into WO_3, MoO_3, and V_2O_5 films forming hydrogen metal bronzes with the general formula $H_xM_yO_z$, where M is a transition metal. This ability of the TMO to arrange great quantities of hydrogen atoms dramatically changing their properties was explored in electrochromism (EC). Interest in this phenomenon was initiated by the pioneering works by Deb [549–551]. Since that

Hydrogen Energy for Beginners
Alexander Gavrilyuk

ISBN 978-981-4364-77-5 (Hardcover), 978-981-4364-78-2 (eBook)
www.panstanford.com

time investigations of the EC in these films have been intensive (see, e.g., [24] and [552] and references therein), owing to the possibility of creating new displays, indicators, and smart windows.

Although photochromism (PC) in the TMO films was the first to be discovered [549–551], its investigations did not receive as much interest as EC because its probable applications were not so clear. Nevertheless, fundamental investigations of PC in the TMO films have continued over years, see reviews [553,554].

For a long time PC was not considered a pronounced effect compared with the EC; the number of inserted guest atoms being much higher in the latter case. First the nature of the PC arising in the MoO_3 thin films, when the oxide surface is excited by photons with energy exceeding the oxide forbidden energy gap, was attributed to the formation of F-centers: oxygen vacancies capture electrons, excited by light from the valence band into the conduction band of the oxide [549,550]; absorption of light arising due to electron optical transitions between these levels and the oxide conduction band. These transitions are caused by photons with energy less than the oxide forbidden energy gap and are observed in the visible and near-infrared ranges.

The hypothesis of F-centers was also applied for similar PC in WO_3 [551] and V_2O_5 thin films [555]. Later it became clear that this model was not suited to explain the variety of experimental data. It was reported that the photochromic sensitivity of MoO_3 films quickly drops down to zero upon lowering the temperature [556], which contradicts the model of a pure electronic mechanism of the color center formation.

Furthermore, it has been shown that the PC in MoO_3 films, as well as in WO_3 and V_2O_5 thin films, can be enhanced drastically via adsorption of specially selected organic molecules on the oxide surface [557]. Thus, the reason of the PC is the injection of hydrogen into TMO films under the action of light, that is, photoinjection of hydrogen (PIH).

Obviously, the inalienable stage of the PIH, under the action of light, is detachment of hydrogen atoms from hydrogen-containing molecules previously adsorbed on the oxide surface. In other words, it is the production of atomic photochemical hydrogen that is then inserted into the oxide structure.

At the present time, it is clear that the traditional PC [549–551,555] also arises due to hydrogen atoms, detached under the action of light from hydrogen-containing molecules (water and organics), that have been uncontrollably adsorbed on the films during preparation or from the ambient [558,559] (the "wild" PC). Obviously, this effect is small compared with the "cultivated" PIH where specially selected hydrogen-donor molecules are adsorbed in great quantities on the oxide surface [557,560–585], which makes it possible to insert approximately the same amounts of hydrogen atoms as by the EC.

The PC arising with the PIH is a unique phenomenon. It is not connected with intrinsic defects of solids but the defects being injected from outside, which makes it possible to achieve a record-breaking value of color centers using the PIH. When hydrogen atoms are inserted into the TMO $H_xM_yO_z$ are formed, which are under intensive long-time investigations due to their various interesting properties [586–590 and references therein]. These bronzes can be prepared by various methods, and it has been reported that x can reach the maximum value of ~0.5 in WO_3 and of ~2 in MoO_3 and V_2O_5 which corresponds to the concentration of the injected hydrogen atoms within the limits 10^{21}–10^{22} cm^{-3} [24,552].

In this chapter the main basic principles and ideas important for understanding of the state of art in the PIH are considered; the details of which can be found in the references. The PIH involves various aspects, including the mechanisms of hydrogen transfer between the adsorbed molecules and the oxide surface, the nature of the drastic changes of the optical parameters in the oxide films, the factors influencing the efficiency of the PIH. The PIH is the crossing point of hydrogen science; it is related to production of atomic photochemical hydrogen, its diffusion, reactions, and behavior inside solids. For this reason it makes to elucidate the PIH here, since the issues to be discussed have much in common to those in other hydrogen processes covered above.

First the mechanism of the PIH will be elucidated since the aims of ongoing investigations, besides the PC as such, involve also the production, storage, and transport of hydrogen for different functions, for example, as a fuel, dopant, or catalyst, which may find applications in science and technology, especially, in hydrogen energetics, biology, and medicine.

11.2 Direct Photoinjection of Hydrogen in Transition Metal Oxide Films

The PIH into the oxide films can be carried out by two methods. The first method is connected with photo-initiated direct detachment of hydrogen atoms from hydrogen donor molecules previously adsorbed on the oxide surface. The mostly full information concerning the mechanism of the PIH can be obtained from consideration of the PIH in highly disordered films with very rough and fractal surface.

11.2.1 The Mechanism of Direct Photoinjection of Hydrogen

To achieve highly disordered fractal surface, the TMO films were evaporated directly in dimethylformamide (DMFA)—$(CH_3)_2NCHO$ vapor. Details of the film preparation are given elsewhere [563,570,572–575,577,581–583,585].

Atomic force microscope images for the MoO_3 films are shown in Figs. 11.1 and 11.2 [582]. The scales along *x* and *y* axes are given in μm, whereas the scale along *z* axis is given in nm. The films were found to be nanocomposites, whereas the absence of long-range order was confirmed by X-ray analyses. For this reason, the films are characterized as "quasi-amorphous".

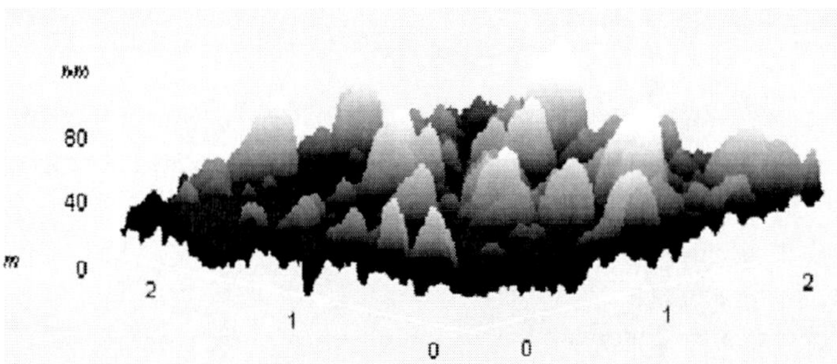

Figure 11.1 Three-dimensional atomic force images of the MoO_3 film surface. Reprinted with permission from Ref. [582], copyright (2007) Taylor & Francis.

The films possess a high porosity with an average of ~100 nm pore diameter and ~200 nm depth, where the adsorbed molecules can freely move, rotate, and bond to the film surface (Figs. 11.1

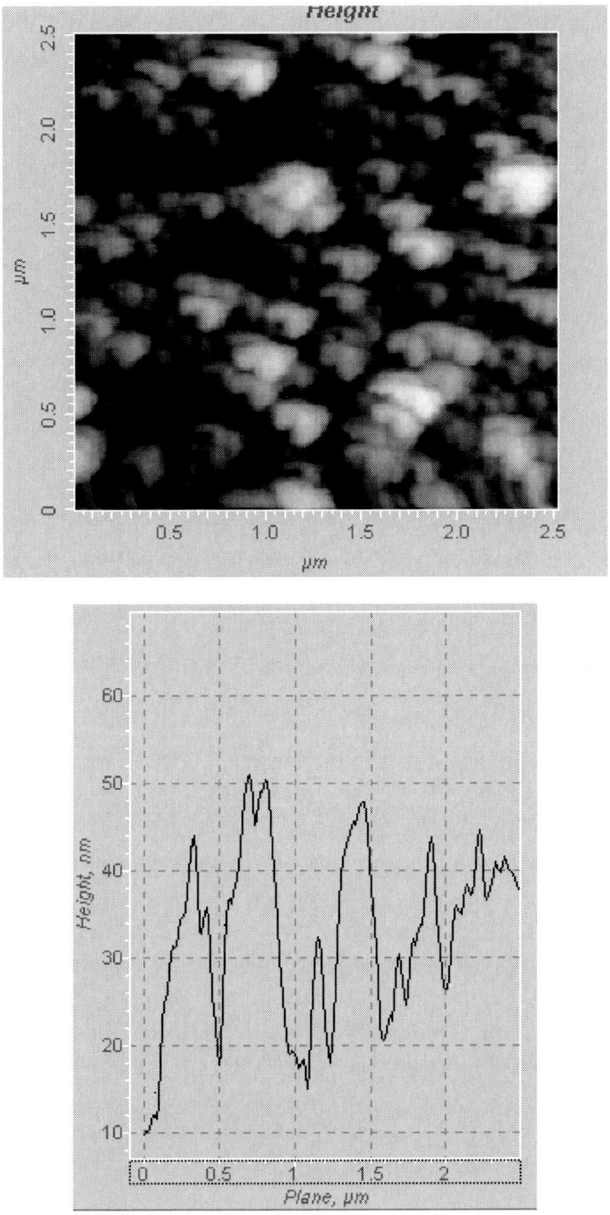

Figure 11.2 Two-dimensional atomic force microscope images of the surface of the MoO_3 films in the *x*–*y* plane (a) and one of the cross-sections in the *x*–*z* plane (b). Reprinted with permission from Ref. [582], copyright (2007) Taylor & Francis.

and 11.2); the surface being essentially heterogeneous and fractal, which predetermines different chemical reactivity for different surface reaction centers. One can observe (Fig. 11.2a) that ~25 nm–diameter clusters have coalesced to form irregular-shaped ~100–200 nm–diameter grains in the MoO_3 films. The investigated films are macroporous with a rough and highly disordered surface.

Similar investigation of the morphology were carried out in WO_3 films prepared under the same technology; all films being also macroporous with a rough and highly disordered fractal surface [582].

The fractal surface with macropores provides additional degrees of freedom for the hydrogen donor molecules, which helps in their adjustment to the surface adsorption centers. The absorbing molecules can find the sites with maximum conformity to their geometrical parameters, which yields the reduction of the energy barriers for hydrogen transfer. The additional degrees of freedom facilitate different types of vibrations for the adsorbed molecules; the fractal surface providing possibilities for fluctuative preparation of the energy barrier for hydrogen transfer.

The scheme elucidating the mechanism of the first stage of the PIH, namely, of dark adsorption for a DMFA molecule on the oxide surface is shown in Fig. 11.3 [581–583]. A surface cation always has in its coordination sphere one or two terminal oxygen atoms O_t with one σ and one π bond (Fig. 11.3) [581–583]. The nucleophilic attack is carried out by a DMFA oxygen atom O_c (carbonyl) in the vicinity of the O_t since the Coulomb repulsion between the O_t and the O_c is smaller compared with that between the bridging oxygen atoms O_{br} and the O_c due to the O_t has only one strong bond. The donor–acceptor (DA) bond is formed by overlapping of vacant *d*-orbitals of the surface cation and *p*-orbital of the O_c with a lone electron pair. The important point is that the *d*-orbital of the surface cation must be vacant. This requirement is fulfilled in d^0 oxides having the maximum valency relative to oxygen (WO_3, MoO_3, V_2O_5, etc). The DA bond couples electronic systems of the oxide and the molecules [581–583].

Proton systems are bonded via hydrogen (H) bonds arising between hydrogen atoms of the molecule and the O_t atoms. To be an effective hydrogen donor these molecules must contain first an oxygen atom and, second, H-C = O_c – or –O_{br}-CH_n- groups, where n = 2, 3 [583]. It is well known that O–H····O bonding is easily achieved, whereas C–H····O-bonding is much less effective owing to the small C–H bond polarity, although, after long discussions, the possibility of the C–H····O-bonding is recognized [32]. The DA-bonding can in turn facilitate the C–H····O-bonding, since the electron density outflow from the oxygen atom of the adsorbed molecule in turn stimulates the electron density outflow to electronegative O_c within the H–C = O_c-group, inducing an additional polarization of the C–H bond, which enhances the inclination for the H-bonding (Fig. 11.3). The C–H···O hydrogen bonding between the oxide surface and organic molecules has already been observed on the surface of nanosized WO_3 particles [591]. Although the H-bond is weak, it plays a very important role determining the direction of the C–H vibration and preparing the system for the hydrogen transfer. Among these two bonds, the main bond is the DA bond, which determines a position of the adsorbed molecules on the oxide surface.

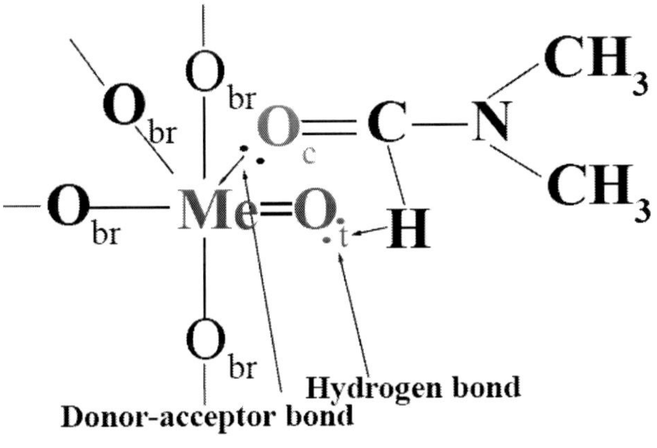

Figure 11.3 Schematic diagram of the bond formation between an adsorbed DMFA molecule and the oxide surface. Reprinted with permission from Ref. [582], copyright (2007) Taylor & Francis.

The formation of the DA bond was confirmed with the help of IR spectroscopy. Figure 11.4 shows the transmission spectra for the C=O absorption band in vapor, liquid, and adsorbed DMFA [583]. The relatively narrow band peaked at 1725 cm^{-1}, attributed to the ν(C=O) stretching vibration in vapor, shits to 1675 cm^{-1} in liquid due to formation of associations [592,593]; the bandwidth being significantly enhanced. The decrease of the ν(C=O) vibration frequency from 1675 cm^{-1} in liquid to 1650 cm^{-1} upon adsorption is an indication of the DA bonding achieved due to a nucleophilic attack of the adsorbed molecule.

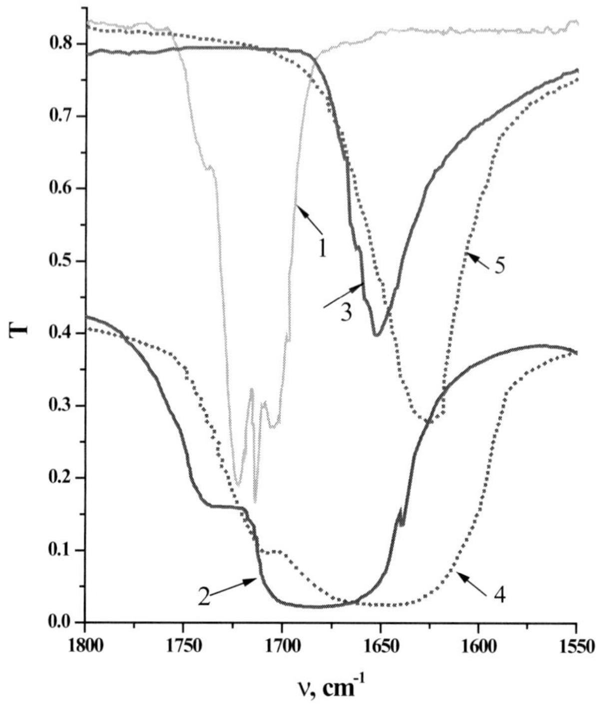

Figure 11.4 Transmission spectra in the vicinity of the C=O stretching vibration frequency for DMFA in various states: 1, gas; 2, liquid; 3, adsorbate; 4, deuterium-containing DMFA, liquid; 5, deuterium-containing DMFA adsorbate. Reprinted with permission from Ref. [582], copyright (2007) Taylor & Francis.

At the same time, the bandwidth for the ν(C=O) is narrower for adsorbed DMFA compared with liquid.

The spectra of the C=O absorption band for DMFA with substitution of all hydrogen atoms for deuterium are also presented. The similar shift from 1650 cm^{-1} in liquid to 1625 cm^{-1} upon adsorption occurs in DMFA-containing deuterium (compare curves 4 and 5 in Fig. 11.4). The O_t atoms are known to be catalytic centers for many reactions of detachment of hydrogen atoms [594,595]. Obviously, definite bond-length conformity should exist to form the hydrogen bond in the adsorbent–adsorbate system. For instance, an O–H bond of length ~1 Å is too short to form the hydrogen bond with the surface double-bonded oxygen, whereas the length of O–C–C–H chain (as in acetone) is too long. For this reason, the chain O–C–H seems to have the most appropriate size and the ability of being adjusted to the oxide surface by varying its bond angles.

To observe hydrogen bonding directly by IR spectroscopy is difficult due to low intensity of the band attributed to the H bond and its non-uniform broadening.

On the other hand, it is possible to obtain evidence of the hydrogen bonding in an indirect way by observing the changes in the frequencies of the M=O bond vibration. In the oxide films, photo-detached hydrogen atoms are often localized near the doubly cation-bonded surface oxygen atoms; this is confirmed by a significant reduction of the W=O vibration frequencies and an enhancement of the band intensities upon the PIH in the WO_3 films [596].

The opposite situation was observed in the MoO_3 films prepared by evaporation in DMFA vapor: the M=O···H-C bond between a DMFA molecule and the O_t atom was broken when a photo-detached proton, released from a DMFA molecule, left its position near the O_t atom and moved to a bridging atom O_{br} (see Fig. 11.3); the Mo=O stretching vibration frequencies being enhanced together with a reduction of the band intensities [596]. Similar effects were observed also in MoO_3 powders upon intercalation of hydrogen atoms [597].

The double (DA and H) bonding yields the dramatic decrease in the energy barrier for the hydrogen transfer between the adsorbed molecule and the oxide surface. Surface cations (Lewis acid centers) ensure perturbation of the molecular electron system, whereas the O_t atoms (Fig. 11.3) play the role of proton detachment centers (Lewis base centers).

The photoreaction mechanism can be understood with the help of a simplified electron energy band diagram for the oxide films [598,599], as shown in Fig. 11.5.

The bonding atomic metal and oxygen orbitals form the valence band, whereas the anti-bonding orbitals form the conduction band. Different types of bonding arise in the TMO: σ-bonds due to overlapping of transition metal hybridized *d*-, *s*-, and *p*-orbitals and *p*-orbitals of an oxygen atom and π-bonds due to overlapping of *d*-cation orbitals and *p*-oxygen orbitals of the doubly cation-bonded oxygen atoms [598,599]. The top of the valence band is formed by states of the oxygen lone-pair electrons, whereas the bottom of the conduction band is formed by vacant *d*-orbitals (non-bonding states). The DA bond between the molecule and oxide surface yields the DA bonding and anti-bonding surface states close to the top of the valence band and bottom of the conduction band, respectively (Fig. 11.5), whereas the hydrogen bond yields perturbation of the non-bonding states of the oxygen lone pair electrons forming a relatively narrow sub-band at the top of the valence band (Fig. 11.5).

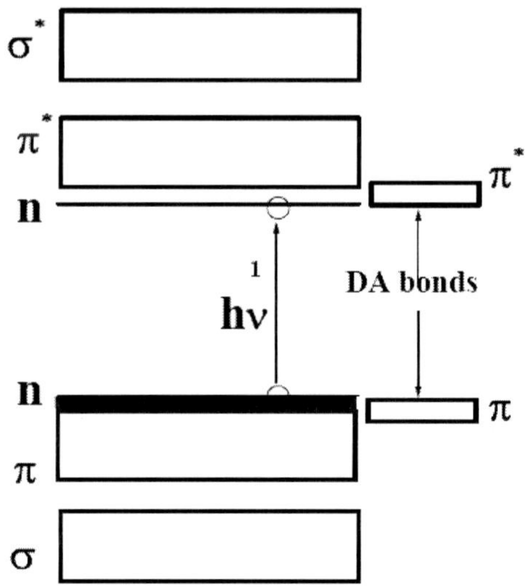

Figure 11.5 The simplified oxide electron energy band diagram. Reprinted with permission from Ref. [582], copyright (2007) Taylor & Francis.

When the surface is illuminated by photons with energy, exceeding the oxide forbidden band gap width, electron-hole pairs are created. If a photogenerated hole reaches the surface, it flows up to the top of the valence band appearing at the O_t, and can be scavenged by an electron of the lone pair of the O_c of the donor molecule, which yields in turn the translocation of the hydrogen bonded proton to the surface oxygen (Fig. 11.3); thus, hydrogen atoms being transferred from the molecule to the surface. The exchange of the hole for the proton makes it impossible recombination of the photogenerated electron; it is localized at a cation forming lower-valency cation.

The reaction may be considered as the hydrogen transfer from the initial U_i to the final U_f potential energy terms along the proton coordinate r with the fixed distance $r_0 = r_{0f} - r_{0i}$ and the salvation parameter q^* corresponding to the initial solvation of the system after appearance of the photogenerated hole in the vicinity of the reaction site (Fig. 11.6); a and b denoting the turning points.

A hydrogen atom can be transferred from the molecule to the oxide surface either in a classical way by overcoming the energy barrier or via tunneling under the barrier. Protons are exceptional particles since they are able to overcome the energy barrier for their transfer not only by classical way but, under appropriate conditions, also by tunneling. ΔE^e is the difference between the ground and excited electronic states at the proton coordinate r^*; the splitting arising due to the possibility of tunneling for the electrons. There is a manifold of localized proton states on the initial side, and a corresponding manifold of states on the final side [600]. Different solvent configurations produce different initial and final energy levels. The scale of the energetic solvent fluctuations is consonant with the proton level separation. For this reason proton can easily find an appropriate excited state on the final state E_f^* to which it can be transferred. The proton transfer may occur between the excited states E_i^* in the initial side and the excited states E_f^* in the final side (Fig. 11.6). The energy released due to the electron-hole recombination can excite the C–H vibration modes or other vibration modes of the oxide and of the adsorbed molecule yielding the changes in the solvent configuration of the proton. The proton is first transferred to the excited state E_i^* and then either tunnels to the final states "cutting" the top of the energy barrier ΔE (Fig. 11.6) or overcomes the barrier in the classical way. Then the relaxation of the solvent configuration follows and the proton occurs in the final ground state E_{of} (Fig. 11.6).

The possibility of the proton tunneling between two potential energy terms separated by the energy barrier causes the splitting of the initial levels for the value E_p, which means that the resonance condition should be fulfilled with an accuracy $\sim E_p$. The transitional configuration q^* ensuring the equalizing of the proton energy levels is characterized by the finite lifetime τ_s during which the proton must tunnel from the initial side to the final side; the reaction of the proton transfer should be adiabatic.

The same holds true also for the adiabatic electron transfer, since the electron should follow in time the motion of the medium dipoles as well as the proton motion. The possibility of tunneling is facilitated here since the electrons tunnel between relatively narrow levels of the molecules into the wide valence band of the oxide (Fig. 11.5). Both levels lay energetically in the same band; fine tuning to the resonance is achieved due to fluctuations in the system.

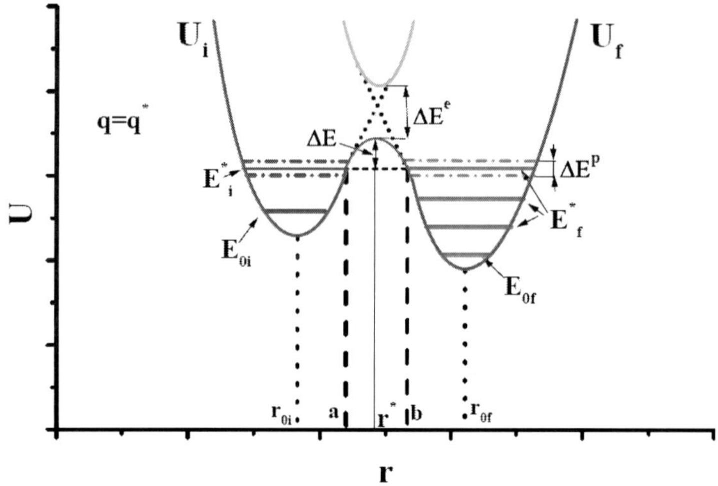

Figure 11.6 Potential energy terms for hydrogen transfer. Reprinted with permission from Ref. [582], copyright (2007) Taylor & Francis.

The probability for the proton transfer from the ground state seems to be low. The C–H···O interaction is weak forming a low-energy hydrogen bond, the C–H···O bond length being ~2–2.5 Å [32]; the probability to eliminate the energy barrier between the initial proton ground state and the final state being is thus very small.

The series of experiments was carried out to determine the reaction mechanism. The reaction rate was calculated by measuring of the optical density D spectra of the films under illumination at various temperatures [582]. A non-zero reaction rate limit was observed in these films, as well as in the WO_3 films obtained by the same technology [581–583], which provides direct evidence for the tunneling character of the reaction [13].

Other types of the oxide films (quasi-amorphous and polycrystalline) prepared at elevated substrate temperatures have been also investigated. In these films the PIH has sometimes a very high rate at room temperature, which it loses quickly with a decrease in temperature; the reaction rate dropping down to a value beyond the sensitivity of our measurements even higher than liquid nitrogen temperatures. For those samples, the plots of the reaction rate versus $1/kT$ exhibited nearly linear dependence of classical Arrhenius type [582].

The other interesting result was a very small kinetic isotope effect inferred only at 20 K obtained with the use of different modification of DMFA [582]. Both the non-zero low-temperature reaction rate limit and the kinetic isotope effect at 20 K show the tunneling character for the hydrogen transfer, whereas the small value for the kinetic isotopic effect shows in turn the small value of the energy barrier for the hydrogen transfer.

The energy barriers differ at different reaction sites of the disordered fractal surface, which yields a distribution for the barrier parameters. Medium dynamics affects the barrier, yielding, sometimes, its total elimination; this process being named in as "fluctuative" preparation of the barrier for chemical reactions [582]. In quantum mechanics, barrier height ΔE and width d determine the reaction probability, and the former depends upon the latter [582]. When the interreagent interactions are strong, the donor–acceptor distance fluctuations are enhanced causing a fluctuation of the barrier height, thereby reducing the effective barrier, sometimes even to zero. To verify whether this model is valid in our case, plots of the reaction rate versus temperature T were made for all samples [582]: all plots exhibited fairly good exponentional dependence, which, according to [13], is an evidence of the fluctuative preparation of the barrier.

The fundamental reasons exist for this reaction mechanism. First, the fluctuative preparation of the barrier is achieved by the topological surface disorder (space fluctuations of the barrier parameters), since the oxide surface is essentially heterogeneous having the variety of reaction sites with different barrier parameters; for a definite number of the centers the highest degree of interaction between the adsorbate and the adsorbent being achieved, which yields the most effective reduction of the barrier. The time-fluctuative preparation of the barrier is achieved by phonon stimulation. At very low temperatures phonon modes excited due to the recombination of the electron photo-excitations provide the time fluctuation of the barrier; in the first turn, the C–H stretching vibration being excited. These fluctuations together with the space fluctuations on the fractal highly disordered surface provide hydrogen transfer at the most low-barrier reaction centers. For centers with higher value of $\sqrt{\Delta E} * d$, the "unfreezing" of other vibration modes adds the contribution to this process at higher temperatures providing the hydrogen transfer. For the reaction cites with the lowest energy barrier, the configuration appropriate for the electron and proton tunneling is achieved even at very low temperatures, whereas for the others it is achieved only at high temperatures. For all films the non-zero low-temperature limit is achieved within the same temperature range ~90 K, which may be is attributed to "unfreezing" of the same intermolecular vibration modes [13].

The oxide structure having very disordered heterogeneous surface with macropores provides high fluctuations of the energy barrier parameters for the hydrogen transfer. It is possible then to make the double-bond interface between the hydrogen donor molecules and the oxide surface on the fractal surface, which reduce the parameters of the energy barrier. A very important point is that the transfer of the electron and proton is carried out to the same oxygen atom, namely, the O_t (Fig. 11.3); the reaction running without charge transfer. The non-uniform broadening of the energy levels due to the high degree of disorder facilitates fulfillment of the requirements for tunneling. In this system, the non-zero low temperature reaction rate limit was inferred. The small kinetic isotope effect and linear dependence of the reaction rate upon temperature showed that the reaction proceeds via tunneling of both quantum particles: the electron and the proton; the reaction being essentially activationless. The reaction mechanism is the fluctuative preparation of the energy barrier: the barrier being eliminated due to unfreezing of the intermolecular vibration modes at higher temperatures.

The experiments showed that the reaction has all the features of the proton-coupled electron transfer (PCET) [600–602], which often occurs as concerted transfer of both particles (electrons and protons). The tunneling PCET on the transition metal oxide surface can be very useful to mimic hydrogen abstraction reactions occurring in biological systems.

11.2.2 Molecules Used as Hydrogen Donors

The organic molecules must satisfy various requirements to be effective hydrogen donors.

Some of these requirements are clear after previous consideration of the mechanism of the PIH.

1. The molecule should contain an oxygen atom or atoms since the structure of electron orbitals on the oxide surface is

inclined to the formation of the DA bond between the oxide surface and the oxygen atom of the molecule.

2. An important requirement for a molecule to be an effective hydrogen donor is to contain H–C=O or –O–CH_n groups (where n = 2, 3) [582,583]. Obviously, definite bond-length conformity should exist to form the hydrogen bond in the adsorbent–adsorbate system.
3. The molecule should have stable radicals after splitting of hydrogen atoms, which usually occurs in compounds having CH_2– or CH_3– groups. Otherwise, the products of the photoreaction can easily recombine decreasing its efficiency. For this reason, water molecules cannot be considered as effective hydrogen donors since the OH^- radical formed when a hydrogen atom is detached from water molecule is very unstable. This radical can easily recombine with a detached hydrogen atom or capture an electron from the arisen lower valency surface cation forming stable OH^- radical. Illumination of the TMO films in water vapor or film preparation by evaporation in a humid atmosphere does not give rise to the noticeable enhancement of photochromic sensitivity. On the other hand, water molecules can be partners for organic (or sometimes inorganic) molecules in many photoreactions, when OH^- radicals can oxidize organic molecules or the products of their decomposition. Water molecules also can facilitate the decomposition of organic molecules, making solvent shells around them [596]. Water molecules also can form a hydrogen bonded layer on the oxide surface, promoting the proton diffusion through it [596].
4. The organic molecules should be able to undergo easy transformations under light illumination forming eventually volatile molecules which can readily desorb from the oxide surface.
5. The organic molecules should be small for effective use of the available oxide surface area. The adsorption of the organic molecules on the TMO surface is usually carried out in one of the two ways. The first way, illumination of the TMO films, is carried out in a sealed, previously evacuated cell filled with organic vapor. This method is especially suitable for volatile substances having a low evaporation temperature (e.g., methanol), since in this case a pressure of several tens

of millibar can be achieved in the sealed cell, which accelerate the adsorption rate and the exchange between methanol and the products of its decomposition. Illumination in vapor is more effective than illumination under a liquid layer of the organic material, since in the first case the cell and the organic substance source can be carefully pumped so as to reduce the oxygen content. The presence of oxygen causes oxidation of hydrogen oxide bronze, decreasing the photochromic sensitivity. Inclination to oxidation increases with hydrogen content in the films and the partial oxygen pressure in the cell.

The other way of organic molecule adsorption is suitable for the organic molecules having rather high evaporation temperature. Due to its technological convenience and other useful properties, DMFA was often used here, as a hydrogen donor. A great amount of DMFA molecules facilitates their capture into the oxide films during evaporation. Being adsorbed on the oxide film surface, non-volatile DMFA molecules remain there indefinitely at room and low temperatures if an access of oxygen is prevented, otherwise the adsorbed DMFA molecules are completely turned into volatile products during several months via surface oxidation reactions. The other important peculiarity of DMFA used in this investigation is the absence of OH groups. The other useful quality of DMFA is the capability of playing the role of the probe molecule due to the presence of the C=O carbonyl group, which makes it possible to observe formation of the DA bond between the adsorbed molecule and the oxide surface by IR spectroscopy. The DA bond arises from a lone electron pair of an oxygen atom of the C=O group; a common orbital being formed due to an overlap of an orbital of the loan electron pair and a vacant *d*-orbital of the surface cation, which yields the C=O bond lengthening, a decrease in the π-bond electron density within the C=O group, a decrease in the bond force constant and, in turn, a decrease of the C=O stretching vibration frequency [603].

The choice of the adsorption method depends also upon the oxide used for the PIH. For example, for WO_3 films both methods are acceptable, whereas for the quasi-amorphous MoO_3 and V_2O_5 films, the adsorption of organic molecules (hydrogen donors) must be carried out immediately after the film preparation, that is, before exposure of the prepared films to air. Otherwise, ambient water molecules adsorb on the oxide surface blocking the adsorption centers intended for the further the hydrogen donor molecules. Since the adsorption energy for water molecules is high enough, they do not exchange for the hydrogen donor molecules during their storage in the vapor of volatile hydrogen donors (methanol or ethanol).

11.3 Changes in the Optical Parameters of the Oxide due to Photoinjection of Hydrogen

11.3.1 Method of Illumination

The thin films were illuminated from a 250 W power mercury lamp with a quartz envelope whose spectrum of irradiation contains a series of bands in the UV range [604]. For the sake of acceleration of the experiments, the full lamp output was often used, with the samples being illuminated through a water filter to get rid of IR components of the radiation. The special measures were undertaken to prevent heating of the samples, with the highest temperature not exceeding 45°C during illumination.

11.3.2 The Nature of the Absorption Arising with the PIH

The PIH in the TMO yields dramatic changes of their characteristics. In this section, the changes of the optical parameters which yield the PC will be considered, although, there are publications concerning the changes of other parameters, among those electrical or structural parameters should be mentioned.

The changes in the optical and electrical parameters arise due to the injected electrons, but the injected protons also could play important role in the changes of the optical parameters, especially in V_2O_5 films. In the "perfect" oxides all cations have maximal valency, which is +6 in WO_3 and MoO_3, and +5 in V_2O_5. The injected electrons are trapped by cations W^{+6}, Mo^{6+}, or V^{5+}, which turn into lower valency cations W^{+6}, Mo^{6+}, or V^{5+}, correspondingly. The general reason for the arising absorption is then the electron transfer between non-isovalent cations under the action of light. For example, for WO_3, it is possible to write:

$$W^{6+}(A) + W^{5+}(B) \leftrightarrow W^{5+}(A) + W^{6+}(B),$$

where A and B neighboring cations.

Meanwhile, to create an adequate model of the absorption appeared to be a difficult task. The nature of color centers arising in TMO upon insertion of hydrogen atoms is one of the most important issues for both the EC and the PC. Although several models have been proposed to describe the absorption arising at insertion of hydrogen atoms (small polaron, intervalence charge transfer, F-center), no one has been able to satisfy the variety of experimental data.

The TMO have strong electron–phonon interaction and are inclined to formation of polarons [24,552]. Thus, when the concentration of the injected electrons is low, the polaron model is valid [608]. But at a high concentration of the injected electrons, the electron–electron interaction must be also taken in account. Very often the PIH is carried out into highly disordered materials, where Anderson localization has a significant income. Various mechanisms of localizations enhance the difficulties in creating of the model. Furthermore one should answer the question whether one type of the absorbing centers or several ones are created at injection of hydrogen atoms.

The author believes that the absorption can be described as a superposition of three bands belonging to different color centers, each described by a Gaussian [573,575,577,579]. There are fundamental physical arguments, which suit the experimental data.

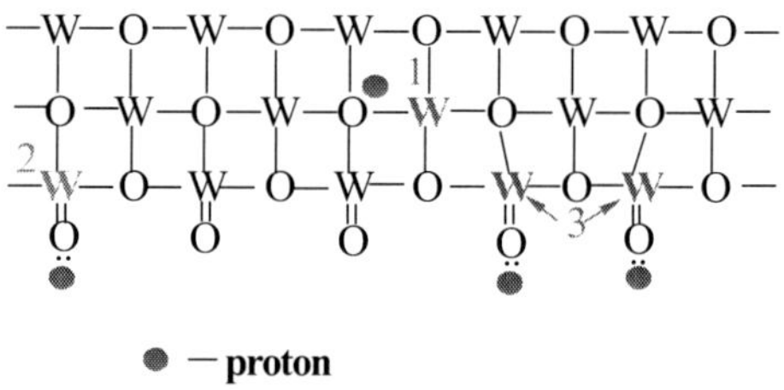

Figure 11.7 Schematic of the centers arising in WO_3 with injection of hydrogen atoms: (1) Bulk paramagnetic cation W^{5+}; (2) Surface paramagnetic cation W^{5+}; (3) Surface center, consisting of two neighboring paired $[W^{5+}W^{5+}]$ cations.

Let us use WO_3 as a model oxide. One of the bands can be assigned to paramagnetic centers localized inside of the oxide bulk (Fig. 11.7, center 1). Besides the bulk center, surface centers can also arise (Fig. 11.7, center 2). Usually surface cations have distorted coordination, sometimes a coordination number being reduced. Oxygen atoms also can have various coordination numbers and the inserted proton can be localized either in the vicinity of terminal or bridging oxygen atoms, giving rise to different configurations of inserted centers. For this reason the coordination of a surface and a bulk low-valency cations can be essentially different, which is reflected in the absorption spectra or in the ESR spectra. It was reported, for example, that the ESR signal arising in the WO_3 or MoO_3 films with the PIH is not elementary and contains two components, belonging to Mo^{5+} or Wo^{5+} cations in different coordination [564,585].

Furthermore, under definite conditions, the energy corresponding to a center with two electrons may be lower than that for two single centers, if the Coulomb repulsion is compensated by lattice distortion. The existence of bipolarons in transition metal compounds was confirmed theoretically and experimentally [606–612]. The pairing of electrons can be either onsite (Anderson bipolarons) or intersite (Heitler–London bipolarons) [606]. When two electrons are localized on one cation, an Anderson bipolaron can arise. In the case of Heitler–London bipolarons, the displacement of neighboring atoms toward each other makes a common bonding orbital where two localized electrons with anti-parallel spins are localized [611]. In TMO, the formation of Heitler–London bipolarons often happens to be energy beneficial compared with that of Andersson bipolarons [607,610]. It has been reported on the diamagnetic bipolarons in non-stoichiometric MoO_3 [613] or in alkaline molybdenum bronzes [614]. The crossover from polarons to bipolarons occurs easier in two-dimensional systems, even for large polarons [615,616]. MoO_3 and V_2O_5 films have a layer structure, which yields to these materials the properties of low-dimensional systems. The formation of the bipolarons was also observed in sodium vanadium bronzes, where it was shown that the Heitler–London bipolarons lay energetically way below the Anderson bipolarons [614]. In the highly disordered oxides, a substantial income of Anderson localization also facilitates the formation of diamagnetic pairs [612]. It is reported on the pairing of the Mo^{5+} paramagnetic centers in MoO_3 milled powder in the temperature range 90–300 K [613]. Thus, favorable conditions are achieved to form the paired diamagnetic centers in highly disordered films. Schematic of the paired center is shown in Fig. 11.7, center 3.

The electronic states appeared due to the formation of the bipolarons lie within the energy band gap; optical electronic transitions between them and the conduction band occur at higher energy, whereas the optical band is much wider than that for unpaired cations. Due to the additional stabilization by stronger lattice distortion, an electron in a bipolaron requires more energy to be transferred to a neighboring undistorted lattice site than a single polaron.

Since the crossover from polarons to bipolarons occurs easier in two-dimensional systems, one would expect their appearance rather on the surface. Due to high income of the Anderson localization, the paired diamagnetic centers can be called "dimers", which reflects the impact of various mechanisms of localization.

The optical density spectra D for various TMO films before and after PIH are in shown Fig. 11.8 along with the fits of the absorption spectra.

A typical absorption band arising with the PIH in quasi-amorphous WO_3 films is shown in Fig. 11.8.

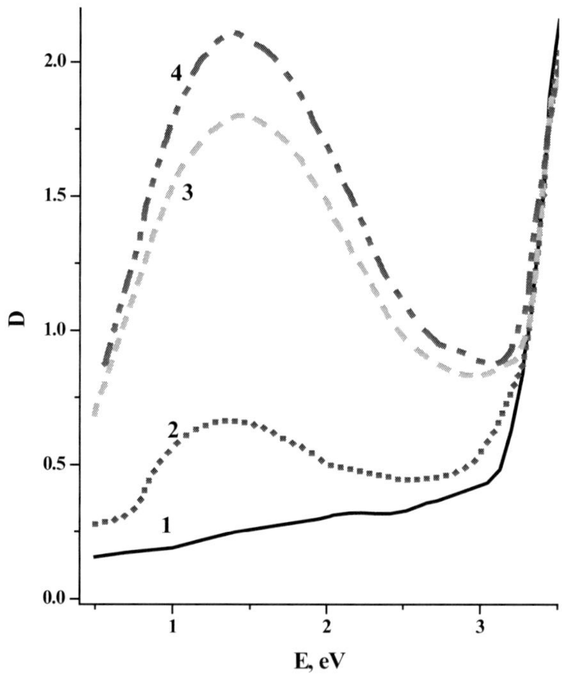

Figure 11.8 Optical density spectra for a quasi-amorphous 0.6 μm thick WO_3 film; (1) before illumination, (2) after illumination, illumination time t = 2 h, (3) t = 9 h, (4) t = 22 h. Illumination under pressure p = 40 mbar pressure of methanol vapor.

One can see that the reason for the PC is the appearance of the wide absorption band peaked at the near-IR range, but spreading into the whole visible range, which yields a blue color into the film.

An example of the arising absorption band and enhancement of the efficiency of the PIH with the use of the special hydrogen donors in polycrystalline WO_3 films evaporated onto substrates heated up to 400°C is presented in Fig. 11.9. These films possess a very high [001] texture [574,576]. For these films the absorption band is noticeably narrower and shifted to lower energy compared with the quasi-amorphous films, which occurs due to a smaller degree of the film disorder.

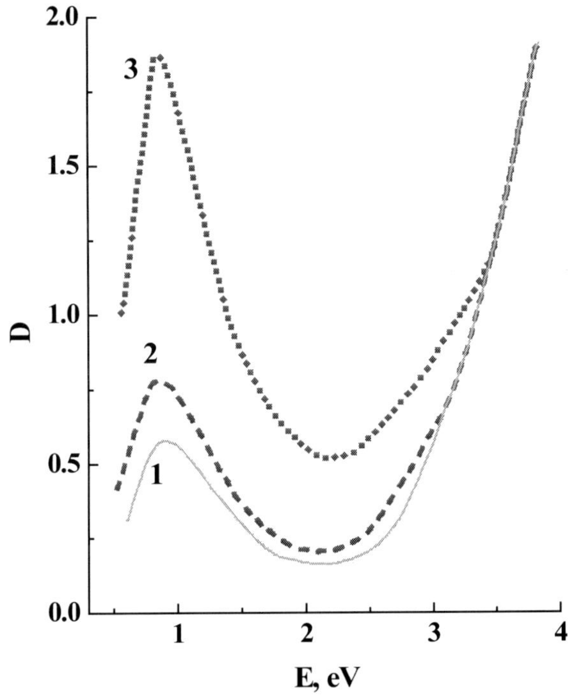

Figure 11.9 The enhancement of the PC in polycrystalline 10 μm thick films. (1) The optical density D spectrum before illumination, (2) after illumination in air, (3) after illumination under a 40 mbar pressure of methanol vapor. Illumination time in both cases was 6 s.

The influence of ordering upon the absorption spectra is illustrated by the fits of the absorption bands arising with the PIH in

WO_3 films prepared under different conditions (Figs. 11.10–11.12) [553]. One can observe that on increasing the film order the income of the bulk centers in the total absorption grows continuously, whereas the share of the surface centers in the total absorption is drastically reduced. However, for WO_3 films, separate bands belonging to different types of centers are not resolved, being detected only by computer simulation.

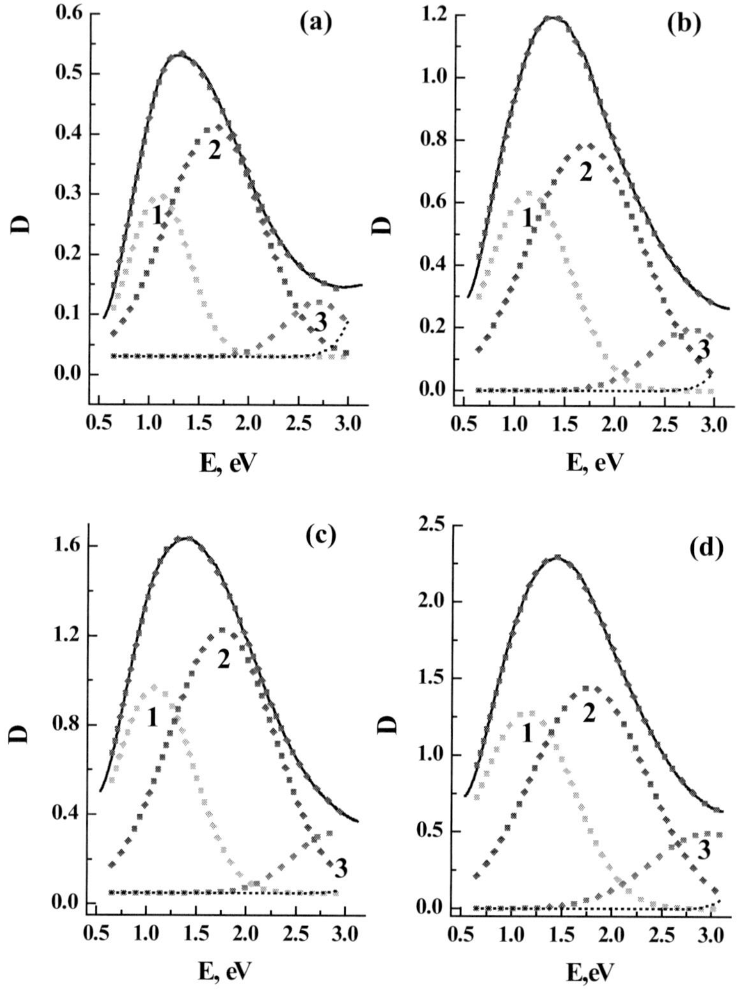

Figure 11.10 Fits of the absorption spectra for quasi-amorphous 0.45 μm-thick WO_3 films prepared by evaporation onto unheated substrates.

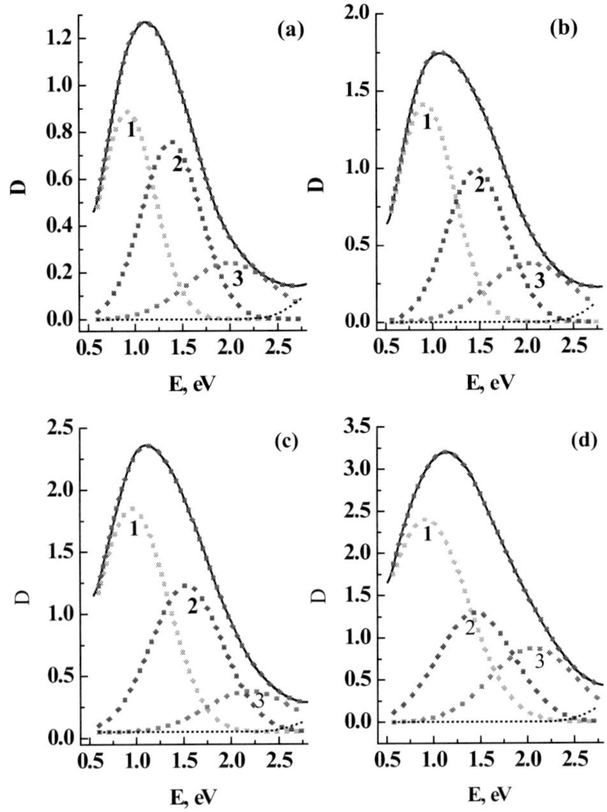

Figure 11.11 Fits of the absorption spectra for polycrystalline 0.6 μm-thick WO_3 films prepared by evaporation onto substrates heated up to 400°C.

On the other hand, these bands are resolved in the MoO_3 films prepared by evaporation in DMFA vapor. The films prepared in one technological run were illuminated under different conditions, either under p = 10^{-2} mbar vacuum, or in air with ~100% relative humidity. The optical density D spectra are shown in Fig. 11.13. Comparing the changes in the optical density spectra caused by the UV-illumination, one can see the great difference. The decomposition of the spectra into three Gaussians (Fig. 11.14) shows that in both cases the same optical bands arise under different condition of the illumination, but the relative income of each type of the centers into the total absorption is different. The fitting of all other spectra gave similar results.

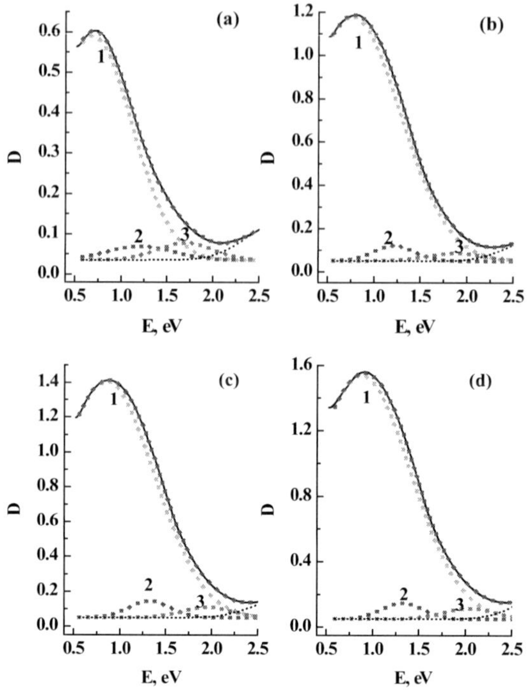

Figure 11.12 Fits of the absorption spectra for polycrystalline 0.4 μm thick WO_3 films prepared by evaporation onto substrates heated up to 200°C and annealed in air at 500°C.

The hydrogen diffusion coefficient for the films with very rough and heterogeneous surface is very sensitive to the type of the adsorbed molecules. When the illumination is carried out under vacuum, the proton surface diffusion is impeded, since DMFA has bad electrolytic properties. This yields high local concentration of the injected color centers. Thus, the favorable conditions for the pairing of the Mo^{5+}_{S} centers are achieved, which simulates the formation of the [$Mo^{5+}Mo^{5+}$] surface diamagnetic centers, and band 3 dominates in the optical density spectra (Figs. 11.13a and 11.14a). When the illumination is carried out in humid air, water molecules adsorbing on the oxide surface, form the net of hydrogen bonds, which serve as specific tunnels for the hydrogen diffusion. IR spectroscopy showed that in the latter case DMFA molecules are quickly oxidized to volatile products and exchanged for water molecules, which facilitates the surface hydrogen diffusion. This yields the higher income for the bands 1 and 2 (Figs. 11.13b and 11.14b). The enhanced mobility of the Mo^{5+} surface centers makes it possible to find the surface sites with smaller energy barriers between the surface and the bulk states, which enhances the probability for the Mo_b^{5+} center formation. The local concentration of the injected surface centers decreases compared with the illumination under vacuum, which in turn reduces the probability for the formation of the $[Mo^{5+}Mo^{5+}]$ surface centers. For this reason, the bands 1 and 2 dominate in the optical density spectra (Figs. 11.13b and 11.14b). The most important achievement of this experiment is that the absorption bands are clearly resolved even without the fitting process.

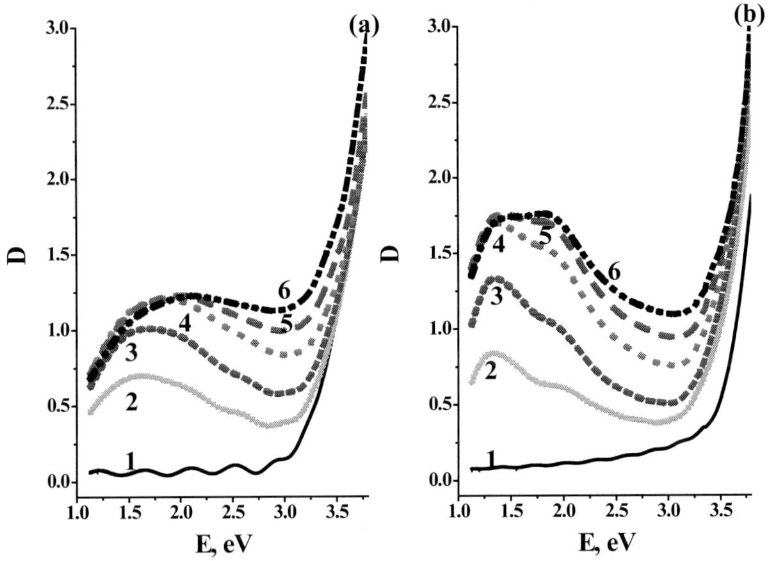

Figure 11.13 The optical density D spectra for the MoO_3 films before and after the illumination: (a) Illumination in vacuum, (b) illumination in humid air, (1) before illumination, (2) after illumination; illumination time t = 1 min, 2 – t = 3 min, 3 – t = 10 min, 4 – t = 20 min, 5 – t = 40 min.

Similar effect can be observed in V_2O_5 films. Figure 11.15 shows the optical density spectra for the V_2O_5 films evaporated in DMFA vapor onto unheated substrates [572]. Besides the spectra in the UV, the visible, and the near-IR band (Fig. 11.15a), Fig. 11.15b contains the IR spectra in the range where the V–O stretching vibrations are revealed.

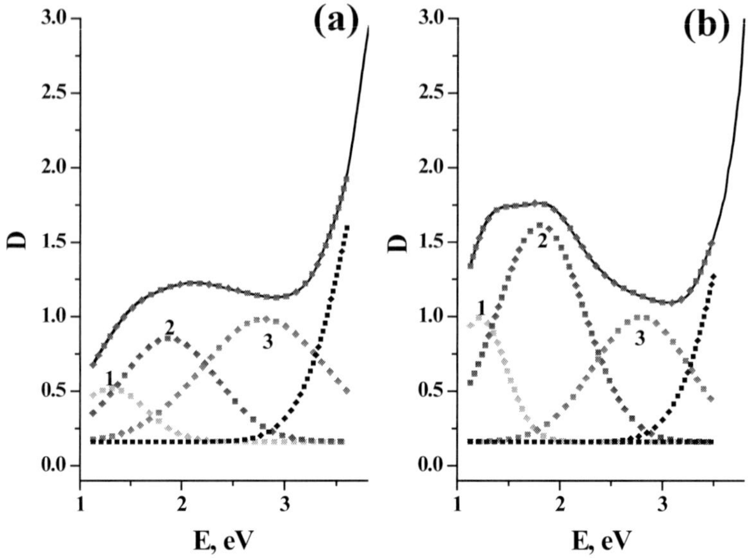

Figure 11.14 Examples of the decomposition of the optical density spectra; the curves 6 in Fig. 11.13a and b being decomposed.

Illumination yields the appearance of a very broad low-intensity band peaked at E = 2.5–2.6 eV. Meanwhile the photochromic V_2O_5 films can be prepared also by splat cooling of V_2O_5 melt by pouring it into water [617]. Figure 11.16 illustrates the PC in these films, illuminated under a 40 mbar pressure of methanol. One can easily see that the arising absorption has a quite different spectrum compared with the films prepared by evaporation in DMFA vapor and illuminated under vacuum (Fig. 11.15a). The reason for such great difference is the same as in the case for the MoO_3 films. Due to the bad overlap of 3d vanadium orbitals, the injected electrons are strongly inclined to localization, especially when the injected protons have difficulties in the surface diffusion being surrounded by DMFA molecules. For this reason predominantly the [$V^{4+}V^{4+}$] paired centers 3 are formed with the PIH in these films. Completely different case occurs in the film obtained by splat cooling. These films contain great amount of adsorbed water which facilitates surface diffusion for the detached protons. Thus, the photogenerated centers diffuse over the surface and penetrate into the bulk. Low inclination to localization yields formation of single electron centers 1 and 2. Figure 11.17 shows some fits which confirm the above.

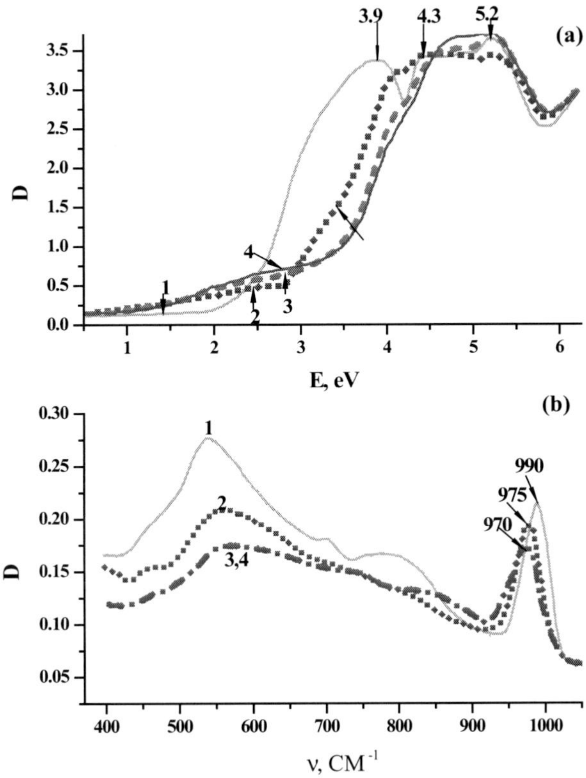

Figure 11.15 Optical density D of spectra for quasi-amorphous 1 μm thick V_2O_5 films evaporated in DMFA vapor before (1) and after illumination under a 10^{-2} mbar vacuum; (2) illumination time $t = 30$ min, (3) $t = 2$ h, and (4) $t = 8$ h [572].

Besides the appearance of the absorption band one can observe the "blue" shift of the fundamental absorption edge in the V_2O_5 films (Figs. 11.15 and 11.16). This shift is especially great in the V_2O_5 films evaporated in DMFA (Fig. 11.15). It is possible to speak about spectral hole "burning" with the PIH in the V_2O_5 films.

Many workers observed the "blue" shift of the fundamental absorption edge upon intercalation of guest atoms in V_2O_5 films [24,552]. There are several reports on the giant shift (~1 eV) achieved either under the action of light [563,572] or electric field [618]. The nature for the shift was attributed to the filling up of the split-off part of the conduction band with the injected electrons [4,5,84], or the Moss–Burstein effect [619,620].

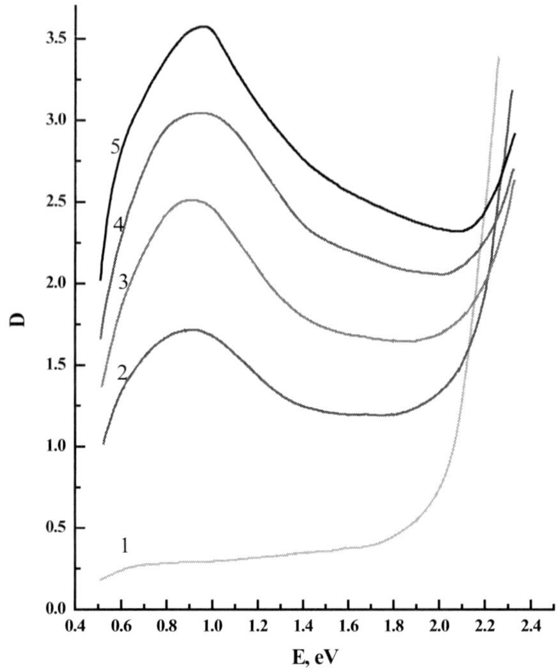

Figure 11.16 Optical density spectra before (1) and after illumination of a 5.5 μm thick V_2O_5 layer prepared by splat cooling of V_2O_5 melt; (2) after illumination, illumination time t = 30 min, (3) t = 1.5 h (4) t = 3 h, and (5) t = 5 h.

Meanwhile one can see (Fig. 11.15a) that the PIH yields total elimination of the absorption in the range 2–4 eV (Fig. 11.15a), and the blue shift achieves of the giant value ~1 eV! One should search the reasons for such great effect in the specific structure and the electron structure of vanadium oxides, which have low extended 3*d* orbitals. V_2O_5 also has a rather complicated structure consisting of zigzag rows consisting of vanadium-oxide pyramids. There are three types of oxygen atoms coordinated by 1, 2, and 3 vanadium atoms, correspondingly [621]. As it was already shown, the oxygen doubly bonded to vanadium (vanadyl) plays a very important role in the adsorption of the hydrogen donor molecules and the reactions of hydrogen transfer. On the other hand the vanadyl bond length is only d = 1.58 Å [621]. The frequency of the stretching vibration of the vanadyl bond is at 990 cm^{-1} (Fig. 11.15b). Due to this the essential π-overlap is achieved between *d*-orbital of the V and *p*-orbital of the vanadyl, but, on the other hand, this overlap can be easily ruined by the V=O bond lengthening.

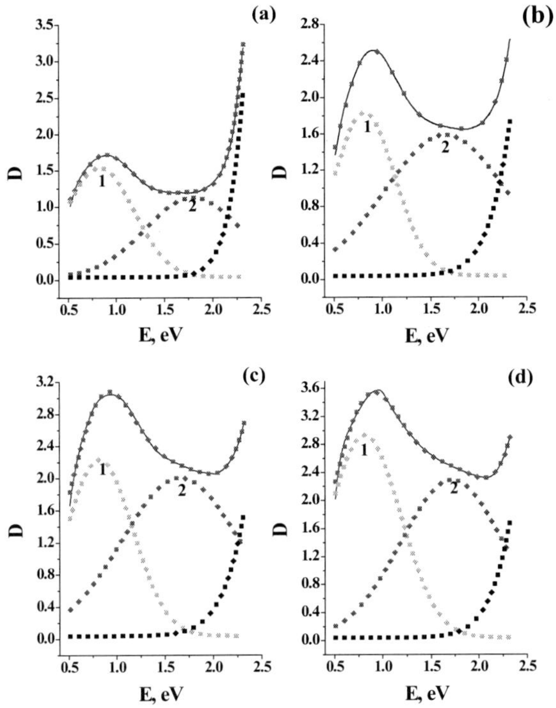

Figure 11.17 Fits of the optical density spectra presented in Fig. 11.16.

The very important effect is the correlation between the shift of the V=O stretching vibration frequency from 990 to 975 cm^{-1} and the "blue" shift of the fundamental absorption edge observed during the PIH (Fig. 11.15b), which is an evidence of the V=O bond lengthening due to the formation of the $V=O\cdots H^+$ bond.

Figure 11.18a shows highly schematically the density of states (DOS) for highly disordered V_2O_5 films, which is based on the pattern energy band diagram discussed before. The bond lengthening reduces drastically the oscillator strength for the V-$O_\pi \rightarrow$ the V-O_{π^*} optical transitions, which yields in turn drastic rearrangement in the energy band diagram (Fig. 11.18). In the initial state, the V and the O_1 orbitals provide the essential π overlap. On insertion of hydrogen, the V–O lengthening due to the $V=O\cdots H^+$ bond formation, diminishes the overlap. Figures 11.18b and c show the changes in the DOS arising with the PIH, which makes clear the reason of the great "blue" shift of the fundamental absorption edge. The diminishing π overlap causes elimination of the resulting π states in the valence bands and π^* states in the conduction band, which results in a decrease in the absorption on increasing *x* (Figs. 11.18b and c).

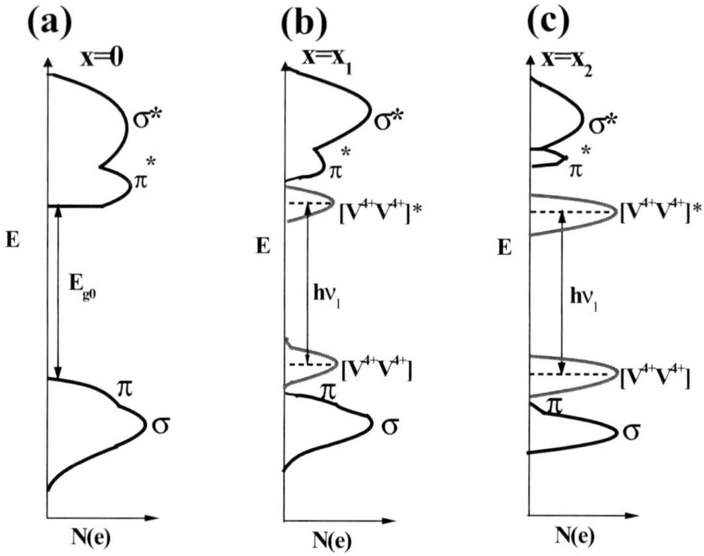

Figure 11.18 Schematic diagram of the density of states in V_2O_5 films.

Thus, the nature of the giant blue shift is spectral hole burning due to elimination of the π states. The overwhelming majority of the injected protons are accommodated near the O_t vanadyl oxygen atoms in the V_2O_5 films evaporated in DMFA vapor, which causes the V = O bond lengthening, deterioration of the π-bonds, and the spectral hole burning. A different situation occurs in the V_2O_5 films prepared by splat cooling; the injected protons here being distributed in the vicinity of various types of oxygen. As a result the shift of the fundamental absorption edge is much smaller in the latter case (compare Figs. 11.15a and 11.16).

It is important to stress that the centers arising in the highly disordered V_2O_5 films are different compared with those in other oxide films (MoO_3, WO_3) or even with the V_2O_5 films prepared by splat cooling. The oscillator strength f was determined for the absorption band peaked at E ~ 2.5 eV from the experiments on Li intercalation in the highly disordered V_2O_5 films [618] where the changes of the optical parameters in highly disordered V_2O_5 films on insertion of Li atoms were investigated since, in the latter case, the absorption spectra of the arising centers are identical to those presented in Fig. 11.16. The amounted value was $f = 0.014$, which is typical for intracenter *d–d* optical transitions [622]. These optical transitions appear due to the displacements of the neighboring V cations toward each other and formation of the common $[V^{4+}V^{4+}]$ bonding and $[V^{4+}V^{4+}]^*$antibonding orbitals. The optical transitions between these orbitals are usually not-allowed, but turn into partly allowed due a low local symmetry of the centers and hybridization of the *d*-orbitals with some other orbitals [622]. These transitions are different compared with those with charge transfer arising in WO_3 and MoO_3 films on insertion of hydrogen atoms, where f being ~0.1 [24,552]. The value of $f \sim 0.01$ and the position of the band maximum at $E \sim 2.5$ eV unambiguously prove formation of the dimers $[V^{4+}V^{4+}]$. The optical transitions observed in the V_2O_5 films prepared by splat cooling are mainly the transitions between non-isovalent cations, thus, having the oscillator strength ~0.1; the absorption band peaking at $E = 0.9$ eV. So, it has been shown that there are different possibilities to control the changes of the optical parameters arising in the TMO films upon insertions of hydrogen atoms. Next section shows the influence of various parameters upon the film photochromic sensitivity.

11.4 Parameters Influencing the Photochromic Sensitivity (Efficiency of the PIH)

The photochromic sensitivity S can be determined as the value reverse to the energy necessary to achieve the photoinduced optical density $D = 1$. Obviously this can be done easier at the wavelength (energy) corresponding to the peak of the band arising under illumination. The value of the optical density at the band maximum $D_{\max}$ and the band half-width W serve also for estimation of the concentration of the injected hydrogen atoms N with the help of Smakula formula [623]:

$$N * f = 0.87 * 10^{17} \frac{n}{(n^2 + 2)^2} \alpha_{\max} W$$

where N is concentration of induced centers in cm^{-3}, f the oscillator strength, n the refractive index, α_{max} the absorption coefficient in the band maximum in cm^{-1}, W the band half-width in eV.

Fig. 11.19a shows the dependences of the maximum photo-induced optical density D_{max} versus illumination time t (characteristic curves) for the WO_3 films prepared by different technologies. These characteristic curves can be used for the presentation of the film sensitivity. With the help of the Smakula formula, the curves in Fig. 11.19a were also recalculated to present the surface concentration N_s of the injected hydrogen atoms (Fig. 11.19b). The value of the oscillator strength was used $f = 0.1$ for the quasi-amorphous films [24,552] and $f = 0.115$ for the polycrystalline films [624], whereas the values of $W = 1.25$ eV were taken for the quasi-amorphous films, $W = 0.9$ for the polycrystalline films prepared at $T_s > 400°C$, and $W = 0.65$ for the films prepared by crystallization of the quasi-amorphous films [579].

11.4.1 Specific Surface Area

Specific surface area is an extensive factor for the photochromic sensitivity. One can easily understand that in the TMO films whose surface area is small (e.g., single crystals or polycrystals with large grain size) the direct PIH is non-effective. For example, WO_3 film prepared by crystallization of amorphous films at $T > 500°C$ for more than 30 min lose practically all photochromic sensitivity. The quasi-amorphous WO_3 film prepared by evaporation to give rise to large surface area also has high photochromic sensitivity. For instance, the quasi amorphous WO_3 film evaporated in a vacuum 10^{-5}–10^{-4} mbar onto an unheated substrate has a specific surface area $\sim 10^5$ cm^2/g [567]. Film evaporation in organic vapor also drastically enhances the surface area, which in turn enhances the photochromic sensitivity [567,576].

11.4.2 Fermi Level Position

The PIH efficiency increases on lowering the Fermi level, since the donor organic molecules play the role of the hydrogen source. For this reason, the PIH efficiency continuously decreases with the concentration of the injected hydrogen atoms, because the Fermi level continuously rises with the concentration of inserted hydrogen species. The presence of oxygen vacancies also raises the Fermi level, reducing the PIH efficiency [567,576].

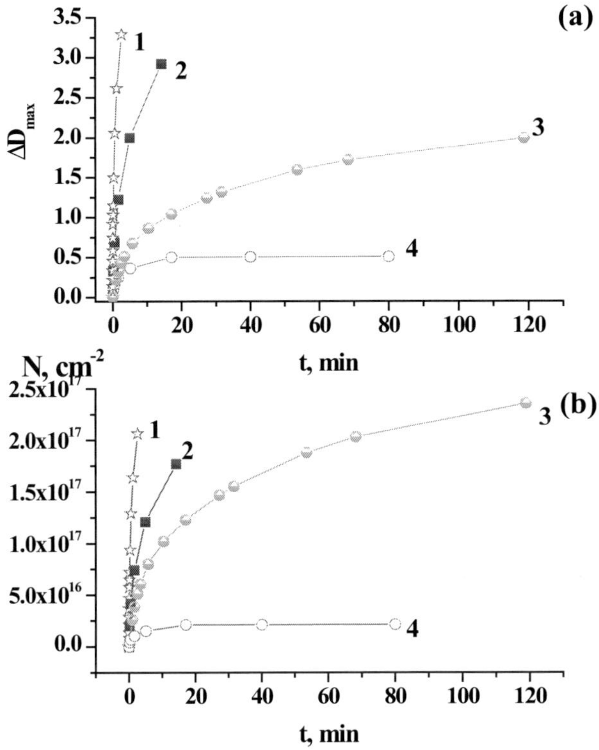

Figure 11.19 Photoinduced optical density ΔD_{max} (a) and surface concentration N_s of injected hydrogen atoms (b) versus illumination time t; (1) for a polycrystalline WO_3 film, texture [001], average grain size 0.5 μm, film thickness d = 11.2 μm; (2) for a polycrystalline WO_3 film, texture [001], average grain size 0.5, film thickness d = 2 μm; (3) for a quasi-amorphous film, film thickness d = 0.7 μm; (4) for a polycrystalline WO_3 film, texture [100], average grain size 15 μm, film thickness d = 2 μm; (2) for a polycrystalline WO_3 film, texture [001], average grain size 15 μm, film thickness d = 2 μm. Illumination was carried out at a 40 mbar pressure of methanol at room temperature.

11.4.3 Proton-Diffusion Coefficient

The photochromic sensitivity is dependent on the proton-diffusion coefficient since diffusion of the injected color centers deeper into the film lowers the Fermi level. On the other hand, diffusion of color centers is determined by the diffusion coefficient $D_H{}^+$ of the slowest particle, that is, proton. However, proton diffusion in the TMO films is a complicated process. Protons can diffuse through the surface and the film bulk. The diffusion through the surface depends on many various parameters, such as (1) pore size, (2) pore shape, (3) character of the pores (open or closed), and (4) state of pores (empty or filled by adsorbed molecules).

For example, water molecules are able to form a partly ordered quasi one-dimensional chain of hydrogen bonds in the layers adjacent to the surface of solids. Proton motion in these layers which are named Bernal–Fowler filaments can be of soliton type [47].

Some alcohols (ethanol and methanol) are also able to form the chain of hydrogen bonds. The Bernal–Fowler filaments formation (or even at least of their fragments) can drastically enhance the hydrogen diffusion coefficient. This possibility in turn depends on whether the pores are open or closed and their size, since it is known that the so-called molecular sieve effects were observed in quasi-amorphous WO_3 films [625]. The film texture can also play an important role since the orientation of the film surface can influence the formation of Bernal–Fowler filaments [47].

The surface reconstruction which is continuously going on in the quasi-amorphous WO_3 films even at room temperature can lead to changes in porosity: some open pores can be transformed to closed ones [625]. This in turn leads to a drop of photochromic sensitivity, possibly due to breaks of Bernal–Fowler filaments. For this reason, as-deposited quasi-amorphous WO_3 films always have a maximum sensitivity which then decreases with time. For polycrystalline WO_3 films, the photochromic sensitivity remains practically unchanged over very long time storage. For the same reason, the photochromic sensitivity for the quasi-amorphous WO_3 films slightly increases and only at long exposures with film thickness when the latter exceeds a value about of 1 μm, that is, the depth of penetration of light (Fig. 11.20b). However, for the crystalline WO_3 films having a grain texture, the film sensitivity increases with thickness even when the latter exceeds dozen of microns (see Fig. 11.20a), since the proton diffusion can occur through the Bernal–Fowler filaments formed near the oriented surface in interlayer grain boundaries. The experiments with both types of the films were carried out using only one line of the lamp emitting profile at E_{max} = 3.4 eV, which

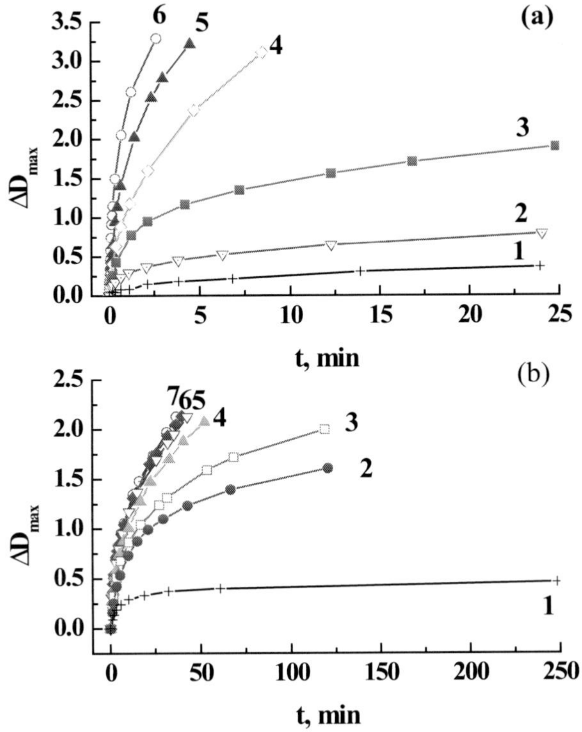

Figure 11.20 Photoinduced optical density ΔD_{max} at the maximum of the optical absorption band versus illumination time t for WO_3 films of various thickness *d*. (a) For polycrystalline WO_3 films, texture [001]; average grain size, 0.5 μm. (1) *d* = 0.35 μm, (2) *d* = 0.65 μm, (3) *d* = 2.1 μm, (4) *d* = 6 μm, (5) *d* = 8.3 μm, (6) *d* = 11 μm. (b) For quasi-amorphous WO_3 films: (1) *d* = 0.2 μm, (2) *d* = 0.47 μm, (3) *d* = 0.7 μm, (4) *d* = 1 μm, (5) *d* = 1.5 μm, (6) *d* = 2.8 μm, *d* = 5.5 μm. Illumination at a 40 mbar pressure of methanol at room temperature.

was discriminated by a filter. The crystalline WO_3 films prepared by evaporation onto a substrate heated to $T > 400°C$ have a very high [001] texture and average grain size about a value of 0.1 μm [576]. They are of a column type being highly oriented in the direction perpendicular to the film surface. A very high degree of the surface orientation facilitates formation of the Bernal–Fowler filaments or, at least, their fragments and enhances the proton-diffusion coefficient. The Bernal–Fowler filaments can be formed by methanol molecules into the grain interlayer boundaries. The proton-diffusion coefficient can be roughly estimated by the formula [626]:

$$D = l^2/4t$$

where l is the film thickness and t is characteristic diffusion time. With values $l \sim 10$ μm and $t \sim 1$ s (Fig. 11.20a), one obtains the value of $D_{H^+} \sim 2.5*10^{-7}$ cm^2/s. Similar estimation for the quasi-amorphous WO_3 film (Fig. 11.20b) with the values $l \sim 3$ μm and $t \sim 1.8*10^3$ s give the value of $D_{H^+} \sim 10^{-11}$ cm^2/s. It should be noticed that the value of D_{H^+} for bulk-type WO_3 is reported to be in the range from 10^{-12} to 10^{-14} cm^2/s [627,628].

Thus, D_{H^+} is a very important factor for photochromic sensitivity. The textured polycrystalline films have a much smaller specific surface area compared with the quasi-amorphous films. But, despite that, this drawback can be overcompensated by very quick diffusion of the injected electron–proton plasma, which can be seen in Fig. 11.19.

On the other hand, the films with different degree of order have different values of the oscillator strength, the refractive index, and the band half-width. For this reason, a pronounced difference can occur between characteristics curves and the dependences of N_s versus illumination time t, which is seen especially clear between the curves.

11.5 Photoinjection of Hydrogen with the help of a Hydrogen Injector

Despite the fact that direct PIH is a very interesting and promising method for changing the TMO film parameters by means of illumination, it is clear that the area of its application is restricted. The PIH can be effective only in films having a great specific surface area, that is, in porous amorphous films and some polycrystalline films with a small grain size. The PIH is really ineffective in single crystals and polycrystalline films with a large grain size. To overcome these restrictions a new method of the PIH was proposed.

For this purpose double-layer structures can be prepared; WO_3 being the upper layer of the double-layer structure, whereas the other layer is a hydrogen acceptor material whose doping with hydrogen atoms should be carried out (Fig. 11.21). Demands of compatibility predetermine the regime of depositing for the WO_3 layer: namely, it must be deposited onto the hydrogen acceptor at room temperature obtaining the quasi-amorphous films.

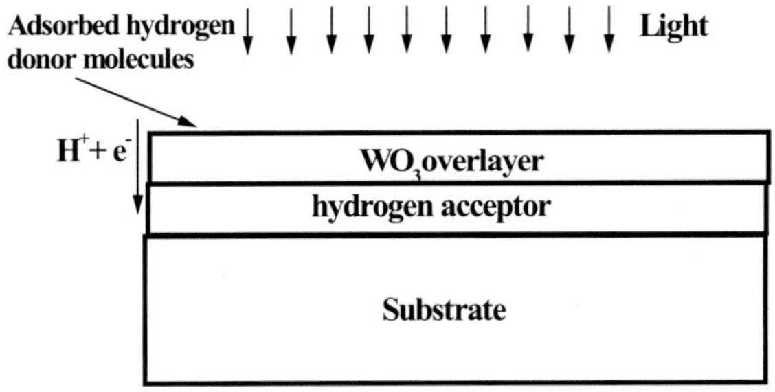

Figure 11.21 A double-layer structure for carrying out the PIH with the help of a hydrogen injector.

These films have very small electron mobility, which ranges within between 0.03 and 0.2 $cm^2/V*sec$ [629]; the electron diffusion coefficient being determined as $D = 0.0025 \pm 0.0006$ cm^2*sec^{-1} [630]. The photogenerated electrons have a great inclination to localization via Anderson, electron–phonon, and electron–electron mechanisms of localization [629].

The electron work function φ for the WO_3 films is $\varphi = 4.3 - 4.9$ eV [630–636]. This is a relatively low value as compared with many other materials.

The electron flow occurs from the WO_3 layer to the hydrogen acceptor when the double-layer structure is produced; the surface of the hydrogen acceptor being negatively charged. This charging prevents injection of the photogenerated electrons into the partner. Thus, the photogenerated electrons have a very small probability to reach the hydrogen acceptor surface. The overwhelming majority of the photogenerated electron-hole pairs recombine. Only small amount of the photogenerated holes reach the grain surface where they can be scavenged by the adsorbed organic molecules (hydrogen donors) and exchanged for protons. The electron conduction in the WO_3 films is variable range hopping [629]. Under continuous illumination, the concentration of the electrons and protons on the WO_3 surface increases, which forces the electron–proton plasma to spread toward the surface of the hydrogen acceptor; the diffusion rate being determined by the diffusion coefficient D of the slowest particle, that is of a proton $D_H{}^+$, which was amounted for the WO_3

films illuminated in methanol vapors as $D_H^+ \sim 10^{-11}$ cm^2/sec [576]. If the WO_3 film is ~1 μm thick, the plasma front will reach the hydrogen acceptor surface in a time:

$$t = d^2/4D$$

where d is film thickness. Thus, the plasma front reaches the partner surface in several minutes after beginning of the illumination.

The consideration presented shows that the WO_3 films have a number of qualities which make it possible to use them for in situ doping of the hydrogen acceptor materials with atomic hydrogen, which in turn yields new interesting effects.

The use of the hydrogen injector (HI) can drastically enhance the number of solids in which the PIH can be achieved solving many scientific and technological problems. It was shown [562], that by the use of the HI, the PIH can be achieved in VO_2 influencing the parameters of the phase transition occurring in this material at 68°C. Direct PIH is impossible in VO_2, since this material has electrons in its *d*-orbital which makes donor–acceptor bond formation impossible between the oxide surface and organic molecules.

The use of the HI makes possible creation of photochromic structures combining the photochromic sensitivity of the quasi-amorphous WO_3 films together with the parameters of the photoinduced state for polycrystalline WO_3 films [566]. The use of the HI makes it possible directly to determine of hydrogen content inserted in the oxide film which is usually a difficult problem. The double-layer film structure consisting of crystalline and amorphous WO_3 layers was prepared. The polycrystalline WO_3 films underwent the heat treatment at 550°C for several hours. Such treatment leads to an increase of the average grain size up to 10 μm, reducing practically to zero the photochromic sensitivity due to a very small specific surface area of the treated films.

The nuclear reaction $^2D(^3He, p)\alpha$ was used to determine deuterium content [624]. Methanol was used as the hydrogen source in which protons were substituted by deuterium atoms. After illumination of the double-layer structure, the quasi-amorphous WO_3 film was etched in a weak alkaline solution and the deuterium content was analyzed by the nuclear reaction. Since the specific surface area of the polycrystalline WO_3 film was very small and the hydrogen background due to adsorption of hydrogen-containing molecules could be neglected, the determined deuterium content was assigned to injected hydrogen which is connected with the optical absorption arising in the film with injection of hydrogen. This also allows determining the value of the oscillator strength for optical transitions in the polycrystalline WO_3 films [598]. The use of the quasi-amorphous WO_3 HI makes it possible to carry out PIH in Ag- and Cu-halide, materials which are basic for investigations of the photographic process [574,584,637]. It was shown that the PIH in these material causes a great enhancement of the so-called "print-out" effect (formation of metallic colloids under illumination). Atomic photochemical hydrogen, detached under the action of light from hydrogen donor molecules, previously adsorbed on the WO_3 surface, migrated through the WO_3 film into the halide film, which provided doping of the halide surface and yielded hydrogen sensitization simultaneous to illumination and yielded the enhancement of PC in the halide films. The atomic hydrogen played the role of a reducing agent and triggered formation of sensitization centers on the halide surface, which in turn facilitated growth of silver clusters and colloids under the action of light. The double-layer structure realizes the idea of two-stage catalysis: first the oxide surface catalyses hydrogen production under the action of light, then the photochemical hydrogen atoms act themselves as catalysts in the photolysis of the halide. The employment of the double-layer structures made it possible to connect two famous photochemical processes: the proton-coupled electron transfer and the photolysis in the silver halides. Also two types of PC have been observed, which made it possible to combine the photochromic in the WO_3 films together with the print-out effect in silver and cuprous halides.

11.6 Pending Application for the Photochromism Induced by Photoinjection of Hydrogen

11.6.1 Photochromic Registration Medium

From the above it is clear that the TMO can be used as photochromic registration mediums for various aims. It was reported on holographic recording in quasi-amorphous MoO_3 films [638]. The TMO films can be used for patterning either directly or using the double-layer structures. The phenomenon can be used for production of nanosized silver and copper particles which in turn can be employed as catalysts in various chemical and biological processes. The examples of the photochromic images are shown in Fig. 11.22.

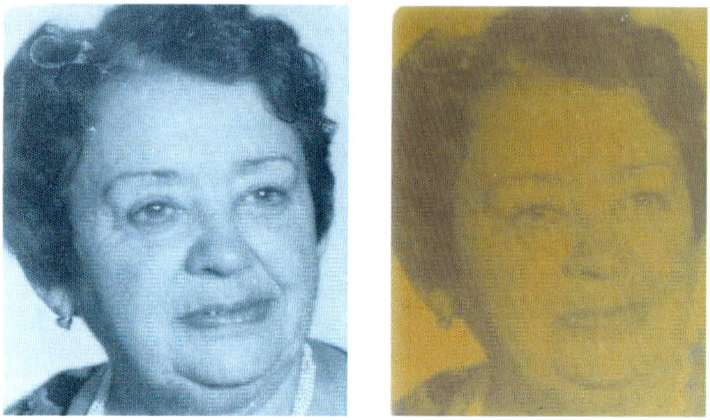

Figure 11.22 Examples of the photochromic images on the MoO_3 (blue) and V_2O_5 (yellow) films.

11.6.2 Registration of X- and γ-Irradiation

Obvious that hydrogen atoms can be split off from the hydrogen donor molecules adsorbed on the TMO surface not only under the action of light but also under high-energy excitation. The splitting occurs very effectively for secondary electrons, which can excite not only the oxide surface but directly the intramolecular bonds, facilitating catalytic transformations and particularly splitting of hydrogen atoms. On the other hand, the contact of the film with ambient oxygen must be excluded since high-energy irradiation produces ozone that can easily oxidize hydrogen bronzes formed under the action of radiation. For this reason all experiments were carried out either under vacuum or with the use of a special hermetic layer [639]. The film of the TMO must have large thickness to absorb large amount of irradiation. Usually, several dozens micron thick MoO_3 films evaporated in DMFA were employed [639]. The films were evaporated onto light and thin substrates (usually of mica) to minimize the changes of the optical parameters of the substrate. Threshold sensitivity was determined as 10^4 R for X-ray irradiation from a X-ray tube with molybdenum anode with voltage of 55 kV and current of 10 mA and 10^6 R for γ irradiation of ^{60}Co [639].

11.6.3 Registration of β-Irradiation

The quasi-amorphous MoO_3 films evaporated in DMFA vapor exhibited a good sensitivity for registration of β-irradiation. A molybdenum plate with titanium surface layer containing tritium was used as a source of β-irradiation. Film thickness ranged from 3 to 4 μm. The experiments were carried in a 10^{-2} mbar vacuum. The sensitivity threshold registered by optical measurements was within the limits 10^{10}–10^{11} β-particles/cm^2 [640].

11.6.4 Visualization of Ionic Beams

The TMO films with adsorbed hydrogen donor molecules can be employed for visualization and registration of various ionic beams; from ionic implantators, linear accelerators, cyclotrons, etc. The series of experiments was carried out with ions of different mass (from protons up to division fragments) and energies (from keV up to GeV). It has been shown that the quasi-amorphous TMO films may be used for this purpose due to relatively high sensitivity and high dynamic range. The edge sensitivity is within the limits from 10^{10} up to 10^{12} particles/cm^2, which depends upon mass and energy of visualized particles [641]. The advantage of these materials is practically linear dependence for the induced optical density against the logarithm of the energy absorbed.

11.7 Conclusions

In this chapter, several important issues concerning the PIH have been considered. They give some basic concepts that are useful for engineering of various photochromic systems based upon photoinjection. The scientific trend concerning the PIH in solids is being developed and there are no doubts that new materials and new ideas will forward research and development in this area.

References

Chapter 1

[1] Komatsu, E., Dunkley, J., Nolta M. R., Bennett C. L., Gold B., Hinshaw, G., et al. (2009) Five-year Wilkinson microwave anisotropy probe observations: Cosmological interpretation, *ApJS*, **180**, 300–376.

[2] Menegoni, E., Galli, S., Bartlett, J. G., Martins, C. J. A. P., and Melchiorri, A. (2009) New constraints on variations of the fine structure constant from CMB anisotropies, *Phys. Rev. D*, **80**, 087302.

[3] Chronology of the universe. http://en.wikipedia.org/wiki/Timeline_of_the_Big_Bang.

[4] Planck epoch. http://en.wikipedia.org/wiki/Planck_epoch.

[5] Cosmic microwave back-ground radiation. http://en.wikipedia.org/wiki/Cosmic_microwave_background_radiation.

[6] Cosmic history. http://en.wikipedia.org/wiki/File:Cosmic_History_020622_b.jpg

[7] Hydrogen line. http://en.wikipedia.org/wiki/Hydrogen_line.

[8] Grochala, W., Edwards, P. (2004) Thermal decomposition of the non-interstitial hydrides for the storage and production of hydrogen, *Chem. Rev.*, **104**, 1283–1315.

[9] Electron orbiting a proton with parallel spins. http://en.wikipedia.org/wiki/File:HydrogenLineAntiParallel.png

[10] Fine and hyperfine structure in hydrogen. http://en.wikipedia. org/wiki/File:Fine_hyperfine_levels.png.

[11] Interstellar gas clouds. http://www.ras.ucalgary.ca/CGPS/press/aas00/pr/pr_14012000/newfilamentdetail4.tif

[12] Bell, R. P. (1973) *The Proton in Chemistry* (Chapman and Hall, London).

[13] Gol'daskii, V. I., Trakhtenberg, L. I., Fleurov, V. N. (1989) *Tunnelling Phenomena in Chemical Physics* (Gordon and Breach Science Publisher, New York).

[14] Pauling, L. (1927) The theoretical prediction of the physical properties of many-electron atoms and ions. Mole refraction, diamagnetic susceptibility, and extension in space, *Proc. R. Soc. A*, 114, 181–211

[15] Hydrogen (H_2) Properties, Uses, Applications: Hydrogen Gas and Liquid Hydrogen. http://www.uigi.com/hydrogen.html.

[16] Tikhonov, V. I., Volkov, A. A. (2002) Separation of water into its ortho and para isomers, *Science*, **296**, 2363.

[17] Milenko, Y. Y., Sibileva, R. M., Strzhemechny, M. A. (1997) Natural ortho-para conversion rate in liquid and gaseous hydrogen, *J. Low Temp. Phys.*, **107**, 77–92.

[18] Ortho-para conversion. http://www.mae.ufl.edu/NasaHydrogen Research/h2webcourse/L11liquefaction2.pdf.

[19] Svadlenak, R. E, Scott, A. B. (1957) The conversion of ortho- to parahydrogen on iron oxide-zinc oxide catalysts, *J. Am. Chem. Soc.*, **79**, 5385–5388.

[20] Hydrogen. http://en.wikipedia.org/wiki/Hydrogen.

[21] Shinitzky, M., Elitzur, A. C. (2006) Ortho–para spin isomers of the protons in the methylene group—Possible implications for protein structure, *Chirality*, **18**, 754–756.

[22] Spin isomers of molecular hydrogen. http://en.wikipedia.org/wiki/File:Spinisomersofmolecularhydrogen. gif

[23] Report from the Commission on Dietary Food Additive Intake. European Union. http://ec.europa.eu/food/fs/sfp/addit_flavor/flav15_en.pdf.

[24] Granqvist, C. G. (1995) *Handbook of Inorganic Electrochromic Materials* (Elsevier, Amsterdam).

[25] Gavrilyuk, A. (2010) Chapter 14: Photochromism in transition metal oxides induced by photoinjection of hydrogen. In: Somani, P. R. (ed.), *Chromic Materials, Phenomena and Their Technological Applications* (Applied Science Innovations Pvt. Ltd., India), pp. 473–508.

Chapter 2

[26] Davydov, A. S. (1979) *Biology and Quantum Mechanics* (Naukova dumka Kiev) (in Russian).

[27] Hankins, D., Woskowitz, J. W., Stillinger, F. H. (1970) Water molecules interactions, *J. Chem. Phys.*, **53**, 4544–4553.

[28] Physics News Preview. The Secret Nature of Hydrogen Bonds. http://swift.cmbi.ru.nl/teach/B2/HTML/hbonds.html.

[29] Isaacs, E. D, Shukla, A. P., Platzman, M., Hamann, D. R., Barbiellini, B., and Tulk, C. A. (1999) Covalency of the hydrogen bond in ice: A direct X-Ray measurement, *Phys. Rev. Lett.*, **82**, 600–603. Bibcode:1999PhRvL.82.600I. Doi:10.1103/PhysRevLett.82.600

[30] Von Brill, R., Tippe, A. (1967) *Gitterparameter von Ice I bei tiefen Temperaturen, Acta Crystallogr.*, **23**, 343–345.

[31] Hamann, D. R. (1997) H_2O hydrogen bonding in density-functional theory, *Phys. Rev. B*, **55**, 10157–10160.

[32] Jeffrey, G. A. (1999) The ups and downs of C-H hydrogen bonds, *J. Mol. Struct.*, **485/486** 293–298.

[33] Desiraju, G. R. (2011) A bond by any other name, *Angew. Chem., Int. Ed.*, **50**, 52–59.

[34] Pauling, L. (1939) *The Nature of the Chemical Bond* (Cornell University Press, Ithaca).

[35] Pimentel, G. C., McClellan, A. L. (1960) *The Hydrogen Bond* (W. H. Freeman, San Francisco).

[36] Hamilton, J W., Ibersd, C. A. (1968) *Hydrogen Bonding in Solids* (W. A. Benjamin, New York).

[37] Jeffrey, G. A., Saenger, W. (1991) *Hydrogen Bonding in Biological Structures* (Springer-Verlag, Berlin).

[38] (2009) *The Nature of the Hydrogen Bond.* In: Gilli, G., Gilli, P. (eds.) (Oxford University Press, Oxford).

[39] Perrin, Ch. L., Nielson, J. B. (1997) "Strong" hydrogen bonds in chemistry and biology, *Annu. Rev. Phys. Chem.*, **48**, 511–544.

[40] Model of hydrogen bonds in water. http://commons.wikimedia.org/wiki/File:3D_model_hydrogen_bonds_in_water.svg

[41] Davydov, A. S. (1986) Solitons in Bioenegetics (Naukova Dumka, Kiev) (in Russian).

[42] Davydov, A. S. (1984) *Solitons in Molecular Systems* (Naukova Dumka, Kiev) (in Russian).

[43] Davydov, A. S. (1991) Solitons and proton motion in hydrogen-bonded molecular structure. In *Mathematics and Its Applications* (Soviet

Series), Chapter 12, *Solitons in molecular systems*, Vol. 61, 301–339 (Springer).

[44] Crystal structure of hexagonal ice. http://en.wikipedia.org/wiki/File:Hex_ice.GIF

[45] Hobbs, P. V. (1974) *Ice Physics* (Oxford, Clarendon Press).

[46] Eigen, M., de Maeyer, L, Spatz, H. C. (1962) *Physics of Ice Crystals* (Colloquium, London).

[47] Antonchenko, V. Ya., Davydov, A. S., Zolotaryuk, A. V. (1983) Solitons and proton motion in ice-like structures, *Phys. Status Solidi (b)*, **115**, 631–640.

[48] Pang, X. F. (2006) The conductivity properties of protons in ice and mechanism of magnetization of liquid water, *Eur. Phys. J. B*, **49**, 5–23.

Chapter 3

[49] Griffiths, D. J. (1987) *Introduction to Elementary Particles* (John Wiley & Sons, New York).

[50] Halzen, F., Martin A. D. (1984). *Quarks and Leptons: An Introductory Course in Modern Particle Physics* (John Wiley & Sons, New York).

[51] Binding energy. http://en.wikipedia.org/wiki/Binding_energy.

[52] Nuclear Fusion in the Sun. http://www.buzzle.com/articles/nuclear-fusion-in-the-sun.html

[53] Bethe, H. A. (1939) Energy production in stars, *Phys. Rev.*, **55**, 434–456.

[54] Krane, K. S. (1987) *Introductory Nuclear Physics* (John Wiley & Sons, New York).

[55] Proton–proton chain reaction. http://en.wikipedia.org/wiki/Proton%E2%80%93proton_chain_reaction.

[56] Christman, J. (2001) "The Weak Interaction". Physnet. Michigan State University. http://physnet2.pa.msu.edu/home/modules/pdf_modules/m281.pdf.

[57] Greiner, W., Müller, B. (2009). *Gauge Theory of Weak Interactions* (Springer-Verlag, Berlin, Heidelberg), p. 2.

[58] Fusion in the Sun. http://en.wikipedia.org/wiki/File:FusionintheSun.svg

[59] Deuterium–tritium_fusion. http://en.wikipedia.org/wiki/File: Deuterium-tritium_fusion.svg

[60] Nuclear fusion. http://en.wikipedia.org/wiki/Nuclear_fusion.

Chapter 4

[61] Smith, A. L. (1997). *Oxford Dictionary of Biochemistry and Molecular Biology* (Oxford: Oxford University Press).

[62] Mullineaux, C. W (1999). The thylakoid membranes of cyanobacteria: Structure, dynamics and function, *Aust. J. Plant Physiol.*, **26**, 671–677.

[63] Chloroplast. http://en.wikipedia.org/wiki/File:Chloroplast.svg

[64] Campbell, N. A., Williamson, B. Heyden, R. J. (2006). *Biology: Exploring Life* (Pearson Prentice Hall, Boston, Massachusetts).

[65] Haddy A. (2010). Oxygen evolution by photosystem II: The Cl^- and Ca^{2+} requirements. http://www.uncg.edu/che/faculty/haddy.html

[66] Knowles, J. R (1980). Enzyme-catalyzed phosphoryl transfer reactions, *Annu. Rev. Biochem.*, **49**, 877–919.

[67] Adenosine triphosphate. http://en.wikipedia.org/wiki/Adenosine_triphosphate

[68] Structure of Adenosine triphosphate. http://en.wikipedia.org/wiki/File:Adenosintriphosphat_protoniert.svg

[69] Törnroth-Horsefield, S., Neutze R. (2008) Opening and closing the metabolite gate", *Proc. Natl. Acad. Sci. USA*, **105** (50): 19565–19566.

[70] Nature's batteries may have helped power early life forms, *Science Daily*, May 25, 2010.

[71] Kalyanasundaram, K., Grätzel M. (2010) Artificial photosynthesis: Biomimetic approaches to solar energy, conversion, and storage, *Curr. Opin. Biotechnol.*, **21**, 298–310.

[72] Nicotinamide adenine dinucleotide. http://en.wikipedia.org/wiki/Nicotinamide_adenine_dinucleotide

[73] Nicotinamide adenine dinucleotide phosphate. http://en.wikipedia.org/wiki/Nicotinamide_adenine_dinucleotide_phosphate.

[74] Structure of nicotinamide adenine dinucleotide. http://en.wikipedia.org/wiki/File:NAD%2B_phys.svg

[75] Reduction and oxidation of the coenzyme NAD. http://en.wikipedia.org/wiki/File:NAD_oxidation_reduction.svg

[76] Light-dependent reactions of photosynthesis at the thylakoid membrane.http://en.wikipedia.org/wiki/File:Thylakoid_membrane.png

[77] Jakubowski, H. Biochemistry, Chapter 8—Oxidation/Phosphorylation D: Photosynthesis: The light reaction. http://biowiki.ucdavis.edu/Biochemistry/Oxidation_and_Phosphorylation/Photosynthesis%3a_the_light_reaction

[78] Blankenship, R. E. *Photosynthetic Antennas and Reaction Centers: Current Understanding and Prospects for Improvement*, Department of Chemistry and Biochemistry Arizona State University, Tempe, AZ 85287–1604. http://bioenergy.asu.edu/photosyn/education/antenna.html.

[79] Blankenship, R. E., Madigan, M. T., Bauer, C. E. (eds.) (1995) *Anoxygenic Photosynthetic Bacteria* (Kluwer Academic Publishers, Dordrecht, The Netherlands).

[80] van Grondelle, R., Dekker, J. P., Gillbro, T., Sundstrom, V. (1994) Energy transfer and trapping in photosynthesis, *Biochim. Biophys. Acta*, **1187**, 1–65.

[81] Klimov, V. V. (1996) Oxidation of water and oxygen release in photosythesis, *Sorosovski Obrazovatelny Zhurnal*, **11**, 9–12 (in Russian).

[82] Photosystem-II. http://en.wikipedia.org/wiki/File:Photosystem-II_2AXT.PNG

[83] Mitchell, P. (1993). David Keilin's respiratory chain concept and its chemiosmotic consequences, *Nobel Lecture, 8 December 1978.* In: Tore Frängsmyr (ed-in-charge), Sture Forsén (ed.), *Nobel Lectures, Chemistry 1971–1980* (World Scientific Publishing Co., Singapore), pp. 295–330.

[84] Danon, A., Stoeckenius, W. (1974) Photophosphorylation in *Halobacterium halobium*, *PNAS*, **71**, 1234–1238.

[85] Oesterhelt, D., Stoeckenius, W. (1973) Functions of a new photoreceptor membrane, *PNAS*, **70**, 2853–2857.

[86] Stoeckenius, W. (1976) The purple membrane of salt-loving bacteria, *SciAm*, **234**, 38–46.

[87] Calvin, M. (1962) The path of carbon in photosynthesis, *Angew. Chem., Int. Ed.*, **1**, 65–75.

[88] Overview of the Calvin Cycle pathway. http://en.wikipedia.org/wiki/File:Calvin-cycle4.svg

[89] Raven, P. H., Ray, F., Evert, S. E. (2005) *Eichhorn Biology of Plants*, 7th Ed. (W. H. Freeman and Company Publishers, New York), 124–127.

Chapter 5

[90] Flavin adenine dinucleotide. http://en.wikipedia.org/wiki/FAD.

[91] Metzler, D. E. (2001) *Biochemistry. The Chemical Reactions of Living Cells*, 2nd Ed. (Harcourt, San Diego).

[92] Flavin mononucleotide. http://en.wikipedia.org/wiki/Flavin_mononucleotide.

[93] Ferguson, S. J., Nicholls, D. Ferguson, S. (2002). *Bioenergetics*, 3rd Ed. (Academic, San Diego).

[94] Mitochondrion of the eukaryotic cell . http://en.wikipedia.org/wiki/File:Animal_mitochondrion_diagram_en_(edit).svg

[95] Bruce, A., Johnson, A., Lewis, J., Raff, M., Roberts, K., Walter, P. (1994). *Molecular Biology of the Cell* (Garland Publishing Inc., New York).

[96] Herrmann, J. M., Neupert, W. (2000) Protein transport into mitochondria, *Curr. Opin. Microbiol.*, **3**, 210–214.

[97] Chipuk, J. E, Bouchier-Hayes, L., Green, D. R. (2006) Mitochondrial outer membrane permeabilization during apoptosis: The innocent bystander scenario, *Cell Death Differ.*, **13**, 1396–1402.

[98] Voet, D., Voet, J. G., Pratt, Ch. W. (2006) *Fundamentals of Biochemistry*, 2nd Ed. (John Wiley and Sons, New York).

[99] King, A., Selak, M. A., Gottlieb, E. (2006) Succinate dehydrogenase and fumarate hydratase: Linking mitochondrial dysfunction and cancer, *Oncogene*, **25**, 4675–4682.

[100] Mitochondrial electron transport chain. http://commons.wikimedia.org/wiki/File:Mitochondrial_electron_transport_chain%E2%80%94Etc4.svg

[101] Kaneko, M. (1993) Artificial photosynthesis as a new energy source, *RIKEN Review (Focused on Light Science and Technology)*, **1**, 41–42.

Chapter 6

[102] Mathews, J., Wang, G. (2009) Metabolic pathway engineering for enhanced biohydrogen production, *Int. J. Hydrogen Energy*, **34**, 7404–7416.

[103] Levin, D. B., Pitt, L., Love, M. (2004) Biohydrogen production: Prospects and limitations to practical application, *Int. J. Hydrogen Energy*, **29**, 173–89.

[104] Nitrogenase. http://en.wikipedia.org/wiki/Nitrogenase.

[105] Goodsell D. (2002) Nitrogenase. *Molecule of the Month* (February). Doi: 10.2210/rcsb_pdb/mom_2002_2

[106] Goodsell D. (2009) Hydrogenase. *Molecule of the Month* (March). Doi: 10.2210/rcsb_pdb/mom_2009_3.

[107] Casalot, L., Rousset, M. (2001) Maturation of the [NiFe] hydrogenases, *Trends Microbiol.*, **9**, 228–336.

[108] *Stigeoclonium* sp. http://upload.wikimedia.org/wikipedia/commons/4/42/Stigeoclonium_sp_zugespitzte_seitenzweige.jpeg

[109] Cyanobacteria. http://en.wikipedia.org/wiki/Cyanobacteria

[110] Gust, D. Biomimetic production of hydrogen. http://www.aps.org/meetings/multimedia/upload/Devens_Gust.pdf.

[111] Ghirardi, M. L., Zhang, L., Lee, J. W., Flynn, T., Seibert, M., Greenbaum, E., Melis A. (2000) Microalgae: A green source of renewable H_2, *Trends Biotechnol.*, **18**, 506–511.

[112] Adams, M. W. W. (1990) The structure and mechanism of iron hydrogenases, *Biochem. Biophys. Acta*, **1020**, 115–145.

[113] Wykoff, D. D., Davies, J. P., Melis, A., Grossman A. R. (1998) The regulation of photosynthetic electron-transport during nutrient deprivation in *Chlamydomonas reinhardtii*, *Plant Physiol.*, **117**, 129–139.

[114] Kosourov, S., Tsygankov, A., Seibert, M., Ghirardi, M. L. (2002) Sustained hydrogen photoproduction by *Chlamydomonas reinhardtii*: Effects of culture parameters, *Biotechnol. Bioeng.*, **78**, 731–740.

[115] Melis, A., Zhang, L., Forestier, M., Ghirardi, M. L., Seibert, M. (2000) Sustained photobiological hydrogen gas production upon reversible inactivation of oxygen evolution in the green alga *Chlamydomonas reinhardtii*, *Plant Physiol.*, **122**, 127–135.

[116] Ghirardi, M. L., Posewitz, M. C., Maness, P.-C., Dubini, A., Yu, J., Seibert, M. (2007) Hydrogenases and hydrogen photoproduction in oxygenic photosynthetic organisms, *Annu. Rev. Plant Biol.*, **58**, 71–91.

[117] Tamagnini, P., Leitao, E., Oliveira, P., Ferreira, D., Pinto, F., Harris, D. J., et al. (2007) Cyanobacterial hydrogenases: Diversity, regulation and applications, *FEMS Microbiol. Rev.*, **31**, 692–720.

[118] Xu, Q., Yooseph, S., Smith, H. O., Venter, C. J. (2005) Development of a novel recombinant cyanobacterial system for hydrogen production from water. In: *Genomics: GTL program projects.* (J. Craig Venter Institute, Rockville, MD), p. 64.

[119] Tamagnini, P., Axelsson, R., Lindberg, P., Oxelfelt, F., Wünschiers, R., Lindblad, P. (2002) Hydrogenases and hydrogen metabolism of cyanobacteria, *Microbiol. Mol. Bio. Rev.*, **66**, 1–20.

[120] Koksharova, O. A., Wolk, C.P. (2002) Genetic tools for cyanobacteria, *Appl. Microbiol. Biotechnol.*, **58**, 123–137.

[121] Miyake, M., Asada, Y. (1997) Direct electroporation of clostridial hydrogenase into cyanobacterial cells. *Biotechnol. Tech.*, **11**, 787–790.

[122] Asada, Y., Koike. Y., Schnackenberg, J., Miyake, M., Uemura, I., Miyake, J. (2000) Heterologous expression of clostridial hydrogenase in the cyanobacterium *Synechococcus PCC7942. Biochim. Biophys. Acta*, **1490**, 269–278.

[123] King, P. W., Posewitz, M. C., Ghirardi, M. L., Seibert, M. (2006) Functional studies of [FeFe] hydrogenase maturation in an Escherichia coli biosynthetic system, *J. Bacteriol.*, **188**, 2163–2172.

[124] Masukawa, H., Mochimaru, M., Sakurai, H. (2002) Disruption of the uptake hydrogenase gene, but not of the bidirectional hydrogenase gene, leads to enhanced photobiological hydrogen production by the nitrogen-fixing cyanobacterium *Anabaena sp. PCC 7120, Appl. Microbiol. Biotechnol.*, **58**, 618–624.

[125] Schopf, J. W. (2000) The fossil record. Tracing the roots of the cyanobacterial lineage. In: Whitton, B. A., Potts, M. (eds.) *The Ecology of Cyanobacteria* (Kluwer Academic Publishers, Dordrecht, The Netherlands).

[126] Hansel, A., Lindblad, P. (1998) Mini-review: Toward optimization of cyanobacteria as biotechnologically relevant producers of molecular hydrogen, a clean energy source, *Appl. Environ. Microbiol.*, **50**, 153–160.

[127] Pinto, F. A. L., Troshina, O., Lindblad, P. (2002) A brief look at three decades of research on cyanobacterial hydrogen evolution, *Int. J. Hydrogen Energy*, **27**, 1209–1215.

[128] Howarth, D. C., Codd, G. A. (1985) The uptake and production of molecular hydrogen by unicellular cyanobacteria, *J. Gen. Microbiol.*, **131**, 1561–1569.

[129] Masukawa, H., Nakamura, K., Mochimaru, M., Sakurai, H. (2001) Photohydrogen production and nitrogenase activity in some heterocystous cyanobacteria. In: *BioHydrogen II* (Elsevier B. V., Amsterdam, The Netherlands), pp. 63–66.

[130] Sveshnikov, D. A., Sveshnikov, N. V., Rao, K. K., Hall, D. O. (1997) Hydrogen metabolism of *Anabaena variabilis* in continuous cultures and under nutritional stress, *FEBS Lett.*, **147**, 297–301.

[131] Arik, T., Gunduz, U., Yucel, M., Turker, L., Sediroglu, V., Eroglu, I. (1996) Photoproduction of hydrogen by *Rhodobacter sphaeroides* OU001. In: Viroglu, T. N., Winter, C. J., Baselt, J. P., Kreysa, G. (eds.), *Proceedings of the 11th World Hydrogen Energy Conference* (Stuttgart, Germany. Frankfurt: Scon & Wetzel GmbH), pp. 2417–2426.

[132] Bolton, J. R. (1996) Solar photoproduction of hydrogen, *Sol. Energy,* **57**, 37–50.

[133] Fedorov, A. S., Tsygankov, A. A., Rao, K. K., Hall, D. O. (1998) Hydrogen photoproduction by Rhodobacter sphaeroides immobilised on polyurethane foam, *Biotechnol. Lett.*, **20**, 1007–1009.

[134] Tsygankov, A. A., Hirata, Y., Miyake, M., Asada, Y., Miyake, J. (1994) Photobioreactor with photosynthetic bacteria immobilized on porous glass for hydrogen photoproduction, *J. Ferment. Bioeng.*, **77**, 575–578.

[135] Tsygankov, A. A., Fedorov, A. S., Laurinavichene, T. V., Gogotov, I. N., Rao K. K., Hall D. O. (1998) Actual and potential rates of hydrogen photoproduction by continuous culture of the purple non-sulphur bacteria Rhodobacter capsulatus, *Appl. Microbiol. Biotechnol.*, **49**, 102–107.

[136] Zurrer, H., Bachofen, R. (1979) Hydrogen production by the photosynthetic bacterium, *Rhodospirillum rubrum*, *Appl. Environ. Microbiol.*, **37**, 789–793.

[137] Fascetti, E., Todini, O. (1995) Rhodobacter sphaeroides RV cultivation and hydrogen production in a one- and two-stage chemostat, *Appl. Microbiol. Biotechnol.*, **22**, 300–305.

[138] Zurrer, H., Bachofen, R. (1982) Aspects of growth and hydrogen production of the photosynthetic bacterium *Rhodospirillum rubrum* in continuous culture, *Biomass*, **2**, 165 –174.

[139] Francou, N., Vignais, P. M. (1984) Hydrogen production by *Rhodopseudomonas capsulata* cells entrapped in carrageenan beads, *Biotechnol. Lett.*, **6**, 639–644.

[140] Vincenzini, M., Materassi, R., Sili, C., Florenzano, G. (1986) Hydrogen production by immobilized cells III. Prolonged and stable H_2

photoevolution by *Rhodopseudomonas palaustris* in light-dark-cycles, *Int. J. Hydrogen Energy*, **11**, 623–626.

[141] Tsygankov, A. A., Fedorov, A. S., Talipova, I. V., Laurinavichene, T. V, Miyake, J., Gogotov, I. N., Rao K. K., Hall D. O. (1998) Application of immobilized phototrophic bacteria for simultaneous waste water treatment and hydrogen photoproduction, *Appl. Biochem. Microbiol.*, **34**, 1–5 (in Russian).

[142] Jouanneau, Y., Lebecque, S., Vignais, P. M. (1984) Ammonia and light effect on nitrogenase activity in nitrogen-limited continuous cultures of Rhodopseudomonas capsulata: Role of glutamate synthetase, *Arch. Microbiol.*, **119**, 326–331.

[143] Willison, J. C., Jouanneau, Y., Michalski, W. P., Colbeau, A., Vignais, P. M. (1983) Nitrogen fixation and H_2 metabolism in the photosynthetic bacterium Rhodopseudomonas capsulate. In: Papageorgiou, G. C., Packer, L. (eds.) *Photosynthetic Prokaryotes: Cell Differentiation and Function* (Elsevier New York), pp. 333–352.

[144] Tsygankov, A. A., Fedorov, A. S., Talipova I. V., Laurinavichene T. V., Miyake, J., Gogotov, I. N. (1998) Use of immobilized phototrophic microorganisms for waste water treatment and simultaneous production of hydrogen, *Appl. Biochem. Microbiol.*, **34**, 362–366 (in Russian).

[145] Champine, J. E., Uffen R. L. (1987) Membrane topography of anaerobic carbon monoxide oxidation in *Rhodocyclus gelatinosus*, *J. Bacteriol.*, **169**, 4784–4789.

[146] Kerby, R. L., Ludden, P. W., Robert, G. P. (1995) Carbon monoxide dependent growth of *Rhodospirillum rubrum*, *J. Bacteriol.*, **177**, 2241–2244.

[147] Uffen, R. L. (1983) Metabolism of carbon monoxide by *Rhodopseudomonas gelatinosa*: Cell growth and properties of the oxidation system, *J. Bacteriol.*, **155**, 956–965.

[148] Wakim, B. T., Uffen, R. L. (1982) Membrane association of the carbon monoxide oxidation system in Rhodopseudomonas gelatinosa, *J. Bacteriol.*, **153**, 571–573.

[149] Maness, P. C., Weaver, P. F. (1997) A potential bioremediation role for photosynthetic bacteria. In: Sikdal, S. K., Irvine, R. L. (eds.) *Bioremediation: Principles and Practice*, vol. II (Technomic Publishing Co, Inc., Lancaster PA).

[150] Uffen, R. L. (1976) Anaerobic growth of a Rhodopseudomonas species in the dark with carbon monoxide ass ole carbon and energy substrate, *Proc. Natl. Acad. Sci. USA*, **73**, 3298–3302.

[151] Watt, A. S., Huang J., Smolinski S., Maness, P. C., Wolfrum, E. J. The water–gas shift pathway in *Rubrivivax gelatinosus* (in preparation).

[152] Dabrock, B., Bahl, H., Gottschalk, G. (1992) Paramenters affecting solvent production by *Clostridium pasteurianum*, *Appl. Environ. Microbiol.*, **58**, 1233–1239.

[153] Melis, T. (2002) Green alga production: Process, challenges, and prospects, *Int. J. Hydrogen Energy*, **27**, 1217–1228.

[154] Polle, J. E. W., Kanakagiri, S., Jin, E., Masuda, T., Melis, A. (2002) Truncated chlorophyll antenna size of the photosystems—a practical method to improve microalgal productivity and hydrogen production in mass culture, *Int. J. Hydrogen Energy*, **27**, 1257–1264.

[155] Gordon, J. (2002) Tailoring optical systems to optimized photobioreactors, *Int. J. Hydrogen Energy*, **27**, 1175–1184.

[156] Laurinavichene, T. V., Tolstygina, I. V., Galiulina, R. R., Ghirardi, M. L., Seibert, M., Tsygankov, AA. (2002) Dilution methods to deprive *Chlamydomonas reinhardtii* cultures of sulphur for subsequent hydrogen photoproduction, *Int. J. Hydrogen Energy*, **27**, 1245–1250.

[157] Tsygankov, A. A., Kosourov S., Seibert, M., Ghirardi, M. L. (2002) Hydrogen photoproduction under continuous illumination by sulphur-deprived, synchronous *Chlamydomonas reinhardtii* cultures, *Int. J. Hydrogen Energy*, **27**, 1139–1144.

[158] Carrasco, C. D., Buettner, J. A., Golden, J. W. (1995) Programmed DNA rearrangement of a cyanobacterial hupL gene in heterocysts, *Proc. Natl. Acad. Sci. USA*, **92**, 791–795.

[159] Lindblad, P., Christensson, K., Lindberg, P., Federov, A., Pinto, F., Tsygankov, A. (2002) Photoproduction of H_2 by wild type Anabaena PCC 1720 and a hydrogen uptake deficient mutant: From laboratory to outdoor culture, *Int. J. Hydrogen Energy*, **27**, 1271–1281.

[160] Ko, I-B., Noike, T. (2002) Use of blue optical filters for suppression of growth of algae in hydrogen producing non-axenic cultures of Rhodobacter sphaeroides RV, *Int. J. Hydrogen Energy*, **27**, 1297–1302.

[161] Kondo, T., Arawaka, M., Wakayama, T., Miyake, J. (2002) Hydrogen production by combining two types of photosynthetic bacteria with different characteristics, *Int. J. Hydrogen Energy*, **27**, 1303–1308.

[162] Lee, C-M., Chen, P-C., Wang, C-C., Tung, Y-C. (2002) Photohydrogen production using purple non-sulfur bacteria with hydrogen fermentation reactor effluent, *Int. J. Hydrogen Energy*, **27**, 1308–1314.

[164] Hoekema, S., Bijmans, M., Janssen, M., Tramper, J., Wijffels, R. H. (2002) A pneumatically agitated Uat-panel photobioreactor with gas recirculation: Anaerobic photoheterotrophic cultivation of a purple non-sulfur bacterium, *Int. J. Hydrogen Energy*, **27**, 1331–1338.

[164] Zhu, H., Ueda, S., Asada, Y., Miyake, J. (2002) Hydrogen production as a novel process of wastewater treatment—studies on tofu waste water with entrapped *R. Sphaeroides* and mutagenesis, *Int. J. Hydrogen Energy*, **27**, 1349–1358.

[165] Das, D., Nejat Veziroglu, T. (2001) Hydrogen production by biological processes: A survey of literature, *Int. J. Hydrogen Energy*, **26**, 13–28.

[166] Kataoka, N., Miya, A., Kiriyama, K. (1997) Studies on hydrogen production by continuous culture of hydrogen-producing anaerobic bacteria. *Proceedings of the Eighth International Conference on Anaerobic Digestion*, vol. 2, Sendai, Japan, 383–390.

[167] Mizuno, O., Dinsdale, R., Hawkes, F. R., Hawkes, D. L., Noike, T. (2000) Enhancement of hydrogen production from glucose by nitrogen gas sparing, *Bioresource Technol.*, **73**, 59–65.

[168] Liang, T.-M., Cheng, S.-S., Wu, K.-L. (2002) Behavioral study on hydrogen fermentation reactor installed with silicone rubber membrane, *Int. J. Hydrogen Energy*, **27**, 1157–1165.

[169] Teplyakov, V. V., Gassanova, L. G., Sostina, E. G., Slepova, E. V., Modigell, M., Netrusov, A. I. (2002) Lab-scale bioreactor integration with active membrane system for hydrogen production: Experience and prospects, *Int. J. Hydrogen Energy*, **27**, 1149–1155.

[170] Chang, J-S., Lee, K-S., Lin, P-J. (2002) BioHydrogen Production with fixed-bed bioreactors, *Int. J. Hydrogen Energy*, 27, 1167–1174.

[171] Microbial electrolysis cell. http://www.rsc.org/chemistryworld/News/2011/February/11021103.asp

[172] Breakthrough for bacterial hydrogen production RSC Advancing the chemical sciences, *Chemistry World*, February 2011.

Chapter 7

[173] Abe, R. (2010) Recent progress on photocatalytic and photoelectrochemical water splitting under visible light irradiation, *J. Photochem. Photobiol. C: Photochem. Rev.*, **11**, 179–209.

[174] Scaife, D. E. (1980) Oxide semiconductors in photoelectrochemical conversion of solar energy, *Sol. Energy*, **25**, 41–54.

[175] Darwent, J. R., Mills, A. (1982) Photo-oxidation of water sensitized by WO_3 powder, *J. Chem. Soc., Faraday Trans.*, 2, **78**, 359–367.

[176] Erbs, W. Desilvestro, J., Borgarello, E., Gratzel, M. (1984) Visible-light-induced oxygen generation from aqueous dispersions of tungsten(VI) oxide. *J. Phys. Chem.*, **88**, 4001–4006.

[177] Sayama, K., Yoshida, R., Kusama, H., Okabe, K., Abe, Y., Arakawa, H. (1997) Photocatalytic decomposition of water into H_2 and O_2 by a two-step photoexcitation reaction using a WO_3 suspension catalyst and an Fe^{3+}/Fe^{2+} redox system, *Chem. Phys. Lett.*, **277**, 387–391.

[178] Kudo, A., Ueda, K., Kato, H. (1998) Photocatalytic O_2 evolution under visible light irradiation on $BiVO_4$ in aqueous $AgNO_3$ solution, *Catal. Lett.*, **53**, 229–230.

[179] Kudo, A., Omori, K., Kato, H. (1999) A novel aqueous process for preparation of crystal form-controlled and highly crystalline $BiVO_4$ powder from layered vanadates at room temperature and its photocatalytic and photophysical properties, *J. Am. Chem. Soc.*, **121**, 11459–11467.

[180] Tokunaga, S., Kato, H., Kudo, A. (2001) Selective preparation of monoclinic and tetragonal $BiVO_4$ with scheelite structure and their photocatalytic properties, *Chem. Mater.*, **13**, 4624–4628.

[181] Yu, J. Q., Kudo, A. (2006) Effects of structural variation on the photocatalytic performance of hydrothermally synthesized $BiVO_4$, *Adv. Funct. Mater.*, **16**, 2163–2169.

[182] Matsumura, M., Saho, Y., Tsubomura, H. (1983) Photocatalytic hydrogen production from solutions of sulfite using platinized cadmium sulfide powder, *J. Phys. Chem.*, **87**, 3807–3808.

[183] Kudo, A. (2006) Development of photocatalyst materials for water splitting. *Int. J. Hydrogen Energy*, **31**, 197–202.

[184] Kudo, A., Nagane, A., Tsuji, I., Kato, H. (2002) H_2 Evolution from aqueous potassium sulfite solutions under visible light irradiation over a novel sulfide photocatalyst $NaInS_2$ with a layered structure, *Chem. Lett.*, **31**, 882–884.

[185] Lei, Z., You, W., Liu, M., Zhou, G., Takata, T., Hara, M., Domen, K., Li, C. (2003) Photocatalytic water reduction under visible light on a novel $ZnIn_2S_4$ catalyst synthesized by hydrothermal method, *Chem. Commun.*, **2003**, 2142–2143.

[186] Zheng, N., Bu, X., Vu, H., Feng, P. (2005) Open-framework chalcogenides as visible-light photocatalysts for hydrogen generation from water, *Angew. Chem., Int. Ed.*, **44**, 5299–5303.

[187] Chen, D., Ye, J. (2007) Photocatalytic H_2 evolution under visible light irradiation on $AgIn_5S_8$ photocatalyst, *J. Phys. Chem. Solids*, **68**, 2317–2320.

[188] Hitoki, G., Ishikawa, A., Kondo, J. N., Hara, M., Domen, K. (2002) Ta_3N_5 as a novel visible light-driven photocatalyst ($\lambda < 600$ nm), *Chem. Lett.*, **31**, 736–738.

[189] Lee, Y., Nukumizu, K., Watanabe, T., Takata, T., Hara, M., Yoshimura, M., Domen, K. (2006) Effect of 10 MPa ammonia treatment on the activity of visible light responsive Ta_3N_5 photocatalyst, *Chem. Lett.*, **35**, 352–354.

[190] Grätzel M. (2001) Photoelectrochemical cells, *Nature*, **414**, 338–344.

[191] Gerisher, H. (1972) Electrochemical techniques for the study of photosensitization, *Photochem. Photobiol.*, **16**, 243–260.

[192] Kiwi, J., Grätzel, M. (1979) Projection, size factors, and reaction dynamics of colloidal redox catalysts mediating light induced hydrogen evolution from water, *J. Am. Chem. Soc.*, **101**, 7214–7217.

[193] Kajiwara, T., Hasimoto, K., Kawai, T., Sakata, T. (1982) Dynamics of luminescence from $Ru(bpy)_3Cl_2$ adsorbed on semiconductor surfaces, *J. Phys. Chem.*, **86**, 4516–4522.

[194] Hashimoto, K., Kawai, T., Sakata, T. (1983) Using dye-sensitized TiO_2 powder, *Nouv. J. Chim.*, **7**, 249–253.

[195] Misawa, H., Sakuragi, H., Usui, Y., Tokumaru, K. (1983) Photosensitizing action of eosin Y for visible light induced hydrogen evolution from water, *Chem. Lett.*, 1021–1023.

[196] Shimizu, T., Iyoda, T., Koide, Y. (1985) Advanced visible-light-induced water reduction with dye-sensitized semiconductor powder catalyst, *J. Am. Chem. Soc.*, **107**, 35–41 .

[197] Mau, A. W.-H., Johansen, O., Sasse, W. H.F. (1985) Xanthene dyes as sensitizers for the photoreduction of water, *Photochem. Photobiol.*, **41**, 503–509.

[198] Heleg, V., Willner, I. (1994) Photocatalysed CO_2-fixation to formate and H_2-evolution by eosin-modified Pd–TiO_2 powders. *J. Chem. Soc. Chem. Commun.*, 2113–2114.

[199] Malinka, E. A., Kamalov, G. L., Vodzinskii, S. V., Melnik, V. I., Zhilina, Z. I. (1995) Hydrogen production from water by visible light using zinc porphyrin-sensitized platinized titanium dioxide, *J. Photochem. Photobiol. A*, **90**, 153–158.

[200] Abe, R., Hara, K., Sayama, K., Domen, K., Arakawa, H. (2000) Steady hydrogen evolution from water on eosin Y-fixed TiO_2 photocatalyst

using a silane-coupling reagent under visible light irradiation, *J. Photochem. Photobiol. A: Chem.*, **137**, 63–69.

[201] Tennakone, K., Bandara, J. (2001) Photocatalytic activity of dye-sensitized tin(IV) oxide nanocrystalline particles attached to zinc oxide particles: Long distance electron transfer via ballistic transport of electrons across nanocrystallites, *Appl. Cat. A: General*, **208**, 335–341.

[202] Kim, Y. I., Atherton, S. J., Brigham, E. S., Mallouk, T. E. (1993) Sensitized layered metal oxide semiconductor particles for photochemical hydrogen evolution from nonsacrificial electron donors, *J. Phys. Chem.*, **97**, 11802–11810.

[203] Saupe, G. B., Mallouk, T. E., Kim, W., Schmehl, R. H. (1997) Visible light photolysis of hydrogen iodide using sensitized layered metal oxide semiconductors: The role of surface chemical modification in controlling back electron transfer reactions, *J. Phys. Chem. B*, **101**, 2508–2513.

[204] Abe, R., Sayama, K., Arakawa, H. (2003) Significant influence of solvent on hydrogen production from aqueous I_3^-/I^- redox solution using dye-sensitized Pt/TiO_2 photocatalyst under visible light irradiation, *Chem. Phys. Lett.*, **379**, 230–235.

[205] Abe, R., Sayama, K., Arakawa, H. (2004) Dye-sensitized photocatalysts for efficient hydrogen production from aqueous I^- solution under visible light irradiation, *J. Photochem. Photobiol. A*, **166**, 115–122.

[206] Abe, R., Shinmei, K., Hara, K., Ohtani, B. (2009) Robust dye-sensitized overall water splitting system with two-step photoexcitation of coumarin dyes and metal oxide semiconductors, *Chem. Commun.*, 3577–3579. Doi: 10.1039/B905935K.

[207] Kudo A., Miseki, Y. (2009) Heterogeneous photocatalyst materials for water splitting, *Chem. Soc. Rev.*, **38**, 253–278.

[208] Kawai, T., Sakata, T. (1979) Hydrogen evolution from water using solid carbon and light energy, *Nature*, **282**, 283- 284.

[209] Kawai, T., Sakata, T. (1979) Production of H_2 and CO from liquid water and carbon using solar energy, *J. Chem. Soc., Chem. Commun.*, 1047–1048. Doi: 10.1039/C39790001047.

[210] Kawai, T., Sakata, T. (1980) Conversion of carbohydrate into hydrogen fuel by a photocatalytic process, *Nature*, **286**, 474–476.

[211] Kawai, T., Sakata, T. (1980) Photocatalytic hydrogen production from liquid methanol and water, *J. Chem. Soc. Chem. Commun.*, 694–695. Doi: 10.1039/C39800000694.

[212] Kawai, T., Sakata, T (1981). Photocatalytic hydrogen production from water by the decomposition of polyvinylchloride, protein, algae, dead insects, and excrement, *Chem. Lett.*, 81–84.

[213] Sakata, T., Kawai, T. (1981) Hydrogen production from biomass and water by photocatalytic processes, *Nouv. J. Chim.*, **5**, 579–581.

[214] Balashev, K. P. Photocatalytic conversion of solar energy, *Sorosovskii Obrazovatelnyi Jhurnal*, N8, 1998, 58–64 (in Russian).

[215] Hara, M., Waraksa, C. C., Lean, J. T., Lewis, B. A., Mallouk, T. E. (2000) Photocatalytic water oxidation in a buffered Tris(2,2′-bipyridyl) ruthenium complex-colloidal IrO_2 system, *J. Phys. Chem. A*, **104**, 5275–5280.

[216] Youngblood, W. J., Lee, S.- H. A., Kobayashi, Y., Pagan, E. A. H., Hoertz, P. G., Moore, T. A., Moore, A. L., Gust, D., Mallouk, T. E. (2009) Photoassisted overall water splitting in a visible light-absorbing dye-sensitized photoelectrochemical cell, *J. Am. Chem. Soc.*, **131**, 926–927.

[217] Brimblecombe, R., Koo, A., Dismukes, G. G., Swieger, G. F., Spiccia, L. (2010) Solar driven water oxidation by a bioinspired manganese molecular catalyst, *J. Am. Chem. Soc.*, **132**, 2892–2894.

[218] Sayama, K., Mukasa, K., Abe, R., Abe, Y., Arakawa, H. (2001) Stoichiometric water splitting into H_2 and O_2 using a mixture of two different photocatalysts and an IO_3^-/I^- shuttle redox mediator under visible light irradiation, *Chem. Commun.*, 2416–2417. Doi: 10.1039/B107673F.

[219] Sayama, K., Mukasa, K., Abe, R., Abe, Y., Arakawa, H. (2002) A new photocatalytic water splitting system under visible light irradiation mimicking a Z-scheme mechanism in photosynthesis, *J. Photochem. Photobiol. A*, **148**, 71–77.

[220] Abe, R., Sayama, K., Sugihara, H. (2005) Development of new photocatalytic water splitting into H_2 and O_2 using two different semiconductor photocatalysts and a shuttle redox mediator IO_3^-/I^-, *J. Phys. Chem. B*, **109**, 16052–16061.

[221] Abe, R., Sayama, K., Domen, K., Arakawa, H. (2001) A new type of water splitting system composed of two different TiO_2 photocatalysts (anatase, rutile) and a IO_3^-/I^- shuttle redox mediator, *Chem. Phys. Lett.*, **344**, 339–344.

[222] Kato, H., Hori, M., Konta, R., Shimodaira, Y., Kudo, A. (2004) Construction of Z-scheme type heterogeneous photocatalysis systems for water splitting into H_2 and O_2 under visible light irradiation, *Chem. Lett.*, **33**, 1348–1349.

[223] Abe, R., Takata, T., Sugihara, H., Domen, K. (2005) Photocatalytic overall water splitting under visible light by TaON and WO_3 with an IO_3^-/I^- shuttle redox mediator, *Chem. Commun.*, 3829–3831.

[224] Kato, H., Sasaki, Y., Iwase, A., Kudo, A. (2007) Role of iron ion electron mediator on photocatalytic overall water splitting under visible light irradiation using Z-scheme systems bull, *Chem. Soc. Jpn.*, **80**, 2457–2464.

[225] Sasaki, Y., Iwase, A., Kato, H., Kudo, A. (2008) The effect of co-catalyst for Z-scheme photocatalysis systems with an Fe^{3+}/Fe^{2+} electron mediator on overall water splitting under visible light irradiation, *J. Catal.*, **259**, 133–137.

[226] Higashi, M., Abe, R., Ishikawa, A., Takata, T., Ohtani, B., Domen, K. (2008) Z-scheme overall water splitting on Modified-TaON photocatalysts under visible light (λ<500 nm), *Chem. Lett.*, **37**, 138–139.

[227] Higashi, M., Abe, R., Teramura, K., Takata, T., Ohtani, B., Domen, K. (2008) Two step water splitting into H_2 and O_2 under visible light by $ATaO_2N$ (A = Ca, Sr, Ba) and WO_3 with IO_3^-/I^- shuttle redox mediator, *Chem. Phys. Lett.*, **452**, 120–123.

[228] Higashi, M., Abe, R., Takata, T., Domen, K. (2009) Photocatalytic overall water splitting under visible light using $ATaO_2N$ (A = Ca, Sr, Ba) and WO_3 in a IO_3^-/I^- shuttle redox mediated system, *Chem. Mater.*, **21**, 1543–1549.

[229] Sasaki, Y., Nemoto, H., Saito, K., Kudo, A. (2009) Solar water splitting using powdered photocatalysts driven by Z-schematic interparticle electron transfer without an electron mediator, *J. Phys. Chem. C*, **113**, 17536–17542.

[230] Tabata, M., Maeda, K., Higashi, M., Lu, D., Takata, T., Abe, R., Domen, K. (2010) Modified Ta_3N_5 powder as a photocatalyst for O_2 evolution in a two-step water splitting system with an iodate/iodide shuttle redox mediator under visible light, *Langmuir*, **26**, 9161–9165.

[231] Maeda, K., Higashi, M., Lu, D., Abe, R., Domen, K. (2010) Efficient nonsacrificial water splitting through two-step photoexcitation by visible light using a modified oxynitride as a hydrogen evolution photocatalyst, *J. Am. Chem. Soc.*, **132**, 5858–5868.

[232] Abe, R., Higashi, M., Domen, K. (2011) Overall water splitting under visible light through a two-step photoexcitation between TaON and WO_3 in the presence of an iodate–iodide shuttle redox mediator, *ChemSusChem*, **4**, 228–237.

[233] Santato, C., Ulmann, M., Augustynski, J. (2001) Photoelectrochemical properties of nanostructured tungsten trioxide films, *J. Phys. Chem. B*, **105**, 936–940.

[234] Kato, H., Kudo, A. (2002) Visible-light-response and photocatalytic activities of TiO_2 and $SrTiO_3$ photocatalysts codoped with antimony and chromium, *J. Phys. Chem. B*, **106**, 5029–5034.

[235] Ishii, T., Kato, H., Kudo, A. (2004) H_2 evolution from an aqueous methanol solution on $SrTiO_3$ photocatalysts codoped with chromium and tantalum ions under visible light irradiation, *J. Photochem. Photobiol. A*, **163**, 181–186.

[236] Konta, R., Ishii, T., Kato, H., Kudo, A. (2004) Photocatalytic activities of noble metal ion doped $SrTiO_3$ under visible light irradiation, *J. Phys. Chem. B*, **108**, 8992–8995.

[237] Hwang, D. W., Kim, H. G., Jang, J. S., Bae, S. W., Ji, S. M., Lee, J. S. (2004) Photocatalytic decomposition of water–methanol solution over metal-doped layered perovskites under visible light irradiation, *Catal. Today*, **93**, 845–850.

[238] Niishiro, R., Kato, H., Kudo, A. (2005) Nickel and either tantalum or niobium-codoped TiO_2 and $SrTiO_3$ photocatalysts with visible-light response for H_2 or O_2 evolution from, *Phys. Chem. Chem. Phys.*, **7**, 2241–2245.

[239] Hwang, D. W., Kim, H. G., Lee, J. S., Kim, J., Le, W., Oh, S. H. (2005) Photocatalytic hydrogen production from water over M-Doped $La_2Ti_2O_7$ (M = Cr, Fe) under visible light irradiation ($\lambda > 420$ nm), *J. Phys. Chem. B*, **109**, 2093–2102.

[240] Nishimoto, S., Matsuda, M., Miyake, M. (2006) Photocatalytic Activities of Rh-doped $CaTiO_3$ under visible light irradiation, *Chem. Lett.*, **35**, 308–309.

[241] Niishiro, R., Konta, R., Kato, H., Chun, W. J., Asakura, K., Kudo, A. (2007) Photocatalytic O_2 evolution of rhodium and antimony-codoped rutile-type TiO_2 under visible light irradiation, *J. Phys. Chem. C*, **111**, 17420–17426.

[242] Shimodaira, Y., Kato, H., Kobayashi, H., Kudo, A. (2007) Investigations of electronic structures and photocatalytic activities under visible light irradiation of lead molybdate replaced with chromium(VI), *Bull. Chem. Soc. Jpn.*, **80**, 885–893.

[243] Ikeda, T., Nomoto, T., Eda, K., Mizutani, Y., Kato, H., Kudo, A., Onishi, H. (2008) Photoinduced dynamics of TiO_2 doped with Cr and Sb, *J. Phys. Chem. C*, **112**, 1167–1173.

[244] J. Yoshimura, Y. Ebina, J. Kondo, K. Domen, A. Tanaka (1993) Visible light-induced photocatalytic behavior of a layered perovskite-type rubidium lead niobate, $RbPb_2Nb_3O_{10}$, *J. Phys. Chem.*, **97**, 1970–1973.

[245] H. G. Kim, D. W. Hwang, J. S. Lee (2004) An undoped, single-phase oxide photocatalyst working under visible light, *J. Am. Chem. Soc.*, **126**, 8912–8913.

[246] S. Kohtani, N. Yamamoto, K. Kitajima, A. Kudo, H. Kato, K. Tokumura, K. Hayakawa, R. Nakagaki (2004) Photodegradation of 4-*n*-nonylphenol and natural estrogens using heterogeneous visible-light-driven $AgNbO_3$ or $BiVO_4$ photocatalyst. Photo/Electrochem, *Photobiol. Environ. Energy Fuel*, 173–184.

[247] Hosogi, Y., Shimodaira, Y., Kato, H., Kobayashi, H., Kudo, A. (2008) Role of Sn^{2+} in the band structure of SnM_2O_6 and $Sn_2M_2O_7$(M = Nb and Ta) and their photocatalytic properties, *Chem. Mater.*, **20**, 1299–1307.

[248] Hosogi, Y., Kato, H., Kudo, A. (2008) Visible light response of $AgLi_{1/3}M_{2/3}O_2$ (M = Ti and Sn) synthesized from layered Li_2MO_3 using molten $AgNO_3$, *Mater. Chem.*, **18**, 647–653.

[249] Kasahara, A., Nukumizu, K., Hitoki, G., Takata, T., Kondo, J. N., Hara, M., Kobayashi, H., Domen, K. (2002) Photoreactions on $LaTiO_2N$ under visible light irradiation, *J. Phys. Chem. A*, **106**, 6750–6753.

[250] Hitoki, G., Takata, T., Kondo, J. N., Hara, M., Kobayashi, H., Domen, K. (2002) An oxynitride, TaON, as an efficient water oxidation photocatalyst under visible light irradiation ($\lambda \leq 500$ nm), *Chem. Commun.*, 1698–1699. Doi: 10.1039/B202393H.

[251] Hitoki, G., Ishikawa, A., Kondo, J. N., Hara, M., Domen, K. (2001) Ta_3N_5 as a novel visible light-driven photocatalyst ($\lambda < 600$ nm), *Chem. Lett.*, **31**, 736–737.

[252] Kasahara, A., Nukumizu, K., Takata, T., Kondo, J. N., Hara, M., Kobayashi, H., Domen, K. (2003) $LaTiO_2N$ as a visible-light (≤600 nm)-driven photocatalyst (2). *J. Phys. Chem. B*, **107**, 791–797.

[253] Hara, M., Nunoshige, J., Takata, T., Kondo, J. N., Domen, K. (2003) Unusual enhancement of H_2 evolution by Ru on TaON photocatalyst under visible light irradiation, *Chem. Commun.*, 3000–3001. Doi: 10.1039/B309935K.

[254] Yamasita, D., Takata, T., Hara, M., Kondo, J. N., Domen, K. (2004) Recent progress of visible-light-driven heterogeneous photocatalysts for overall water splitting, *Solid State Ionics*, **172**, 591–595.

[255] Liu, M., You, W., Lei, Z., Zhou, G., Yang, J, Wu, G., Ma, G., Luan, G., Takata, T., Hara, M., Domen, K., Li, C. (2004) Water reduction and oxidation on Pt–Ru/$Y_2Ta_2O_5N_2$ catalyst under visible light irradiation, *Chem. Commun.*, 2192–2193. Doi: 10.1039/B407892F

[256] Ishikawa, A., Takata, T., Kondo, J. N., Hara, M., Kobayashi, H., Domen, K. (2002) Oxysulfide $Sm_2Ti_2S_2O_5$ as a stable photocatalyst for water

oxidation and reduction under visible light irradiation ($\lambda \leq 650$ nm), *J. Am. Chem. Soc.*, **124**, 13547–13553.

[257] Ishikawa, A., Yamada, Y., Takata, T., Kondo, J. N., Hara, M., Kobayashi, H., Domen, K. (2003) Novel synthesis and photocatalytic activity of oxysulfide $Sm_2Ti_2S_2O_5$, *Chem. Mater.*, **15**, 4442–4446.

[258] Ishikawa, A., Takata, T., Matsumura, T., Kondo, J. N., Hara, M., Kobayashi, H., Domen, K. (2004) Oxysulfides $Ln_2Ti_2S_2O_5$ as stable photocatalysts for water oxidation and reduction under visible-light irradiation, *J. Phys. Chem. B*, **108**, 2637–2642.

[259] Ogisu, K., Ishikawa, A., Teramura, K., Toda, K., Hara, M., Domen, K. (2007) Lanthanum–Indium oxysulfide as a visible light driven photocatalyst for water splitting, *Chem. Lett.*, **36**, 854–855.

[260] Maeda, K., Takata, T., Hara, M., Saito, N., Inoue, Y., Kobayashi, H., K. Domen (2005) GaN:ZnO Solid solution as a photocatalyst for visible-light-driven overall water splitting, *J. Am. Chem. Soc.*, **127**, 8286–8287.

[261] Maeda, K., Teramura, K., Lu, D., Takata, T., Saito, N., Inoue, Y., Domen, K. (2006) Photocatalyst releasing hydrogen from water, *Nature*, **440**, 295–297.

[262] Lee, Y., Terashima, H., Shimodaira, Y., Teramura, K., Hara, M., Kobayashi, H., Domen, K., Yashima, M. (2007) Zinc germanium oxynitride as a photocatalyst for overall water splitting under visible light, *J. Phys. Chem. C*, **111**, 1042–1048.

[263] Tsuji, I., Kato, H., Kudo, A. (2005) Visible-light-induced H_2 evolution from an aqueous solution containing sulfide and sulfite over a ZnS–$CuInS_2$–$AgInS_2$ solid-solution photocatalyst, *Angew. Chem., Int. Ed.*, **44**, 3565.

[264] Li, G., Kako, T., Wang, D., Zou, Z., Ye, J. (2007) Composition dependence of the photophysical and photocatalytic properties of $(AgNbO_3)_{1-x}(NaNbO_3)_x$ solid solutions, *J. Solid State Chem.*, **180**, 2845–2850.

[265] Irie, H., Maruyama, Y., Hashimoto, K. (2007) Ag^+ and Pb^{2+}-Doped $SrTiO_3$ photocatalysts. A correlation between band structure and photocatalytic activity, *J. Phys. Chem. C*, **111**, 1847–1852.

[266] Bard, A. J. (1979) Photoelectrochemistry and heterogeneous photo-catalysis at semiconductors, *J. Photochem.*, **10**, 59–75.

[267] Sayama, K., Yoshida R., Kusama, H., Okabe, K., Abe, Y., Arakawa, H. (1997) Photocatalytic decomposition of water into H_2 and O_2 by a two-step photoexcitation reaction using a WO_3 suspension catalyst and an Fe^{3+}/Fe^{2+} redox system, *Chem. Phys. Lett.*, **277**, 387–391.

[268] Tennakone, K., Pushpa, S. (1985) Water photodecomposition with two reversible and separable photosystems, *J. Chem. Soc., Chem. Commun.*, 1435–1437. Doi: 10.1039/C39850001435.

[269] Tennakone, K., Wickramanayake, S. (1986) Water photolysis with copper(I) chloride, *J. Chem. Soc., Faraday Trans.*, 2, **82**, 1475–1479.

[270] Tennakone, K., Tantrigoda, R., Abeysinghe, S., Punchihewa, S., Fernando, C. A. N. (1990) Water photolysis via reversible oxidation and reduction between MnO_2 and MnO_4^{2-}, *J. Photochem. Photobiol. A: Chem.*, **52**, 43–46.

[271] Ohno, T., Saito, S., Fujihara, K., Matsumura, M. (1996) Photocatalyzed production of hydrogen and iodine from aqueous solutions of iodide using platinum-loaded TiO_2 powder, *Bull. Chem. Soc. Jpn.*, **69**, 3059–3064.

[272] Ohno, T., Fujihara, K., Saito, S., Matsumura, M. (1997) Forwarding reversible photocatalytic reactions on semiconductor particles using an oil/water boundary, *Solar Energy Mat. Solar Cells*, **45**, 169–174.

[273] Ohno, T., Haga, D., Fujihara, K., Kaizaki, K., Matsumura, M. (1997) Unique Effects of iron(III) ions on photocatalytic and photoelectrochemical properties of titanium dioxide, *J. Phys. Chem. B*, **101**, 6415–6419.

[274] Ohno, T., Nakabeya, K., Fujihara, K., Matsumura, M. (1998) Relay of positive holes from photoirradiated Pt-loaded TiO_2 particles in an aqueous phase to *t*-butylhydroquinone in an oil phase, *J. Photochem. Photobiol. A: Chem.*, **117**, 143–147.

[275] Fujihara, K., Ohno, T., Matsumura, M. (1998) Splitting of water by electrochemical combination of two photocatalytic reactions on TiO_2 particles, *J. Chem. Soc., Faraday Trans.*, **94**, 3705–3709.

[276] Ohno, T., Fujihara, K., Sarukawa, K., Matsumura, M. (1999) Splitting of water by combining two photocatalytic reactions through a quinone compound dissolved in an oil phase, *Zeischrift fur Physikalische Chemie, Bd.*, **213**, 165–175.

[277] Bamwenda, G. R., Sayama, K., Arakawa, H. (1999) The effect of selected reaction parameters on the photoproduction of oxygen and hydrogen from a WO_3–Fe^{2+}–Fe^{3+} aqueous suspension, *J. Photochem. Photobiol. A: Chem.*, **122**, 175–183.

[278] Abe, R., Sayama, K., Domen, K., Arakawa, H. (2001) A new type of water splitting system composed of two different TiO_2 photocatalysts (anatase, rutile) and a IO_3^-/I^- shuttle redox mediator, *Chem. Phys. Lett.*, **344**, 339–344.

[279] Fujishima, A. Honda. K. (1972) Electrochemical photolysis of water at a semiconductor electrode, *Nature*, **238**, 37–38.

[280] Aharon-Shalom, E., Heller, A. (1982) Efficient p-InP(Rh-H alloy) and p-InP(Re-H alloy) hydrogen evolving photocathodes. *J. Electrochem. Soc.*, **129**, 2865–2866.

[281] Heller, A. (1984) Hydrogen-evolving solar cells, *Science*, **223**, 1141–1148.

[282] Khaselev, O., Turner, J. A. (1998) A monolithic photovoltaic-photoelectrochemical device for hydrogen production via water splitting, *Science*, **280**, 425–427.

[283] Lewis, N. S., Nocera, D. G. (2006) Chemical challenges in solar energy utilization, *Proc. Natl. Acad. Sci. USA*, **103**, 15729–15735.

[284] Lewis, N. S. (2007) Towards cost-effective solar energy use, *Science*, **315**, 798–801.

[285] Santato, C., Ulmann, M., Augustynski, J. (2001) Photoelectrochemical properties of nanostructured tungsten trioxide films, *J. Phys. Chem. B*, **105**, 936–940.

[286] Santaro, C., Odziemkowski, M., Ulmann, M., Augustynski, J. (2001) Crystallographically oriented mesoporous WO_3 films: synthesis, characterization, and applications, *J. Am. Chem. Soc.*, **123**, 10639–10649.

[287] Wang, H., Lindgren T., He, J., Hagfeldt, A., Lindquist, S. E. (2000) Photoelectrochemistry of nanostructured WO_3 thin film electrodes for water oxidation: mechanism of electron transport, *J. Phys. Chem. B*, **104**, 5686–5696.

[288] Amano, F., Li, D., Ohatani, B. (2010) Fabrication and photoelectrochemical property of tungsten (VI) oxide films with a flake-wall structure, *Chem. Commun.*, **46**, 2769–2771.

[289] Sayama, K., Nomura, A., Zou, Z., Abe, R., Abe, Y., Arakawa, H. (2003) Photoelectrochemical decomposition of water on nanocrystalline $BiVO_4$ film electrodes under visible light, *Chem. Commun.*, 2908–2909.

[290] Sayama, K., Nomura, A., Arai, T., Sugita, T., Abe, R., Yanagida, M., Oi, T., Iwasaki, Y., Abe, Y. Sugihara, H. (2006) Photoelectrochemical decomposition of water into H_2 and O_2 on porous $BiVO_4$ thin-film electrodes under visible light and significant effect of Ag Ion treatment, *J. Phys. Chem. B*, **110**, 11352–11360.

[291] Long, M., Cai, W., Kisch, H. (2008) Visible light induced photoelectrochemical properties of n-$BiVO_4$ and n-$BiVO_4$/p-Co_3O_4, *J. Phys. Chem.* C, **112**, 548–554.

[292] Su, J., Guo, L., Yoriya, S., Grimes, C. A. (2010) Aqueous growth of pyramidal shaped $BiVO_4$ nanowire arrays and structural

characterization: Application to photoelectrochemical water splitting, *Cryst. Growth Des.*, **10**, 856–861.

[293] Iwase, A. Kudo, A. (2010)Photoelectrochemical water splitting using visible-light-responsive $BiVO_4$ fine particles prepared in an aqueous acetic acid solution, *J. Mater. Chem.*, **20**, 7536–7542.

[294] Chatchai, P., Murakami, Y., Kishioka, S., Nosaka, A., Nosaka, Y. (2009) Efficient photocatalytic activity of water oxidation over $WO_3/BiVO_4$ composite under visible light irradiation, *Electrochim. Acta*, **54**, 1147–1152.

[295] Chatchai, P., Kishioka, S., Murakami, Y., Nosaka, A., Nosaka, Y. (2010) Enhanced photoelectrocatalytic activity of $FTO/WO_3/BiVO_4$ electrode modified with gold nanoparticles for water oxidation under visible light irradiation, *Electrochim. Acta*, **55**, 592–596.

[296] Kennedy, J. H., Frese, K. W. (1978) Photo-oxidation of Water at α-Fe_2O_3 Electrodes, *J. Electrochem. Soc.*, **125**, 709–714.

[297] Shinar, R., Kennedy, J. H. (1982) Photoactivity of doped α-Fe_2O_3 electrodes, *Solar Energy Mater.*, **6**, 323–325.

[298] Sanchez, C., Hendewerk, M., Sieber, K. D., Somorjai, G. A. (1986) Synthesis, bulk, and surface characterization of niobium-doped Fe_2O_3 single crystals, *J. Solid State Chem.*, **61**, 47–55.

[299] Sanchez, C., Sieber, K. D., Somorjai, G. A. (1988) The photoelectrochemistry of niobium doped α-Fe_2O_3, *J. Electroanal. Chem.*, **252**, 269–290.

[300] Duret, A., Grätzel, M. (2005) Visible light-induced water oxidation on mesoscopic α-Fe_2O_3 films made by ultrasonic spray pyrolysis, *J. Phys. Chem. B*, **109**, 17184–17191.

[301] Kay, A., Cesar, I., Grätzel, M. (2006) New benchmark for water photo-oxidation by nanostructured α-Fe_2O_3 films, *J. Am. Chem. Soc.*, **128**, 15714–15721.

[302] Shwarsctein, A. K., Hu, Y. S., Forman, A. J., Stucky, G. D., McFarland, E. W. (2008) Electrodeposition of α-Fe_2O_3 doped with Mo or Cr as photoanodes for photocatalytic water splitting, *J. Phys. Chem., C*, **112**, 15900–15907.

[303] Zhong, D. K., Gamelin, D. R. (2010) Photoelectrochemical water oxidation by cobalt catalyst ("Co–Pi")/α-Fe_2O_3 composite photoanodes: Oxygen evolution and resolution of a kinetic bottleneck, *J. Am. Chem. Soc.*, **132**, 4202–4207.

[304] Brillet, J., Grätzel, Sivula, M. K. (2010) Decoupling feature size and functionality in solution-processed, porous hematite electrodes for solar water splitting, *Nano Lett.*, **10**, 4155–4160.

[305] Kitano, M., Takeuchi, M., Masaoka, M., Thomas, J. M., Anpo, M. (2007) Photocatalytic water splitting using Pt-loaded visible light-responsive TiO_2 thin film photocatalysts, *Catal. Today*, **120**, 133–138.

[306] Kitano, M., Iyatani, K., Tsujimaru, K., Matsuoka, M, Takeuchi, M., Ueshima, M., Thomas, J. M., Anpo, M. (2008)The effect of chemical etching by HF solution on the photocatalytic activity of visible light-responsive TiO_2 thin films for solar water splitting, *Top. Catal.*, **49**, 24–31.

[307] Abe, R. Takata, T., Sugihara, H., Domen, K. (2005) The use of $TiCl_4$ treatment to enhance the photocurrent in a TaON photoelectrode under visible light irradiation, *Chem. Lett.*, **34**, 1162–1163.

[308] Nakamura, R., Tanaka, T., Nakato, Y. (2005) Oxygen photoevolution on a tantalum oxynitride photocatalyst under visible-light irradiation: How does water photo-oxidation proceed on a metal–oxynitride surface? *J. Phys. Chem. B*, **109**, 8920–8927.

[309] Thivet, C. L. P., Ishikawa, A., Ziani, A., Gendre, L. L., Yoshida, M., Kubota, J., Tessier, F., Domen K. (2009) Photoelectrochemical properties of crystalline perovskite lanthanum titanium oxynitride films under visible light, *J. Phys. Chem. C*, **113**, 6156–6162.

[310] Abe, R. Higashi, M., Domen, K. (2010) Facile fabrication of an efficient oxynitride TaON photoanode for overall water splitting into H_2 and O_2 under Visible Light Irradiation, *J. Am. Chem. Soc.*, **132**, 11828–11829.

[311] Nishimura, N., Raphael, B., Maeda, K., Gendre, L. L., Abe, R., Kubota, J. Domen, K. (2010) Effect of $TiCl_4$ treatment on the photoelectrochemical properties of $LaTiO_2N$ electrodes for water splitting under visible light, *Thin Solid Films*, **518**, 5855–5859.

[312] Yokoyama, D., Hashiguchi, H., Maeda, K., Minegishi, T., Takata, T., Abe, R., Kubota, J., Domen, K. (2011) Ta_3N_5 photoanodes for water splitting prepared by sputtering, *Thin Solid Films*, **519**, 2087–2092.

[313] Hashiguchi, H., Maeda, K., Abe, R., Ishikawa, A., Kubota, J., Domen, K. (2009) Photoresponse of GaN:ZnO electrode on FTO under visible light irradiation, *Bull. Chem. Soc. Jpn.*, **82**, 401–407.

[314] Marudachalam, M., Birkmire, R. W., Hichri, H., Schultz, J. M., Swartzlander A., Al-Jassim, M. M. (1997) Phases, morphology, and diffusion in $CuIn_xGa_{1-x}Se_2$ thin films, *J. Appl. Phys.*, **82**, 2896–2905.

[315] Valverde, S. F., Regil, E. O., Alvardo, R. V., Noriega, R. R., Feria, O. S. (1997) Mass spectrometry quantification of hydrogen produced under illumination of p-$CuInSe_2$ and modified surfaces, *Int. J. Hydrogen Energy*, **22**, 581–584.

[316] Bodnar, I.V, Ivanov, V. A., Gaisin, V. A. (2001) Spectral properties of $CuGaxIn_{1-x}Se_2$ solid solutions, *J. Appl. Spectrosc.*, **68**, 1016–1019.

[317] Leisch, J. E., Bhattacharya, R. N., Teeter, G., Turner, J. A. (2004) Preparation and characterization of $Cu(In,Ga)(Se,S)_2$ thin films from electrodeposited precursors for hydrogen production, *Sol. Energy Mater. Sol. Cells*, **81**, 249–259.

[318] Nian, J. N. Hu, C. C., Teng, H. (2008) Electrodeposited p-type Cu_2O for H_2 evolution from photoelectrolysis of water under visible light illumination, *Int. J. Hydrogen Energy*, **33**, 2897–2903.

[319] Hu, C. C., Nian, J. N., Teng, H. (2008) Electrodeposited p-type Cu_2O as photocatalyst for H_2 evolution from water reduction in the presence of WO_3, *Sol. Energy Mater. Sol. Cells*, **92**, 1071–1076.

[320] Yamane, S., Kato, N., Kojima, S., Imanishi, A., Ogawa, S., Yoshida, N., Nonomura, S., Nakato, Y. (2009) Efficient solar water splitting with a composite "*n*-Si/*p*-CuI/*n*-*i*-*p* a-Si/*n*-*p* GaP/RuO_2" semiconductor electrode, *J. Phys. Chem. C*, **113**, 14575–14581.

[321] Zhu, F., Hu, J., Matulionis, I., Deutsch, T., Gaillard, N., Kunrath, A., Miller, E., Madan, A. (2009) Amorphous silicon carbide photoelectrode for hydrogen production directly from water using sunlight, *Phil. Mag.*, **89**, 2723–2739.

[322] Kameyama, T., Osaki, T., Okazaki, K., Shibayama, T., Kudo, A., Kuwabata, S., Torimoto, T. (2010) Preparation and photoelectrochemical properties of densely immobilized Cu_2ZnSnS_4 nanoparticle films, *J. Mater. Chem.*, **20**, 5319–5324.

[323] Yokoyama, D., Minegishi, T., Maeda, K., Katayama, M., Kubota, J., Yamada, A., Konaga, M., Domen, K. (2010) Photoelectrochemical water splitting using a $Cu(In,Ga)Se_2$ thin film, *Electrochem. Commun.*, **12**, 851–853.

[324] Ida, S., Yamada, K., Matsunaga, T., Hagiwara, H., Matsumoto, Y., Ishihara, T. (2010) Preparation of p-type $CaFe_2O_4$ photocathodes for producing hydrogen from water, *J. Am. Chem. Soc.*, **132**, 17343–17345.

[325] Zukalová, M., Procházka, J., Zukal, A., Ho Yum, J., Kavan, L. (2008). Structural parameters controlling the performance of organized mesoporous TiO_2 films in dye sensitized solar cells, *Inorg. Chim. Acta*, 361, 656–662.

[326] Masatai, Y., Hodes, G. (1997) Size quantization in electrodeposited CdTe nanocrystalline films, *J. Phys. Chem. B*, **101**, 2685–2690.

[327] Yamaguti K., Sato, S. (1985) Photolysis of water over metallized powdered titanium dioxide, *J. Chem. Soc., Faraday Trans.*, 1, **81**, 1237–1246.

[328] Sayama K., Arakawa, H. (1997) Effect of carbonate salt addition on the photocatalytic decomposition of liquid water over Pt–TiO_2 catalyst, *J. Chem. Soc., Faraday Trans.*, **93**, 1647–1654.

[329] Nozik, J. (1978) Photoelectrochemistry: Applications to Solar Energy Conversion, *Annu. Rev. Phys. Chem.*, **29**, 189–222.

[330] Mavroides, J. G., Kafalas J. A., Kolesar D. F. (1976) Photoelectrolysis of water in cells with $SrTiO_3$ anodes, *Appl. Phys. Lett.*, **28**, 241–243.

[331] Ellis, A. B., Kaiser, S. W., Wrighton, M. S. (1976) Semiconducting potassium tantalate electrodes. Photoassistance agents for the efficient electrolysis of water, *J. Phys. Chem.*, **80**, 1325–1328.

[332] Domen, K., Naito, S., Soma, S., Onishi, M., Tamaru, K. (1980) Photocatalytic decomposition of water vapour on an NiO–$SrTiO_3$ catalyst, *J. Chem. Soc., Chem. Commun.*, 543–544. Doi: 10.1039/C39800000543.

[333] Domen, K., Naito, S., Onishi, T., Tamaru, T., Soma, M. (1982) Study of the photocatalytic decomposition of water vapor over a nickel(II) oxide-strontium titanate ($SrTiO_3$) catalyst, *J. Phys. Chem.*, **86**, 3657–3661.

[334] Domen, K., Naito, S. Onishi, T., Tamaru, K. (1982) Photocatalytic decomposition of liquid water on a NiO-$SrTiO_3$ catalyst, *Chem. Phys. Lett.*, **92**, 433–434.

[335] Domen, K., Kudo, A., Onishi, T., Kosugi, N., Kuroda, H. (1986) Photocatalytic decomposition of water into hydrogen and oxygen over nickel(II) oxide-strontium titanate ($SrTiO_3$) powder. 1. Structure of the catalysts, *J. Phys. Chem.*, **90**, 292–295.

[336] Domen, K., Kudo A., Ohnishi, T. (1986) Mechanism of photocatalytic decomposition of water into H_2 and O_2 over NiO-$SrTiO_3$, *J. Catal.*, **102**, 92–98.

[337] Kudo, A., Tanaka, A., Domen, K., Onishi, T. (1988) The effects of the calcination temperature of $SrTiO_3$ powder on photocatalytic activities, *J. Catal.*, **111**, 296–301.

[338] Kudo, A. Tanaka, A. K. Domen, K. Maruya, K. Aika and Onishi, T. Photocatalytic decomposition of water over NiO-$K_4Nb_6O_{17}$ catalyst, *J. Catal.*, 1988, 111, 67–76.

[339] Kudo, A., Sayama, K., Tanaka, A., Asakura, K., Domen, K., Maruya, K., and Onishi, T. (1989) Nickel-loaded $K_4Nb_6O_{17}$ photocatalyst in the decomposition of H_2O into H_2 and O_2: Structure and reaction mechanism, *J. Catal.*, **120**, 337–352.

[340] Kudo, A., Kato, H. (2000) Effect of lanthanide-doping into $NaTaO_3$ photocatalysts for efficient water splitting, *Chem. Phys. Lett.*, **331**, 373–377.

[341] Kato, H., Asakura, K., Kudo, A. (2003) Highly efficient water splitting into H_2 and O_2 over lanthanum-doped $NaTaO_3$ photocatalysts with high crystallinity and surface nanostructure, *J. Am. Chem. Soc.*, **125**, 3082–3089.

[342] Iwase, A., Kato, H., Okutomi, H., Kudo, A. (2004) Formation of surface nano-step structures and improvement of photocatalytic activities of $NaTaO_3$ by doping of alkaline earth metal ions, *Chem. Lett.*, **33**, 1260–1261.

[343] Yamakata, A., Ishibashi, T., Kato, H., Kudo, A., Onishi, H. (2003) Photodynamics of $NaTaO_3$ catalysts for efficient water splitting, *J. Phys. Chem. B*, **51**, 14383–14387.

[344] Kudo, A., Kato, H., Nakagawa, S. (2000) Water splitting into H_2 and O_2 on new $Sr_2M_2O_7$ (M = Nb and Ta) photocatalysts with layered perovskite structures factors affecting the photocatalytic activity, *J. Phys. Chem. B*, **104**, 571–575 .

[345] Dare-Edwards, M. P., Goodenough, J. P., Hamnett A., Nicholson, N. D. (1981) Photoelectrochemistry of nickel (II) oxide, *J. Chem. Soc., Faraday Trans.*, 2, **77**, 643–661.

Chapter 8

[346] Natural gas. http://www.naturalgas.org/overview/background.asp.

[347] HFCIT Hydrogen Production: Natural Gas Reforming. U. S. Department of Energy. http://www1.eere.energy.gov/hydrogenandfuelcells/production/natural_gas.html.

[348] http://en.wikipedia.org/wiki/Hydrogen_production

[349] Kanan M. W., Nocera· D. G. (2008) in situ formation of an oxygen evolving catalyst in neutral water containing phosphate and Co^{2+}, *Science*, **321**, 1072–1075.

[350] Nocera solar system fuel cell. http://peswiki.com/index.php/Image:Nocera_solar_system_fuel_cell_diagram_jp70.jpg

[351] Fown. (2001) A. Dr. Nocera's Invention May be the Future of Alternative Energy. But it's Not a Leaf!, Clean Technica. http://cleantechnica.com/2011/04/07/dr-noceras-invention-may-be-the-future-of-alternative-energy-but-its-not-a-leaf/.

Chapter 9

[352] Fuel cell. Wikipedia. http://en.wikipedia.org/wiki/Fuel_cell.

[353] Fuel cell block diagram. http://en.wikipedia.org/wiki/File:Fuel_Cell_Block_Diagram.svg

[354] Kakati, B. K., Deka, D. (2007) Effect of resin matrix precursor on the properties of graphite composite bipolar plate for PEM fuel cell, *Energy Fuels*, **21**, 1681–1687.

[355] Heitner-Wirguin, C. (1996) Recent advances in perfluorinated ionomer membranes: Structure, properties and applications, *J. Membr. Sci.*, **120**, 1–33.

[356] Mauritz, K. A., Moore R. B. (2004) State of Understanding of Nafion, *Chem. Rev.*, **104**, 4535–4586.

[357] Proton exchange mebrane fuel cell. http://en.wikipedia.org/wiki/File:PEM_fuelcell.svg

[358] Nafion. http://en.wikipedia.org/wiki/Nafion#cite_note-Heitner-Wirguin1996-1.

[359] Nafion polymer. http://en.m.wikipedia.org/wiki/File:Nafion2.svg

[360] Hsu, W. Y., Gierke, T. D. (1983) Ion transport and clustering in nafion perfluorinated membranes, *J. Membr. Sci.*, **13**, 307–326.

[361] Schmidt-Rohr, K., Chen, Q. (2008) Parallel cylindrical water nanochannels in Nafion fuel-cell membranes, *Nat. Mater.*, **7**, 75–83.

[362] Chen, Q., Schmidt-Rohr, K. (2004) 19F and 13C NMR signal assignment and analysis in a perfluorinated ionomer (Nafion) by two-dimensional solid-state NMR, *Macromolecules*, **37**, 5995–6003.

[363] Chen, Q., Schmidt-Rohr, K. (2007) Backbone dynamics of the Nafion Ionomer Studied by 19F–13C solid-state NMR, *Macromol. Chem. Phys.*, **208**, 2189–2203.

[364] Chu, B., Wu, C., Buck, W. (1989) Light-scattering characterization of poly(tetrafluoroethylene).2.PTFE in perfluorotetracosane-molecular-weight distribution and solution properties, *Macromolecules*, **22**, 831–837.

[365] Rosi-Schwartz, B., Mitchell, G. R. (1996) Extracting force fields for disordered polymeric materials from neutron scattering data, *Polymer*, **37**, 1857–1870.

[366] Tian, N., Zhou, Z.-Y., Sun, S.-G., Ding, Y., Wang, Z. L. (2007). Synthesis of tetrahexahedral platinum nanocrystals with high-index facets and high electro-oxidation activity, *Science*, **316**, 732–735.

[367] Stamenkovic, V. R., Fowler, B., Mun, B. S., Wang, G., Ross, P. N., Lucas, C. A., Marković. N. M. (2007) Improved oxygen reduction activity on $Pt_3Ni(111)$ via increased surface site availability, *Science*, **315**, 493–497.

[368] Hoogers Boca Raton, G. (ed.) (2003) *Fuel Cell Technology Handbook* (CRC Press, Florida).

[369] Wang, C., Daimon, H., Onodera, T., Koda, T. Sun, S. (2008) A general approach to the size- and shape-controlled synthesis of platinum nanoparticles and their catalytic reduction of oxygen, *Angew. Chem., Int. Ed.*, **47**, 3588–3591.

[370] Alayoglu, S., Nilekar, A. U., Mavrikakis, M., Eichhorn, B. (2008) Ru-Pt core-shell nanoparticles for preferential oxidation of carbon monoxide in hydrogen, *Nat. Mater.*, **7**, 333–338.

Chapter 10

[371] Hydrogen storage. Wikipedia http://en.wikipedia.org/wiki/Hydrogen_storage.

[372] Hydrogen Storage Technologies Roadmap. Department of Energy, Washington, 2005. http://www.doe.gov.

[373] Sing, K. S. W., Everett, D. H., Haul, Raw, Moscou, L., Pierotti, R. A., Rouquérol, J., Siemieniewska, T. (1985) Reporting physisorption data for gas/solid systems with special reference to the determination of surface area and porosity, *Pure Appl. Chem.*, **57**, 603–619.

[374] Broom, D. P. (2011) *Hydrogen Storage Materials. The Characterisation of Their Storage Properties* (Springer-Verlag, London).

[375] Linares-Solano, F. Surez- Garcia, D. Lozano-Castelló, M. Jorda-Beneyto, D. Cazorla-Amorós. (2005) Hydrogen Storage on Carbon Materials, H_2 Workshop 2005, Bad Honnef, Germany 23–28 October 2005.

[376] Rouquerol, F., Rouquerol, J., Sing, K. (1999) *Adsorption by Powders and Porous Solids: Principles, Methodology and Applications* (Academic Press, London).

[377] Thomas, K. M. (2009) Adsorption and desorption of hydrogen on metal–organic framework materials for storage applications: Comparison with other nanoporous materials, *Dalton Trans.*, 1487–1505.

[378] Rowsell, J. L. C., Yaghi, O. M. (2004) Metal–organic frameworks: A new class of porous materials, *Microporous Mesoporous Mater.*, **73**, 3–14.

[379] Kitagawa, S. Kitaura, R., Noro, S.-I. (2004) Funktionale poröse Koordinationspolymere. *Angew. Chem.*, **116**, 2388–2430.

[380] Kitagawa, S., Kitaura, R., Noro S.-I. (2004) Functional Porous Coordination Polymers. *Angew. Chem., Int. Ed.*, **43**, 2334–2375.

[381] Janiak, C. (2003) Engineering coordination polymers towards applications. *Dalton Trans.*, 2781–2804.

[382] Rowsell, J. L. C., Yaghi, O. M. (2005) Strategies for hydrogen storage in metal–organic frameworks. *Angew. Chem., Int. Ed.*, **44**, 4670 –4679.

[383] Yaghi, O. M., O'Keeffe, M., Ockwig, N. W., Chae, H. K., Eddaoudi, M., Kim, J. (2003) Reticular synthesis and the design of new materials, *Nature*, **423**, 705–714.

[384] Chae, H. K., Siberio-Perez, D. Y., Kim, J., Go, Y., Eddaoudi, M., Matzger, A. J., O'Keeffe, M., Yaghi, O. M. (2004) A route to high surface area, porosity and inclusion of large molecules in crystals, *Nature*, **427**, 523–527.

[385] Rowsell, J. L. C, Millward, A. R., Park, K. S., Yaghi, O. M. (2004) Hydrogen sorption in functionalized metal–organic frameworks, *J. Am. Chem. Soc.*, **126**, 5666–5667.

[386] Rzepka, M., Lamp, P., de la Casa-Lillo, M. A. (1998) Physisorption of hydrogen on microporous carbon and carbon nanotubes, *J. Phys. Chem. B*, **102**, 10894–10898.

[387] Large pore of MOF-5 with yellow ball. http://en.wikipedia.org/wiki/File:IRMOF-1_wiki.png

[388] http://en.wikipedia.org/wiki/Metal-organic_framework

[389] Pan, L., Sander, M. B., Huang, X., Li, J., Smith, M., Bittner, E., Bockrath, B., Johnson, J. K. (2004) Microporous metal organic materials: Promising candidates as sorbents for hydrogen storage, *J. Am. Chem. Soc.*, **126**, 1308–1309.

[390] Dybtsev, D. N., Chun, H., Yoon, S. H., Kim, D., Kim K. (2004) Microporous manganese formate: A simple metal–organic porous material with high framework stability and highly selective gas sorption properties, *J. Am. Chem. Soc.*, **126**, 32–33.

[391] Kesanli, B., Cui, Y., Smith, M. R., Bittner, E. W., Bockrath, B. C., Lin, W. (2005) Highly interpenetrated metal-organic frameworks for hydrogen storage, *Angew. Chem., Int. Ed.*, **44**, 72–75.

[392] Roswell, J. L. C., Millward, A. R., Park, K. S., Yaghi, O. M. (2004) Hydrogen sorption in functionalized metal–organic frameworks, *J. Am. Chem. Soc.*, 126, 5666–5667.

[393] Li, H., Eddaoudi, M., Groy, T. L., Yaghi, O. M. (1998) Establishing microporosity in open metal–organic frameworks: gas sorption isotherms for Zn(BDC) (BDC = 1,4-benzenedicarboxylate), *J. Am. Chem. Soc.*, **120**, 8571–8572.

[394] Rosi, N. L., Eddaoudi, M., Kim, J., O'Keeffe, M., Yaghi, O. M. (2002) Infinite secondary building units and forbidden catenation in metal-organic frameworks, *Angew. Chem.*, **114**, 294–297.

[395] Kitaura, R., Onoyama, G., Sakamoto, H., Matsuda, R., Noro, S.-I., Kitagawa, S. (2004) Immobilization of a metallo Schiff Base into a microporous coordination polymer, *Angew. Chem.*, **116**, 2738–2741.

[396] Forster, P. M., Eckert, J., Chang, J.-S., Park, S.-E., Srey, G. F., Cheetham, A. K. (2003) Hydrogen adsorption in nanoporous nickel (II) phosphates, *J. Am. Chem. Soc.*, **125**, 1309–1312.

[397] Crystal structure of MIL-53 with yellow balls. http://commons.wikimedia.org/wiki/File:MIL-53ht.png

[398] Hübner, O., Glöss, A., Fichtner, M., Klopper, W. (2004) On the interaction of dihydrogen with aromatic systems, *J. Phys. Chem. A*, **108**, 3019–3023.

[399] Benard, P., Chahine, R. (2001) Determination of the adsorption isotherms of hydrogen on activated carbons above the critical temperature of the adsorbate over wide temperature and pressure ranges, *Langmuir*, **17**, 1950–1955.

[400] Schimmel, H. G., Kearley, G. J., Nijkamp, M. G., Visser, C. T., de Jong, K. P., Mulder, F. M. (2003) Hydrogen adsorption in carbon nanostructures: Comparison of nanotubes, fibers, and coals, *Chem. Eur. J.*, **9**, 4764–4770.

[401] Sha, X., Jackson, B. (2004) The Location of adsorbed hydrogen in graphite nanostructures, *J. Am. Chem. Soc.*, **126**, 13095–13099.

[402] Eddaoudi, M., Kim, J., Rosi, N., Vodak, D., Wachter, J., O'Keeffe, M., Yaghi, O. M. (2002) Systematic design of pore size and functionality in isoreticular MOFs and their application in methane storage, *Science*, **295**, 469–472.

[403] Rosi, N. L., Eckert, J., Eddaoudi, M., Vodak, D. T., Kim, J., O'Keeffe, M., Yaghi, O. M. (2003) Hydrogen storage in microporous metal–organic frameworks, *Science*, **300**, 1127–1129.

[404] Kubota, Y., Takata, M., Matsuda, R., Kitaura, R., Kitagawa, S., Kato, K., Sakata, M., Kobayashi, T. C. (2005) Direct observation of hydrogen molecules adsorbed onto a microporous coordination polymer, *Angew. Chem., Int. Ed.*, **44**, 920–923.

[405] Anderson, C.-R., Coker, D. F., Eckert, J., Bug, A. L. R. (1999) Computational study of molecular hydrogen in zeolite Na-A. I. Potential energy surfaces and thermodynamic separation factors for ortho and para hydrogen, *J. Chem. Phys.*, **111**, 7599–7613.

[406] McKinnon, J. A., Coker, D. F., Eckert, J., Bug, A. L. R. (2001), Computational study of molecular hydrogen in zeolite Na-A. II. Density of rotational states and inelastic neutron scattering spectra, *J. Chem. Phys.*, **114**, 10137–10150.

[407] Vishnyakov, A., Ravikovitch, P. I., Neimark, A. V., Rlow, M. B., Wang, Q. M. (2003) Nanopore structure and sorption properties of Cu-BTC metal–organic framework, *Nano Lett.*, **3**, 713–718.

[408] Skoulidas, A. I. (2004) Molecular dynamics simulations of gas diffusion in metal–organic frameworks: Argon in CuBTC, *J. Am. Chem. Soc.*, **126**, 1356–1357.

[409] Düren, T., Sarkisov, L., Yaghi, O. M., Snurr, R. Q. (2004) Design of new materials for methane storage, *Langmuir*, **20**, 2683–2689.

[410] Heine, T., Zhechkov, L. Seifert, G. (2004) Hydrogen storage by physisorption on nanostructured graphite platelets, *Phys. Chem. Chem. Phys.*, **6**, 980–984.

[411] Hamel, S., Côté, M. (2004) First-principles study of the rotational transitions of H_2 physisorbed over benzene, *J. Chem. Phys.*, **121**, 12618–12625.

[412] T. Sagara, J. Klassen, E. Ganz (2004) Computational study of hydrogen binding by metal–organic framework-5, *J. Chem. Phys.*, **121**, 12543–12547.

[413] Inelastic neutron scattering. http://en.wikipedia.org/wiki/Inelastic_neutron_scattering.

[414] Eckert, J., Kubas, G. J. (1993).Barrier to rotation of the dihydrogen ligand in metal complexes, *J. Phys. Chem.*, **97**, 2378–2384.

[415] I. F. Silvera (1980) The solid molecular hydrogens in the condensed phase: Fundamentals and static properties, *Rev. Mod. Phys.*, **52**, 393–452.

[416] Dihydrogen complex. http://en.wikipedia.org/wiki/Dihydrogen_complex.

[417] National Institute of Standards and Technology. http://www.nist.gov/public_affairs/techbeat/tb2008_0401.htm#hydrogen.

[418] Wong-Foy, A. G., Matzger, A. J., Yaghi, O. M. (2006) Exceptional H_2 saturation uptake in microporous metal–organic frameworks, *J. Am. Chem. Soc.*, **128**, 3494–3495.

[419] Yan, Y., Lin, X., Yang, S., Blake, A. J., Dailly, A., Champness, N. R., Hubberstey, P., Schröder, M. (2009) Exceptionally high H_2 storage by a metal–organic polyhedral framework, *Chem. Commun.*, 1025–1027.

[420] Zeolite. http://en.wikipedia.org/wiki/File:Zeolite-ZSM-5-3D-vdW.png.

[421] Vitillo, J. G., Ricchiardi, G., Spoto, G., Zecchina, A. (2005) Theoretical maximal storage of hydrogen in zeolitic frameworks, *Phys. Chem. Chem. Phys.*, 7, 3948–3954.

[422] van den Berg, A. W. C., Bromley, S. T., Wojdel, J. C., Jansen, J. C. (2006) Adsorption isotherms of H_2 in microporous materials with SOD structure: A grand canonical Monte Carlo study, *Microporous Mesoporous Mater.*, **87**, 235–242.

[423] Song, M. K., No, K. T. (2007) Molecular simulation of hydrogen adsorption in organic zeolite, *Catal. Today*, **120**, 374–382.

[424] Fraenkel, D., Shabtai, J. (1977) Encapsulation of hydrogen in molecular sieve zeolites, *J. Am. Chem. Soc.*, **99**, 7074–7076.

[425] Efstathiou, A. M., Suib, S. L., Bennett, C. O. (1990) Encapsulation of molecular hydrogen in zeolites at 1 atm, *J. Catal.*, **123**, 456–462.

[426] Weitkamp, J., Fritz, M., Ernst, S. (1995) Zeolites as media for hydrogen storage. *Int. J. Hydrogen Energy*, **20**, 967–970.

[427] Garrone, E., Bonelli, B., Otero Areán, C. (2008) Enthalpy-entropy correlation for hydrogen adsorption on zeolites, *Chem. Phys. Lett.*, **456**, 68–70.

[428] Otero Areán, C., Turnes Palomino, G., Llop Carayol, M. R. (2007) Variable temperature FT-IR studies on hydrogen adsorption on the zeolite (Mg, Na)-Y, *Appl. Surf. Sci.*, **253**, 5701–5704.

[429] Felderhoff, M., Weidenthaler, C., von Helmolt, R., Eberle, U. (2007) Hydrogen storage: The remaining scientific and technological challenges, *Phys. Chem. Chem. Phys.*, **9**, 2643–2653.

[430] van den Berg, A. W. C., Otero Areán, C. (2008) Materials for hydrogen storage: Current research trends and perspectives, *Chem. Commun.*, 668–681. Doi: 10.1039/B712576N.

[431] Anderson, P. A. (2008) Storage of hydrogen in zeolites, In: Walker G. (ed.) *Solid-State Hydrogen Storage: Materials And Chemistry* (Woodhead Publishing, Cambridge).

[432] Yürüm, Y., Taralp, A., Veziroglu, T. N. (2009) Storage of hydrogen in nanostructured carbon materials, *Int. J. Hydrogen Energy*, **34**, 3784–3798.

[433] Rzepka, M., Lamp, P., de la Casa-Lillo, M. A. (1998) Physisorption of hydrogen on microporous carbon and carbon nanotubes, *J. Phys. Chem. B*, **102**, 10894–10898.

[434] Jordá-Beneyto, M., Suárez-García, F., Lozano-Castelló, D., Cazorla-Amorós, D., Linares-Solano, A. (2007) Hydrogen storage on chemically activated carbons and carbon nanomaterials at high pressures, *Carbon*, **45**, 293–303.

[435] Zlotea, C., Moretto, P., Steriotis, T. (2009) A round robin characterisation of the hydrogen sorption properties of a carbon based material, *Int. J. Hydrogen Energy*, **34**, 3044–3057.

[436] Jordá-Beneyto, M., Lozano-Castelló, D., Suárez-García, F., Cazorla-Amorós, D., Linares-Solano, Á. (2008) Advanced activated carbon monoliths and activated carbons for hydrogen storage, *Microporous Mesoporous Mater.*, **112**, 235–242.

[437] Bénard, P., Chahine, R., Chandonia, P. A, Cossement, D., Dorval-Douville, G., Lafi, L., Lachance, P., Paggiaro, R., Poirier, E. (2007) Comparison of hydrogen adsorption on nanoporous materials, *J. Alloys Compd.*, **446–447**, 380–384.

[438] Kroto, H. (2003). Symmetry, space, stars and C60, In: Grenthe I. (ed.) *Nobel Lectures, Chemistry 1996–2000* (World Scientific, Singapore) pp. 44–79.

[439] Kroto, H. W., Heath, J. R., O'Brien, S. C., Curl, R. F., Smalley, R. E. (1985) C_{60}: Buckminsterfullerene, *Nature*, **318**, 162–163.

[440] Fullerene C60. http://upload.wikimedia.org/wikipedia/commons/8/82/Fullerene-C60.png

[441] Iijima, S. (1991) Helical microtubules of graphite carbon, *Nature*, **354**, 56–58.

[442] Carbon nanotubes. http://en.wikipedia.org/wiki/Carbon_nanotube.

[443] Züttel, A., Orimo, S. (2002) Hydrogen in nanostructured, carbon-related, and metallic materials, *MRS Bull.*, **27**, 705–711.

[444] Belle Dumé. Graphene-oxide framework packs in hydrogen. http://physicsworld.com/cws/article/news/42171.

[445] Andre Geim, Random walk to grapheme, Nobel Lecture. http://www.nobelprize.org/nobel_prizes/physics/laureates/2010/geim_lecture.pdf.

[446] Layered graphene sheets could solve hydrogen storage issues. http://www.nist.gov/ncnr/graphene_031610.cfm.

[447] Burress, J. W., Gadipelli, S. Ford, J., Simmons, J. M., Zhou, W., Juldirim., T. (2010) Graphene oxide framework materials: Theoretical predictions and experimental results, *Angew. Chem., Int. Ed.*, **49**, 8902–8904.

[448] Yang, Z., Xia, Y., Mokaya, R. (2007) Enhanced hydrogen storage capacity of high surface area zeolite-like carbon materials, *J. Am. Chem. Soc.*, **129**, 1673–1679.

[449] Hu, Q., Lu, Y., Meisner, G. P. (2008) Preparation of nanoporous carbon particles and their cryogenic hydrogen storage capacities, *J. Phys. Chem. C*, **112**, 1516–1523.

[450] Gogotsi, Y., Dash, R. K., Yushin, G., Yildirim, T., Laudisio, G., Fischer, J. E. (2005) Tailoring of nanoscale porosity in carbide-derived carbons for hydrogen storage, *J. Am. Chem. Soc.*, **127**, 16006–16007.

[451] Gogotsi, Y., Portet, C., Osswald, S., Simmons, J. M., Yildirim, T., Laudisio, G., Fischer, J. E. (2009) Importance of pore size in high-pressure hydrogen storage by porous carbons, *Int. J. Hydrogen Energy*, **34**, 6314–6319.

[452] Züttel, A. (2007) Hydrogen storage and distribution systems, *Mitig. Adapt. Strat. Gl.*, **12**, 343–365.

[453] Oriani, R. A. (1994) A Brief Survey of Useful Information about Hydrogen in Metals. In: International Symposium on Cold Fusion and Advanced Energy Sources. Belarusian State University, Minsk, Belarus (Fusion Information Center, Salt Lake City).

[454] Alefeld G., Volkl, J. (eds.) (1978) *Hydrogen in Metals*, vols. 1 and 2 (Springer-Verlag, Berlin).

[455] McNabb A. M., Foster, P. K. (1963) A new analysis of the diffusion of hydrogen in iron and ferritic steels, *Trans. Metall. Soc. AIME*, **227**, 618–627.

[456] Oriani, R. A. (1970) The diffusion and trapping of hydrogen in steel, *Acta Metall.*, **18**, 147–157.

[457] Li, J. C. M. (1978) Physical chemistry of some microstructural phenomena, *Metall. Trans. A*, **9A**, 1353–1380.

[458] Adams B. D., Chen. A. (2011) The role of palladium in hydrogen economy, *Materials Today*, **14**, 282–289.

[459] Cser, L., Krexner, G., Prem, M., Sharkov, I.,Török, G. (2005) Neutron holography of metal–hydrogen systems, *J. Alloy. Compd.*, **404–406**, 122–125.

[460] Schlapbach, L., Zuttel, A. (2001) Hydrogen-storage materials for mobile applications, *Nature*, **414**, 353–358.

[461] Kishore, S., Nelson, J. A., Adair, J. H., Eklund, P. C. (2005) Hydrogen storage in spherical and platelet palladium nanoparticles, *J. Alloys Compd.*, **389**, 234–242.

[462] Yamauchi, M., Ikeda, R., Kitagawa, H., Takata, M. (2008) Nanosize effects on hydrogen storage in palladium, *J. Phys. Chem. C*, **112**, 3294–3299.

[463] Kobayashi, H., Yamauchi, M., Kitagawa, H., Kubota, Y., Kato, K., Takata, M. (2008) On the nature of strong hydrogen atom trapping inside Pd Nanoparticles, *J. Am. Chem. Soc.*, **130**, 1828–1829.

[464] Gabrielli, C., Grand, P. P., Lasia, A., Perrot, H. (2004) Investigation of hydrogen adsorption-absorption into thin palladium films, *J. Electrochem. Soc.*, **151**, A1925–1936.

[465] C. Gabrielli, P. P. Grand, A. Lasia, and H. Perrot. (2004) Investigation of hydrogen adsorption and absorption in palladium thin films, *J. Electrochem. Soc.*, **151**, A1937–A1942.

[466] B. Łosiewicz, L. Birry, A. (2007) Lasia Effect of adsorbed carbon monoxide on the kinetics of hydrogen electrosorption into palladium, *J. Electroanal. Chem.*, **611**, 26–34.

[467] Duncan, H., Lasia, A. (2008) Separation of hydrogen adsorption and absorption on Pd thin films, *Electrochim. Acta*, **53**, 6845–6850.

[468] Nakatsuji, H., Hada, M. (1985)Interaction of a hydrogen molecule with palladium, *J. Am. Chem. Soc.*, **107**, 8264–8266.

[469] Bartczak, W. M., and Stawowska, J. (2004) Interaction of dihydrogen with transition metal (Pd, Ni, Ag, Cu) clusters, *Struct. Chem.*, **15**, 447–459.

[470] Zaluski, L., Zaluska, A., Tessier, P., Ström-Olsen, J. O., Schulz, R. (1995) Catalytic effect of Pd on hydrogen absorption in mechanically alloyed Mg_2Ni, $LaNi_5$ and FeTi, *J. Alloys Compd.*, **217**, 295–300.

[471] Şennik, E., Kılınç, N., Öztürk Z. Z. (2010) Temperature-dependent H_2 gas-sensing properties of fabricated Pd nanowires using highly oriented pyrolytic graphite, *J. Appl. Phys.*, **108** 054317(7 pages).

[472] Antolini, E. (2009) Palladium in fuel cell catalysis, *Energy Environ. Sci.*, **2**, 915–931.

[473] Sealy, C. (2008) The problem with platinum, *Mater Today*, **11**, 65–68.

[474] Service, R. F. (2007) Platinum in fuel cells gets a helping hand, *Science*, **315**, 172–173.

[475] Mejdell, A. L., Jøndahl, M., Peters, T. A., Bredesen, R., Venvik, H. J. (2009) Effects of CO and CO_2 on hydrogen permeation through a 3 μm Pd/Ag 23 wt.% membrane employed in a microchannel membrane configuration., *Sep. Purif. Technol.*, **68**, 178–184.

[476] Li, H., Xu, H. (2008) Study of *n* value and α/β palladium hydride phase transition within the ultra-thin palladium composite membrane, *J. Membr. Sci.*, **324**, 44–49.

[477] Delogu, F. (2010) Smooth size effects in Pd and PdH_x nanoparticles, *J. Phys. Chem. C*, **114**, 18085–18090.

[478] Yamauchi, M., Kitagawa, I., Kobayashi, H. (2009) Hydrogen storage mediated by Pd and Pt nanoparticles, *ChemPhysChem*, **10**, 2566–2576.

[479] Eastman, J. A., Thompson, L. J., Kestel, B. J. (1993) Narrowing of the palladium-hydrogen miscibility gap in nanocrystalline palladium, *Phys. Rev. B*, **48**, 84–92.

[480] Sun, Y., Tao, Zh., Chen, J., Herricks, T., Xia, Y. (2004) Ag Nanowires coated with Ag/Pd alloy sheaths and their use as substrates for reversible absorption and desorption of hydrogen, *Am. Chem. Soc.*, **126**, 5940–5941.

[481] Pozzo, M., Alfe, D. (2009) Hydrogen dissociation and diffusion on transition metal (= Ti, Zr, V, Fe, Ru, Co, Rh, Ni, Pd, Cu, Ag)-doped Mg(0001) surfaces, *Int. J. Hydrogen Energy*, **34**, 1922–1230.

[482] Shao, H., Liu, T., Wang, Y., Xu, H., Li, X. (2008) Preparation of Mg-based hydrogen storage materials from metal nanoparticles, *J. Alloys Compd.*, **465**, 527–533.

[483] Gross, A. F., Ahn, C. C., Atta, S. L. V., Liu, P., Vajo, J. J. (2009) Fabrication and hydrogen sorption behaviour of nanoparticulate MgH_2 incorporated in a porous carbon host, *Nanotechnology*, **20**, 204005.

[484] Niemann, M. U., Srinivasan, S. S., Phani, A. R., Kumar, A., Goswami, D. Y., Stefanakos, E. K. (2008) Nanomaterials for hydrogen storage applications: A review, *J. Nanomater.*, article ID 950967. Doi: 10.1155/2008/950967

[485] Zaluska, A., Zaluski, L. Ström–Olsen, J. O. (1999) Nanocrystalline magnesium for hydrogen storage, *J. Alloys Compd.*, **288**, 217–225.

[486] Bérubé, V., Radtke, G., Dresselhaus, M., Chen, G. (2007) Size effects on the hydrogen storage properties of nanostructured metal hydrides: A review, *Int. J. Energy Res.*, **31**, 637–663.

[487] Favier, F., Walter, E. C., Zach, M. P., Benter T., Penner, R. M. (2001) Hydrogen sensors and switches from electrodeposited palladium, *Mesowire Arrays Science*, **293**, 2227–2231.

[488] van Vucht J. H. N., Kuijpers F. A., Bruning H. C. A. M. (1970) Reversible room-temperature absorption of large quantities of hydrogen by intermetallic compounds, *Philips Res. Rep.*, **25**, 133–140.

[489] Griessen R., Riesterer T. (1988) Heat of formation models. In: Schlapbach, L. (ed.) *Topics in Applied Physics*, vol. 67: Hydrogen in intermetallic compounds I. Surface and dynamic properties, applications (Springer-Verlag, Berlin).

[490] Luo, S., Clewley, J. D., Flanagan, T. B., Bowman, R. C. Jr., Wade, L. A. (1998) Further studies of the isotherms of $LaNi_{5-x}Sn_xH$ for x = 0–0.5, *J. Alloys Compd.*, **267**, 171–181.

[491] Sandrock, G. (1999) A panoramic overview of hydrogen storage alloys from a gas reaction point of view, *J. Alloys Compd.*, **293–295**, 877–888.

[492] Bowman, R. C. Jr., Luo C. H., Ahn C. C., Witham, C. K., Fultz, B. (1995) The effect of tin on the degradation of $LaNi_{5-y}Sn_y$ metal hydrides during thermal cycling, *J. Alloys Compd.*, **217**, 185–192.

[493] Chandra, D., Reilly, J. J., Chellappa, R. (2006) Metal hydrides for vehicular applications: The state of the art, *JOM*, **58**(2), 26–32.

[494] Feng F, Geng M, Northwood DO (2001) Electrochemical behaviour of intermetallic-based metal hydrides used in Ni/metal hydride (MH) batteries: A review, *Int. J. Hydrogen Energy*, **26**, 725–734.

[495] Töpler, J., Feucht, K. (1989) Results of a test fleet with metal hydride motor cars., *Z. Phys. Chem. NF*, **164**, 1451–1461.

[496] Bowman, R. C. Jr., Fultz, B. (2002) Metallic hydrides I: Hydrogen storage and other gas-phase applications, *MRS Bull.*, **27**, 688–693.

[497] Sakintuna, B., Lamari-Darkrim, F., Hirscher, M. (2007) Metal hydride materials for solid hydrogen storage: A review, *Int. J. Hydrogen Energy*, **32**, 1121–1140.

[498] Nomura, K., Akiba, E. (1995) H_2 Absorbing-desorbing characterization of the Ti-V-Fe alloy system, *J. Alloys Compd.*, **231**, 513–517.

[499] Cho, S-W., Shim, G., Choi, G-S., Park, C-N., Yoo, J-H., Choi, J. (2007) Hydrogen absorption desorption properties of $Ti_{0.32}Cr_{0.43}V_{0.25}$ alloy, *J. Alloys Compd.*, **430**, 136–141.

[500] Huot J., Liang G., Schulz, R. (2001) Mechanically alloyed metal hydride systems, *Appl. Phys. A*, **72**, 187–195.

[501] Zaluska A., Zaluski L., Ström-Olsen J. O. (2001) Structure, catalysis, and atomic reactions on the nano-scale: A systematic approach to metal hydrides for hydrogen storage, *Appl. Phys. A*, **72**, 157–165.

[502] Gross, A. F., Ahn C. C., Van Atta S. L., Liu, P., Vajo, J. J. (2009) Fabrication and hydrogen sorption behavior of nanoparticulate MgH_2 incorporated in a porous carbon host, *Nanotechnology*, **20,** 204005.

[503] Aguey-Zinsou, K.-F., Ares-Fernández, J.-R. (2008) Synthesis of colloidal magnesium: A near room temperature store for hydrogen, *Chem. Mater.,* **20**, 376–378.

[504] Barkhordarian, G., Klassen, T., Bormann, R. (2006) Catalytic mechanism of transition-metal compounds on Mg hydrogen sorption reaction, *J. Phys. Chem.*, **110**, 11020–11024.

[505] Wagemans, R. W. P., van Lenthe J. H., de Jongh P. E., van Dillen A. J., de Jong K. P. (2005) Hydrogen storage in magnesium clusters: Quantum chemical study, *J. Am. Chem. Soc.*, **127**, 16675–16680.

[506] Suryanarayana, C. (2001) Mechanical alloying and milling, *Prog. Mater. Sci.,* **46**, 1–184.

[507] Suryanarayana, C., Koch, C. C. (2000) Nanocrystalline materials—current research and future directions. *Hyperfine Interact.,* **130**, 5–44.

[508] Suryanarayana, C. (2002) The structure and properties of nanocrystalline materials: Issues and concerns, *JOM*, **54**, 24–27.

[509] Pundt, A. (2004) Hydrogen in nano-sized metals, *Adv. Eng. Mater.*, **6**, 11–21.

[510] Pundt, A., Kirchheim, R. (2006) Hydrogen in metals: Microstructural aspects, *Annu. Rev. Mater. Res.*, **36**, 555–608.

[511] Bérubé, V., Radtke, G., Dresselhaus, M., Chen, G. (2007) Size effects on the hydrogen storage properties of nanostructured metal hydrides: A review, *Int. J. Energy. Res.,* **31**, 637–663.

[512] Maeland, A. J., Tanner, L. E., Libowitz, G. G. (1980) Hydrides of metallic glass alloys, *J. Less Common Met.,* **74**, 279–285.

[513] Orimo, S., Fujii, H. (1998) Effects of nanometer-scale structure on hydriding properties of Mg–Ni alloys: A review, *Intermetallics*, **6**, 185–192.

[514] Zaluski, L., Zaluska, A., Tessier, P., Ström-Olsen, J. O., Schulz, R. (1995) Effects of relaxation on hydrogen absorption in Fe-Ti produced by ball-milling, *J. Alloys Compd.*, **227**, 53–57.

[515] Bououdina, M., Fruchart, D., Jacquet, S., Pontonnier, L., Soubeyroux, J. L. (1999) Effect of nickel alloying by using ball milling on the absorption properties of TiFe, *Int. J. Hydrogen Energy*, **24**, 885–890.

[516] Fujii, H., Munehiro, S., Fujii, K., Orimo, S. (2002) Effect of mechanical grinding under Ar and H_2 atmospheres on structural and hydriding properties in $LaNi_5$, *J. Alloys Compd.*, **330–332**, 747–751.

[517] Santos, S. F., Costa, A. L. M., de Castro J. F. R., dos Santos, D. S., Botta W. J., Ishikawa, T. T. (2004) Mechanical and reactive milling of a TiCrV BCC solid solution, *J. Metastable Nanocryst. Mater.*, **20–21**, 291–296.

[518] Huot, J., Enoki, H., Akiba, E. (2008) Synthesis, phase transformation, and hydrogen storage properties of ball-milled $TiV_{0.9}Mn_{1.1}$. *J. Alloys Compd.*, **453**, 203–209.

[519] Takeichi, N., Senoh, H., Takeshita, H. T., Oishi, T., Tanaka, H., Kiyobayashi, T., Kuriyama, N. (2004) Hydrogenation properties and structure of Ti-Cr alloy prepared by mechanical grinding, *Mater. Sci. Eng.: B*, **108**, 100–104.

[520] Singh, B. K., Shim, G., Cho, S-W. (2007) Effects of mechanical milling on hydrogen storage properties of $Ti_{0.32}Cr_{0.43}V_{0.25}$ alloy, *Int. J. Hydrogen Energy*, **32**, 4961–4965.

[521] Singh, A., Singh, B. K., Davidson, D. J., Srivastava, O. N. (2004) Studies on improvement of hydrogen storage capacity of AB_5 type: $MmNi_{4.6}Fe_{0.4}$ alloy, *Int. J. Hydrogen Energy*, **29**, 1151–1156.

[522] Bowman, R. C. Jr. (1988) Preparation and properties of amorphous hydrides, *Mater. Sci. Forum*, **31**, 197–228.

[523]Eliaz, N., Eliezer, D. (1999) An overview of hydrogen interaction with amorphous alloys, *Adv. Perform. Mater.*, **6**, 5–31.

[524] Fukai, Y. (2005) *The Metal-Hydrogen System. Basic Bulk Properties*, 2nd Ed. (Springer, Berlin).

[525] Tsai, A. P (2008) Icosahedral clusters, icosaheral order and stability of quasicrystals—a view of metallurgy, *Sci. Technol. Adv. Mater.*, **9**, 013008.

[526] Bindi, L., Steinhardt, P. J., Yao, N., Lu, P. J. (2009) Natural quasicrystals, *Science*, **324**, 1306–1309.

[527] Takasaki, A., Kelton, K. F. (2006) Hydrogen storage in Ti-based quasicrystal powders produced by mechanical alloying, *Int. J. Hydrogen Energy*, **31**, 183–190.

[528] Bystrzycki, J., Polanski, M., Malka, I. E., Komuda, A. (2009) Hydriding properties of Mg-Al-Zn quasicrystal powder produced by mechanical alloying, *Z. Kristallogr.*, **224**, 105–108.

[529] Bogdanovic′, B., Schwickardi, M. (1997) Ti-doped alkali metal aluminium hydrides as potential novel reversible hydrogen storage materials. *J. Alloys Compd.*, **253–254**, 1–9.

[530] Chen, P., Xiong, Z., Luo, J., Lin, J., Tan, K. L. (2002) Interaction of hydrogen with metal nitrides and imides, *Nature*, **420**, 302–304.

[531] Schüth, F., Bogdanović, B., Felderhoff, M. (2004) Light metal hydrides and complex hydrides for hydrogen storage, *Chem. Commun.*, 2249–2258.

[532] Orimo, S., Nakamori, Y., Eliseo, J. R., Züttel, A., Jensen, C. M. (2007) Complex hydrides for hydrogen storage, *Chem. Rev.*, **107**, 4111–4132.

[533] Jensen, C., Yang, Y., Chou, M. Y. (2008) Alanates as hydrogen storage materials. In: (ed) Walker, G. *Solid-State Hydrogen Storage: Materials and Chemistry* (Woodhead Publishing, Cambridge).

[534] Hino, S., Ichikawa, T., Ogita, N., Udagawa, M., Fujii, H. (2005) Quantitative estimation of NH_3 partial pressure in H_2 desorbed from the Li-N-H system by Raman spectroscopy, *Chem. Commun.*, 3038–3040. Doi: 10.1039/B503356J.

[535] Uribe, F. A., Gottesfeld, S., Zawodzinski, T. A. Jr. (2002) Effect of ammonia as potential fuel impurity on proton-exchange membrane fuel cell performance, *J. Electrochem. Soc.*, **149**, A293–A296.

[536] Züttel, A., Rentsch, S., Fischer, P., Wenger, P., Sudan Mauron, P., Emmenegger, C. (2003) Hydrogen storage properties of $LiBH_4$, *J. Alloys Compd.*, **356–357**, 515–520.

[537] Walker, G. (2008) Multicomponent hydrogen storage systems. In: Walker, G. (ed.) *Solid-State Hydrogen Storage: Materials and Chemistry* (Woodhead Publishing, Cambridge).

[538] Yvon, K. (2003) Hydrogen in novel solid-state metal hydrides, *Z. Kristallogr.*, **218**, 108–116.

[539] Orimo S, Fujii H. (2001) Materials science of Mg-Ni-based new hydrides, *Appl. Phys. A*, **72**, 167–186.

[540] Blomqvist, H., Rönnebro, E., Noréus, D., Kujii, T. (2002) Competing stabilisation mechanisms in Mg_2NiH_4, *J. Alloys Compd.*, **330–332**, 268–270.

[541] Häussermann, U., Blomqvist, H., Noréus, D. (2002) Bonding and stability of the hydrogen storage material Mg_2NiH_4, *Inorg. Chem.*, **41**, 3684–3692.

[542] Mao, W. L., Koh, C. A., Sloan, E. D. (2007) Clathrate hydrates under pressure, *Phys. Today*, **60**(10):42–47.

[543] Mao, W. L., Mao, H., Goncharov, A. F., Struzhkin, V. V., Guo, Q., Hu, J., Shu, J., Hemley, R. J., Somayazulu, M., Zhao, Y. (2002) Hydrogen clusters in clathrate hydrate, *Science*, **297**, 2247–2249.

[544] Struzhkin, V. V., Militzer, B., Mao, W. L., Mao, H. K., Hemley, R. J. (2007) Hydrogen storage in molecular clathrates, *Chem. Rev.*, **107**, 4133–4151.

[545] Florusse, L. J., Peters, C. J., Schoonman, J., Hester, K. C., Koh, C. A., Dec, S. F., Marsh K. N., Sloan, E. D. (2004) Stable low-pressure hydrogen clusters stored in a binary clathrate hydrate, *Science*, **306**, 469–471.

[546] Lee, H., Lee, J., Kim, D. Y., Park, J., Seo, Y-T., Zeng, H., Moudrakovski, I. L., Ratcliffe, C. I., Ripmeester, J. A. (2005) Tuning clathrate hydrates for hydrogen storage, *Nature*, **434**, 743–746.

[547] Plechkova, N. V., Seddon K. R. (2008) Applications of ionic liquids in the chemical industry, *Chem. Soc. Rev.*, **37**, 123–150.

[548] Olivier-Bourbigou, H., Magna, L., Morvan, D. (2010) Ionic liquids and catalysis: Recent progress from knowledge to applications, *Appl. Catal. A*, **373**, 1–56.

Chapter 11

[549] Deb, S. K., Chopoorian, J. A. (1966) Optical properties and color-center formation in thin films of molybdenum trioxide, *J. Appl. Phys.*, **37**, 4818–4825.

[550] Deb, S. K. (1968) Physical properties of a transition metal oxide: Optical and photoelectric properties of single crystals and thin film molybdenum trioxide, *Proc. R. Soc., A*, **304**, 211–231.

[551] Deb, S. K. (1973) Optical and photoelectric properties and colour centres in thin films of tungsten trioxide, *Philos. Mag.*, **27**, 801–822.

[552] Granqvist, C. G. (2000) Electrochromic tungsten oxide films: Review of progress 1993–1998, *Sol. Energy Mater. Sol. Cells*, **60**, 201–262.

[553] He, T., Yao, J. N. (2006) Photochromism in composite and hybrid materials based on transition-metal oxides and polyoxometalates, *Prog. Mater. Sci.*, **51**, 810–879.

[554] He, T., Yao, J. N. (2004) Photochromism in transition-metal oxides, *Res. Chem. Intermed.*, **30**, 459–488.

[555] Gavrilyuk, A. I., Reinov, N. M., Chudnovskii, F. A. (1979) Photo- and thermochromism in V_2O_5 films, *Sov. Tech. Phys. Lett.*, **5**, 514–515.

[556] Wagner, H. J. Driessen, P. Schwerdfeger, C. F. (1979) EPR of Mo^{5+} in amorphous MoO_3 films, *J. Non-Cryst. Solids*, **34**, 335–339.

[557] Gavrilyuk, A. I., Zakharchenya, B. P., Chudnovskii, F. A. (1980) Photochromism in WO_3 thin films, *Sov. Tech. Phys. Lett.*, **6**, 512–513.

[558] Bechinger, C., Herminghaus, S., Leiderer, P. (1994) Photoinduced doping of thin amorphous WO_3 films, *Thin Solid Films*, **239**, 156–160.

[559] Bechinger, C. Wirth, E. Leiderer, P. (1996) Photochromic coloration of WO_3 with visible light, *Appl. Phys. Lett.*, **68**, 2834–2836.

[560] Gavrilyuk, A. I., Prochvatilov, V. G., Chudnovskii, F. A. (1982) Structural changes during electrochromic and photochromic processes in tungsten trioxide, *Sov. Phys. Solid State*, **24**, 558–563.

[561] Fleish, T. H. Mains, G. J. (1982) An XPS study of the UV reduction and photochromism of MoO_3 and WO_3, *J. Chem. Phys.*, **76**, 780–786.

[562] Gavrilyuk, A. I., Lanskaya, T. G., Mansurov, A. A., Chudnovskii, F. A. (1984) Photoinjection of hydrogen in VO_2-WO_3 heterostructure. *Sov. Phys. Solid State*, **26**, 117–120.

[563] Gavrilyuk, A. I., Mansurov, A. A., Chudnovskii, F. A. (1984) Photoinjection of hydrogen in amorphous MoO_3 and V_2O_5 films, *Sov. Tech. Phys. Lett.*, **10**, 292–293.

[564] Zikmach, P. D. (1985) Electrochromism and the localization of charge carriers in tungsten trioxide. Ph.D. Thesis (University of Latvia, Riga, in Russian).

[565] Gavrilyuk, A. I. Gusinskii, G. M., Mansurov, A. A., Rassadin, L. A., Chudnovskii, F. A. (1986) Investigation of photoinjection of hydrogen into transition metal oxides by the nuclear reaction $^2D(^3He, p)\ \alpha$, *Sov. Phys. Solid State*, **28**, 1147–1149.

[566] Gavrilyuk, A. I., Lanskaya, T. G., Chudnovskii, F. A. (1987) Photospillover of hydrogen in a film heterostructure consisting of polycrystalline and amorphous layers of tungsten trioxide, *Sov. Phys. Tech. Phys.*, **32**, 964–966.

[567] Gavrilyuk, A. I., Sekushin N. A. (1990) *Electrochromism and Photochromism in Tungsten and Molybdenum Oxides (*Nauka, Leningrad) (in Russian).

[568] Yao, J. N., Loo, B. H. Fujishima, A. (1990) Study of the photochromic and electrochromic properties of MoO_3 thin films, *Berichte der Bunsengesellshaft/ Phys. Chem. Chem. Phys.*, **94**, 13–17.

[569] Yao, J. N., Loo, B. H., Hashimoto, K., Fujishima, A. (1991) Photochromic characteristics of mixed WO_3 – MoO_3 thin films in alcohol vapors, *Berichte der Bunsengesellschaft/Phys. Chem. Chem. Phys.*, **95**, 554–556.

[570] Chudnovskii, F. A., Shaefer, M. D., Gavrilyuk, A. I., Reifenberger, R. (1992) A study of the morphology of photochromic and thermochromic MoO_3 amorphous films using an atomic force microscope, *Appl. Surf. Sci.*, **62**, 145–149.

[571] U. Tritthart, A. I. Gavrilyuk, W. Gey (1996) Weak localization and electron-electron interaction in amorphous H_xWO_3, *Czech J. Phys.*, **46**(5) (suppl.), 2495–2496.

[572] Gavrilyuk, A. (1997) Nature of photochromism in amorphous V_2O_5 thin films, *Proc. SPIE*, **2968**, 195–200.

[573] Tritthart, U., Gavrilyuk, A., Gey, W. (1998) Low temperature photochromism in quasi-amorphous MoO_3-Films, *Solid State Commun.*, **105**, 653–657.

[574] Gavrilyuk, A. I. (1998) Photoinjection of hydrogen in solids, *Ionics*, **4**, 372–382.

[575] Tritthart, U., Gey, W., Gavrilyuk, A. I. (1998) Low temperature coloration of WO_3 and MoO_3 thin films, *Ionics*, **4**, 299–308.

[576] Gavrilyuk, A. I. (1999) Photochromism in WO_3 thin films, *Electrochim. Acta*, **44**, 3027–3037.

[577] Tritthart, U., Gey, W., Gavrilyuk, A. (1999) Nature of the optical absorption band in amorphous $HxWO_3$ films, *Electrochim. Acta*, **44**, 3039–3049.

[578] Tritthart, U. (1999) Wasserstoffdotierte WO_3-Filme. Elektrische Transporteigenschaften—Metall-Isolator-Übergang. Ph.D. Thesis (Technical University of Braunschweig, Braunschweig) (in German).

[579] Gavrilyuk, A. I. (2000) Nature of the defects induced by photoinjection of hydrogen in transition metal oxides. In: Borstel, G., Krumins, A., Millers D. (eds.), *Defects and Surface-Induced Effects in Advanced Perovskytes*, **77**, 427–438 (NATO Science Series Sub-Series: 3, High Technology, vol. 77, Kluwer Academic Publishers, Dordrecht, Boston, London), pp. 427–438.

[580] Gavrilyuk, A. I. (2001) Photoinjection of hydrogen in solids: Electron-proton plasma doping in transition metal oxides (WO_3, MoO_3, V_2O_5) and drastic enhancement of the print-out effect in silver and cuprous halides (AgCl, AgI, $RbAg_4I_5$, CuCl) Conference paper on 2001 International Semiconductor Device Research Symposium, Washington, December 5–7, *IEEE, Symposium Proceedings* (2001), 55–58.

[581] Gavrilyuk, A. I., Tritthart, U., Gey, W. (2006) Photo-stimulated hydrogen atom transfer quantum reactions in WO_3 and MoO_3 thin films, *Philos. Mag. Lett.*, **86**, 205–214.

[582] Gavrilyuk, A., Tritthart, U., Gey, W. (2007) Photo-stimulated proton-coupled electron transfer in quasi-amorphous WO_3 and MoO_3 thin films, *Philos. Mag.*, **87**, 4519–4553.

[583] Gavrilyuk, A. I. (2007) Proton coupled electron transfer tunneling reactions in WO_3 and MoO_3 nanostructured films, *J. Phys.: Conf. Ser.*, **93,** Article ID 012023.

[584] Gavrilyuk, A. I. (2007) Hydrogen photosensitization, photochromism, and surface-enhanced infrared absorption in AgI thin films, *Phys. Rev. B*, **75**, art. no. 195412.

[585] Gavrilyuk, A. I., Afanasiev, M. M. (2009) ESR study of photoinjection of hydrogen in nanostructured MoO_3 thin films. *Sol. Energy Mater. Sol. Cells*, **93**, 280–288.

[586] Claridge, D. A., Dickens, P. G., Goodenough, J. B. (1989) NMR relaxation times and proton motion in H_xWO_3, *Appl. Phys. A*, **49**, 65–68.

[587] Chippindale A. M, Dickens P. G. (1987) The thermochemistry of the hydrogen vanadium bronzes $H_xV_2O_5$, *Solid State Ionics*, **23**, 83–188.

[588] Dickens, P. G. Birtill, J. J. (1978) Hydrogen molybdenum bronzes, *J. Electron. Mater.*, **7**, 679–686.

[589] Sotani, N., Eda, K., Sadamatu, M., Takagi, S. (1989) Preparation and characterization of hydrogen molybdenum bronzes, H_xMoO_3, *Bull. Chem. Soc. Jpn.*, **62**, 903–907.

[590] Dickens, P. G., Crouch-Baker, S., M. T. Weller. (1986) Hydrogen insertion in oxides, *Solid State Ionics*, **18–19** (Part 1), 89–97.

[591] Chang, H.-C., Jiang, J.-C., Tsai, W.-C., Chen, G.-C., Lin, S. H. (2006) The role of charge-enhanced C–H···O interactions in gel-like mixtures prepared from ionic liquids and tungsten(VI) oxide nanoparticles, *Chem. Phys. Lett.*, **427**, 310–316.

[592] Brand, J. C. D., Eglinton G. (1965) *Applications of Spectroscopy to Organic Chemistry* (Oldbourne Press, London).

[593] Loriner, R. L., Fuson, R. S., Curtin, D. Y., et al. (1980) *The Systematic Identification of Organic Compounds* (John Wiley & Sons, New York).

[594] Morrison, S. R. (1977) *The Chemical Physics of Surface* (Plenum Press, New York).

[595] Yin, X., Han, H., Endou, A., Kubo, M., Teraishi, K., Chatterjee, A., Miyamoto, A. (1999) Reactivity of lattice oxygens present in V_2O_5 (010): A periodic first-principles investigation, *J. Phys. Chem. B*, **10**, 1263–1269.

[596] Gavrilyuk, A. I. (2002) Photoinjection of hydrogen in solids. Habilitation thesis (A. F. Ioffe Physical Technical Institute, St Petersburg) (in Russian).

[597] Sotani, I., Shimada, I., Suzuki, T., Eda, K., Kunitomo M. (1998) Proton insertion compound with tunnel structure. *Solid State Ionics*, **113–115**, 377–385.

[598] Goodenough, J. B. (1971) Metallic oxides. *Prog. Sol. St. Chem.*, **5**, 145–399.

[599] Sienko, M. J. (1963) Electric and magnetic properties of the tungsten and vanadium bronzes. In: Ward, R. (ed.) *Non-stoichiometric Compounds*, vol. 39 (ACS Publications, Washington DC), pp. 224–236.

[600] Cukier, R. I., Nocera, D. G. (1998) Proton-coupled electron transfer, *Annu. Rev. Phys. Chem.*, **49**, 337–369.

[601] Mayer, J. M. (2004) Proton-coupled electron transfer: A Reaction Chemist's View, *Annu. Rev. Phys. Chem.*, **55**, 363–390.

[602] Huynh, M. H., Meyer, T. (2007) Proton-coupled electron transfer, *Chem. Rev.*, **107**, 5004–5064.

[603] Drago, R. (1977) *Physical Methods in Chemistry*, vol. 1 (W. B. Saunders, Philadelphia).

[604] Mills, A., Hunte, S. Le (1997) An overview of semiconductor photocatalysis, *J. Photochem. Photobiol. A*, **108**, 1–35.

[605] Shirmer, O. F. (1980) Small polaron aspects of defects in oxide materials, *J. de Physique Colloque*, **C6**, C479–484.

[606] Anderson, P. W. (1975) Model for the electronic structure of amorphous semiconductors, *Phys. Rev. Lett.*, **34**, 953–955.

[607] Mott, N. F. (1974) *Metal-Insulator Transitions* (Taylor & Francis Ltd., London).

[608] Mott, N. F., Davis, E. A. (1979) *Electron Processes in Non-Crystalline Materials*, 2nd Ed. (Clarendon Press, Oxford).

[609] Alexandrov, A. S., Mott, N. F. (1995) *Polarons and Bipolarons* (World Scientific, Cambridge, Loughborough).

[610] Chakraverty, B. K., Sienko, M. J., Bonnerot, J. (1978) Low-temperature specific heat and magnetic susceptibility of nonmetallic vanadium bronzes, *Phys. Rev. B*, **17**, 3781–3789.

[611] Marezio, M., McWhan, D. B. Dernier, P. D., Remeika, J. P. (1973) Structural aspects of the metal-insulator transition in V_4O_7, *J. Sol. St. Chem.*, **6**, 419–429.

[612] P. W. Anderson. (1958) Absence of diffusion in certain random lattices, *Phys. Rev.*, **109**, 1492–1505.

[613] Mestl, G., Verbruggen, N. F. D., Bosch, E., Knozinger, H. (1996) Mechanically Activated MoO_3. 5. Redox Behavior, *Langmuir*, **12**, 2961–2968.

[614] Sato, M., Fujishita, H., Sato, S., Hoshino, S. (1985) Neutron inelastic scattering and X-ray structural study of the charge-density-wave state in $K_{0.3}MoO_3$, *J. Phys. C Solid State Phys.*, **18**, 2602–2614.

[615] Verbist, G., Peeters, F. M., Devreese, J. T. (1990) The stability region of bipolaron formation in two and three dimensions in relation to high-Tc, *Solid State Commun.*, **76**, 1005–1007.

[616] Verbist, G., Peeters, F. M., Devreese, J. T. (1991) Large bipolarons in two and three dimensions, *Phys. Rev. B*, **43**, 2712–2720.

[617] Gavrilyuk, A. I., Lanskaya, T. G. (1994) Photochromism in thin V_2O_5 layers prepared by the sol-gel technology, *Tech. Phys. Lett.*, **20**, 219–221.

[618] Talledo, A., Granqvist, C. G. (1995) Electrochromic vanadium–pentoxide–based films: Structural, electrochemical, and optical properties, *J. Appl. Phys.*, **77**, 4655–4666.

[619] Moss, T. S. (1954) The interpretation of the properties of indium antimonide, *Proc. Phys. Soc. London, Sect. B*, **76**, 775–782.

[620] Burstein, E. (1954) Anomalous optical absorption limit in InSb, *Phys. Rev.*, **93**, 632–633.

[621] Bachmann, H. G., Ahmed, F. R., Barnes, W. H. (1961) The crystal structure of vanadium pentoxide, *Z. Kristallorg.*, **115**, 110–131.

[622] Bersuker, B. (1996) *Electronic Structure and Properties of Transition Metal Compounds: Introduction to the Theory* (John Wiley & Sons, New York).

[623] Dexter, D. L. (1956) Absorption of light by atoms in solids, *Phys. Rev.*, **101**, 48–55.

[624] Gavrilyuk, A. I., Gusinskii, G. M., Lanskaya, T. G. (1994) Determination of the oscillator strength of an optical transition for color centers in WO_3 thin films, *Tech. Phys. Lett.*, **20**, 295–297.

[625] Schlotter, P., Pickelman, L. (1982) The xerogel structure of thermally evaporated tungsten oxide layers, *J. Electron. Mater.*, **11**, 207–236.

[626] Faughnan, B. W., Crandall, R. S. (1980) Electrochromic displays based on WO_3 films. In: Pankove, J. I. (ed.) *Displays* (Springer-Verlag, Berlin).

[627] M. S. Whittingham, R. A. Huggins. In: Van Gool, W. (ed.), Fast ion transport in solids. *Proceedings of the NATO Sponsored Adv. Study Inst.*

on Fast Ion Transport in Solids, Belgrate, Italy, September 5–15, 1972, Elsevier, 1973, p. 645.

[628] Bludska, J., Jakubec, I. (1996) Insertion of hydrogen into hexagonal tungsten bronzes $A_{0.3}WO_3$ (A = K, NH_4, and Cs), *Z. Phys. Chem.*, **194**, 69–72.

[629] Crandall, R. S., Faughnan, B. W. (1977) Electronic transport in amorphous H_3WO_3, *Phys. Rev. Lett.*, **39**, 232–235.

[630] Crandall, R. S., Faughnan, B. W. (1975) Measurement of the diffusion coefficient of electrons in WO_3 films, *Appl. Phys. Lett.*, **26**, 120–121.

[631] He, T., Ma, Y., Cao, Y. A., Yang, W. S. (2002) Improved photochromism of WO_3 thin films by addition of Au nanoparticles, *Phys. Chem. Chem. Phys.*, **4**, 1637–1639.

[632] Bechinger, C., Herminghaus, S., Leiderer, P. (1994) Photoinduced doping of thin amorphous WO_3 films, *Thin Solid Films*, **239**, 156–160.

[633] Fujii, M., Kawai, T., Nakamatsu, H., Kawai, S. (1983) Photoinduced Colour Transformation in WO_3 Particles, *J. Chem. Soc., Chem. Commun.*, **1983**, 1428–1429.

[634] Kubo, T., Nishikitani, Y., Kuroda, N., Matsuno, M. (1996) Band edge disorder producing Urbach's tail and enhanced photochromic efficiency caused by density of states inside band gap in tungsten oxide films, *Jpn. J. Appl. Phys.*, **35**, 5361–5366.

[635] Shiyanovskaya, I., Hepel, M. (1999) Bicomponent WO_3/TiO_2 films as photoelectrodes, *J. Electrochem. Soc.*, **146**, 243–249.

[636] Tennakone, K., Heperuma, O. A., Bandara, J. M. S., Kiridena, W. C. B. (1992) TiO_2 and WO_3 semiconductor particles in contact: Photochemical reduction of WO_3 to the non-stoichiometric blue form, *Semicond. Sci. Technol.*, **7**, 423–424.

[637] Gavrilyuk, A. (1997). Photoinjection of hydrogen in Ag- and Cu-halides, In: Silinsh, E. A., Medvid, A., Lusis, A. R., Ozols, A. O. (eds.), *Optical Organic and Semiconductor Inorganic Materials, Proc. SPIE*, **2968**, 213–218.

[638] Tubbs, R. (1974) MoO_3 layers—optical properties, colour centres, and holographic recording, *Phys. Status Solidi A*, **21**, 253–260.

[639] Gavrilyuk, A. I., Mansurov, A. A., Razikov, A. H., Chudnovskii, F. A., Shaver, I. K. (1987) Color center formation in x- and γ-irradiated amorphous MoO3 films, *Sov. Phys. Tech. Phys.*, **31**, 585–586.

[640] Gavrilyuk, A. I., Motorny, A. V., Razikov, A. H., Teslenko, S. P., Chudnovskii, F. A., Shaver, I. K. (1985) Visualization of beta irradiation of tritium

with the help of amorphous molybdenum trioxide films, *Pis'ma Zh. Tekh. Fiz.*, **11**, 1089–1093 (in Russian).

[641] Antropov, A. E., Afonin, O. F., Gavrilyuk, A. I., Zakharchenya, B. P., Chudnovskii, F. A., Shaver, I. K. (1978) Use of transition-metal oxide films to study ion-beam profiles, *Sov. Tech. Phys. Lett.*, **4**, 224–225.

Index